OPTIMAL DESIGN AND RELATED AREAS IN OPTIMIZATION AND STATISTICS

Springer Optimization and Its Applications

VOLUME 28

Aims and Scope
Optimization has been expanding in all directions at an astonishing rate during the last few decades. New algorithmic and theoretical techniques have been developed, the diffusion into other disciplines has proceeded at a rapid pace, and our knowledge of all aspects of the field has grown even more profound. At the same time, one of the most striking trends in optimization is the constantly increasing emphasis on the interdisciplinary nature of the field. Optimization has been a basic tool in all areas of applied mathematics, engineering, medicine, economics and other sciences.

The *Springer Series in Optimization and Its Applications* publishes undergraduate and graduate textbooks, monographs and state-of-the-art expository works that focus on algorithms for solving optimization problems and also study applications involving such problems. Some of the topics covered include nonlinear optimization (convex and nonconvex), network flow problems, stochastic optimization, optimal control, discrete optimization, multiobjective programming, description of software packages, approximation techniques and heuristic approaches.

OPTIMAL DESIGN AND RELATED AREAS IN OPTIMIZATION AND STATISTICS

Edited By

LUC PRONZATO
CNRS/Université de Nice Sophia Antipolis, France

ANATOLY ZHIGLJAVSKY
Cardiff University, UK

Springer

Editors

Luc Pronzato
CNRS/Université de Nice Sophia Antipolis
Laboratoire 13S
Bât Euclide, Les Algorithmes
2000 route des Lucioles
BP 121, 06903 Sophia-Antipolis cedex
France
pronzato@i3s.unice.fr

Anatoly Zhigljavsky
Cardiff University
School of Mathematics
Senghennydd Road
CF24 4AG, Cardiff
United Kingdom
zhigljavskyaa@cf.ac.uk

ISBN: 978-0-387-79935-3 e-ISBN: 978-0-387-79936-0
DOI: 10.1007/978-0-387-79936-0

Library of Congress Control Number: 2008940068

Mathematics Subject Classifications (2000): 90C25, 90C30, 090C46, 90C90, 62K05, 62J02, 62F15, 60E15, 13P10

Printed on acid-free paper

springer.com

This volume is dedicated to Henry P. Wynn
on the occasion of his sixtieth birthday

Henry P. Wynn (born 19 February 1945)

Henry Wynn has been involved in some of the major advances in experimental design since 1970, stemming from the work in optimal experimental design of Jack Kiefer and collaborators. His first major result concerned the algorithm for constructing optimal designs, which is now often referred to as the Fedorov–Wynn Algorithm.[1] In joint work with Kiefer he was one of the first to study optimal design for correlated observations[2] and helped to edit

[1] Wynn, H.P. (1970). The sequential generation of D-optimum experimental design. *Annals of Mathematics and Statistics*, **41**, 1655–1664.

[2] Kiefer, J. and Wynn, H.P. (1981). Optimum balanced block and Latin square designs for correlated observations. *Annals of Statistics*, **9**, 737–757.

Kiefer's collected works. He was also a member of the team headed by Jerome Sacks, which wrote the now highly cited paper on computer experiments[3] and continues to be active in the area. An interest in the application of information theory, which started with the introduction of Maximum Entropy Sampling, led to joint work with Sebastiani.[4]

An early paper on multiple comparisons,[5] written with Bloomfield while they were both Ph.D. students, and further work with late Robert Bohrer led to a long-standing collaboration with Daniel Naiman, through which they introduced the concept of a discrete tube.[6] Notable was their use of ideas from computational geometry and topology, which paralleled work on continuous tube theory by Naiman and by Robert Adler and co-workers. Tube theory gives tight inclusion–exclusion (Boole-Bonferroni-Fréchet) bounds for the probability of unions of events. The observation that the close connection between discrete tubes, unions of orthants, monomial ideals and Hilbert functions could be used to get tight systems reliability bounds led to papers.[7,8] This work is continuing with Eduardo Sáenz-de-Cabezón (presented at MEGA2007) using the latter's new algorithm for minimal free resolutions.

Henry Wynn was intimately involved with the introduction of robust engineering methodologies into Europe following his regular participation in a series of National Science Foundation funded workshops in the late 1980s led by Jerome Sacks. This work has led to publications, particularly in the engineering literature, and extensive collaboration with industrial partners, notably Fiat CRF, well funded by both UK and EU grants. Much of this work is joint with Bates.[9]

Initial joint work with Anatoly Zhigljavsky led to a three-way long-term collaboration with Zhigljavsky and Luc Pronzato into the applications of dynamical systems to search and optimization. The novel idea is that a careful renormalization can convert certain search and optimization algorithms into dynamical systems, and rates of convergence can then be translated into rates

---

[3] Sacks, J., Welch, W.J., Mitchell, T.J. and Wynn, H.P. (1989). Design and analysis of computer experiments. *Statistical Science*, **4**, 409–435.

[4] Sebastiani, P. and Wynn, H.P. (2000). Maximum entropy sampling and bayesian experimental design. *Journal of Royal Statistical Society*, **B62**, 145–157.

[5] Wynn, H.P. and Bloomfield, P. (1971). Simultaneous confidence bands in regression analysis. *Journal of Royal Statistical Society*, **B33**, 202–217.

[6] Naiman, D.Q. and Wynn, H.P. (1997). Abstract tubes, improved inclusion–exclusion identities and inequalities and importance sampling. *Annals of Statistics*, **25**, 1954–1983.

[7] Naiman, D. and Wynn, H.P. (2001). Improved inclusion–exclusion inequalities for simplex and orthant arrangements. *Journal of Inequalities in Pure and Applied Mathematics*, **2**, 1–16.

[8] Giglio, G. and Wynn, H.P. (2004). Monomial ideals and the Scarf complex for coherent systems in reliability. *Annals of Statistics*, **32**, 1289–1311.

[9] Bates, R.A., Buck, R.J., Riccomagno, E. and Wynn, H.P. (1996). Experimental design for large systems. *Journal of Royal Statistical Society*, **B58**, 77–94.

of expansion of such systems as measured by Kolmogorov-Shannon entropy. The first tranche of work is summarized in the monograph[10] and includes an improvement over the celebrated Golden Section (Fibonacci) line search algorithm, which they have called the GS4 algorithm. Other joint work, begun there, is a detailed study of the attractors of algorithms of steepest descent type.[11] An important insight has been the link with certain classes of optimal experimental design algorithm, which stems from the fact that both classes of algorithms can be interpreted as updating of measures, spectral measure in the case of renormalized steepest descent.

Following an initial paper on the application of Gröbner bases to experimental design,[12] Wynn and Pistone collaborated with Riccomagno in a monograph,[13] which was an early contribution to the field of "Algebraic Statistics". With others, particularly the Genoa group led by Lorenzo Robbiano, they staged the successful series of GROSTAT workshops every year from 1998 to 2003. The field is growing rapidly with major contributions in the USA by Persi Diaconis, Berndt Sturmfels, Stephen Fienberg and a strong cadre of young researchers.

A long-standing collaboration with Giovagnoli began with work on group invariant orderings (majorization and its generalizations)[14] and widened to include the introduction of D-ordering.[15] Their new work includes the study of measures of agreement and a duality theory for generalized Lorenz ordering and integral stochastic orderings.

Henry Wynn is a professor of statistics at the London School of Economics and leads a research group. From 2001 to 2005 he was also a scientific co-director of EURANDOM, the international stochastics institute based at Eindhoven Technical University (TUE). He has a B.A. in honours mathematics from the University of Oxford and a Ph.D. in mathematical statistics from Imperial College, London. Following a period as a lecturer and then as a reader at Imperial College he became a professor of mathematical statistics at City University, London, in 1985 and Dean of Mathematics from 1987 to 1995. At City University he co-founded the Engineering Design Centre, for which he was a co-director. He moved, in 1995, to the University of Warwick

---

[10] Pronzato, L., Wynn, H.P. and Zhigljavsky, A.A. (2000). *Dynamical Search.* Chapman & Hall, Boca Raton.

[11] Pronzato, L., Wynn, H.P. and Zhigljavsky, A.A. (2006). Asymptotic behaviour of a family of gradient algorithms in $\mathbb{R}^d$ and Hilbert spaces. *Mathematical Programming*, **107**, 409–438.

[12] Pistone, G. and Wynn, H.P. (1996). Generalised confounding with Gröbner bases, *Biometrika* **83**, 653–666.

[13] Pistone, G., Riccomagno, E. and Wynn, H.P. (2001). *Algebraic Statistics.* Chapman & Hall, Boca Raton.

[14] Giovagnoli, A. and Wynn, H.P. (1985). G-majorization with applications to matrix orderings. *Linear Algebra and its Applications*, **67**, 111–135.

[15] Giovagnoli, A. and Wynn, H.P. (1995). Multivariate orderings. *Statistics & Probability Letters*, **22**, 325–332.

as founding director of the Risk Initiative and Statistical Consultancy Unit (RISCU), which he helped build as a leading centre of its kind, well supported by a range of research grants. He holds a number of honours, including the Guy Medal in Silver from the Royal Statistical Society. He claims not to have wholly discarded a radical enthusiasm that led him to San Francisco in 1966, to Paris in support of the students in 1968 and to stand against the council candidate to become the President of the Royal Statistical Society at the age of 32.

Preface

The present volume is a collective monograph devoted to applications of the optimal design theory in optimization and statistics. The chapters reflect the topics discussed at the workshop *"W-Optimum Design and Related Statistical Issues"* that took place in Juan-les-Pins, France, in May 2005. The title of the workshop was chosen as a light-hearted celebration of the work of Henry Wynn. It was supported by the Laboratoire I3S (CNRS/Université de Nice, Sophia Antipolis), to which Henry is a frequent visitor.

The topics covered partly reflect the wide spectrum of Henry's research interests. Algorithms for constructing optimal designs are discussed in Chap. 1, where Henry's contribution to the field is acknowledged. Steepest-ascent algorithms used to construct optimal designs are very much related to general gradient algorithms for convex optimization. In the last ten years, a significant part of Henry's research was devoted to the study of the asymptotic properties of such algorithms. This topic is covered by Chaps. 2 and 3. The work by Alessandra Giovagnoli concentrates on the use of majorization and stochastic ordering, and Chap. 4 is a hopeful renewal of their collaboration. One of Henry's major recent interests is what is now called algebraic statistics, the application of computational commutative algebra to statistics, and he was partly responsible for introducing the experimental design sub-area, reviewed in Chap. 5. One other sub-area is the application to Bayesian networks and Chap. 6 covers this, with Chap. 7 being strongly related. Chapters 8 and 9 focus on nonlinear regression, a topic with strong links to both design and optimization.

We hope that the volume will be of interest to both the specialist in the areas covered and also tempt the non-expert. Although several papers are in the nature of a review they all contain substantial new material and are essentially the beginning of new fields needing a continuing research effort.

The editors are grateful to Rebecca Haycroft for her help in the preparation of the volume.

Sophia Antipolis, France — *Luc Pronzato*
Cardiff, UK — *Anatoly Zhigljavsky*

List of Contributors

Peter E. Caines
Department of Electrical
and Computer Engineering
McGill University
Montreal, Canada
peterc@cim.mcgill.ca

Rob Deardon
Department of Mathematics
and Statistics, University of Guelph
Guelph, ON, Canada
rdeardon@uoguelph.ca

Alessandra Giovagnoli
Department of Statistical Sciences
University of Bologna
Via Belle Arti 41
Bologna 40126, Italy
alessandra.giovagnoli@unibo.it

Rebecca Haycroft
School of Mathematics
Cardiff University
Senghennydd Road
Cardiff CF24 4AG, UK
haycroftRJ@cf.ac.uk

Alexandr Ivanov
Kyiv Polytechnic Institute
National Technical University
37 Peremogy Avenue
03056 Kyiv, Ukraine
Ivanov@paligora.kiev.ua

Nikolai Leonenko
School of Mathematics
Cardiff University
Senghennydd Road
Cardiff CF24 4AG, UK
LeonenkoN@cardiff.ac.uk

Johnny Marzialetti
Department of Statistical Sciences
University of Bologna
Via Belle Arti 41, Bologna 40126
Italy
marzialetti@stat.unibo.it

Andrej Pázman
Department of Applied Mathematics
and Statistics
Faculty of Mathematics
Physics and Informatics
Comenius University
84248 Bratislava, Slovakia
pazman@center.fmph.uniba.sk

Giovanni Pistone
Department of Mathematics
Politecnico di Torino
Corso Duca degli Abruzzi, 24
10129 Torino
Italy
giovanni.pistone@polito.it

Luc Pronzato
Laboratoire I3S, CNRS – UNSA
Les Algorithmes – Bât. Euclide B
2000 route des Lucioles
B.P. 121
06903 Sophia Antipolis
France
pronzato@i3s.unice.fr

Eva Riccomagno
Dipartimento di Matematica
Università degli Studi di
Genova
Via Dodecaneso 35
16149 Genova
Italia
riccomag@dima.unige.it

Maria Piera Rogantin
Dipartimento di Matematica
Università degli Studi di
Genova
Via Dodecaneso 35
16149 Genova
Italia
rogantin@dima.unige.it

Jim Q. Smith
Department of Statistics
The University of Warwick
Gibbet Hill Road
Coventry CV4 7AL, UK
J.Q.Smith@warwick.ac.uk

Ben Torsney
Department of Statistics
University of Glasgow
15 University Gardens
Glasgow G12 8QW UK
bent@stats.gla.ac.uk

Henry P. Wynn
London School of Economics
and Political Science
Houghton Street
London WC2A 2AE UK
h.wynn@lse.ac.uk

Anatoly Zhigljavsky
Cardiff University
School of Mathematics
Senghennydd Road
Cardiff CF24 4AG UK
zhigljavskyAA@cf.ac.uk

Contents

1

W-Iterations and Ripples Therefrom

B. Torsney

Summary. The focus of this contribution is on algorithms for constructing optimal designs. Henry Wynn, one of the earliest contributors in this field, inspired David Silvey to point me in the direction of further algorithmic developments. I first heard of this topic at a seminar David gave at University College London in the autumn of 1970 in which he spoke of Henry's work (he had been Henry's external examiner). Henry also attended this seminar! The rest is history. In this chapter, Henry's work and ripples therefrom will be explored.

1.1 Introduction

The content of the chapter is as follows: in Sect. 1.2, the optimal design problem is reviewed and a general optimal weights problem defined; in Sects. 1.4 and 1.5, algorithms are described, including Henry's method, a multiplicative iteration and steepest ascent; in Sect. 1.6, a new approach to determining optimal designs, exact or approximate, is proposed based on transforming design points on a finite interval to proportions of that interval, so that methods for determining optimal weights can be exploited.

1.2 Optimal Design and an Optimization Problem

1.2.1 Optimal Design

We first summarize optimal design theory. It is necessary to exploit this tool if, to obtain an observation on a response variable y, we must first choose a "value" for a vector $\underline{x}$ of design variables from some design space $X \subset \mathbb{R}^1$. To be precise we assume the following general model:

$$y \sim p(y|\underline{x}, \underline{\beta}), \;\; \underline{x} \in X \;\text{ where }\; X \subset \mathbb{R}^1, \underline{\beta} \subset \mathbb{R}^k.$$

The exact design problem is then that of deciding upon the number $n(\underline{x})$ of observations that should be taken at $\underline{x}$, given the total number, say N, of runs.

L. Pronzato, A. Zhigljavsky (eds.), *Optimal Design and Related Areas in Optimization and Statistics*, Springer Optimization and its Applications 28,
DOI 10.1007/978-0-387-79936-0_1, © Springer Science+Business Media LLC 2009

This is a difficult problem to solve and one that needs to be solved for each value of N. The approximate design problem, a simpler problem than that of the exact design, entails the calculation of the proportion $p(\underline{x})$ of observations that are required to be taken at $\underline{x}$. For a discrete set $X = \{\underline{x}_1, \underline{x}_2, \ldots, \underline{x}_J\}$, p must define a discrete probability distribution (design measure), say

$$\begin{array}{cc} \underline{x} & \underline{x}_1\ \underline{x}_2 \cdots \underline{x}_J \\ p(\underline{x}) & p_1\ p_2 \cdots p_J \end{array}$$

where p_j define design weights and thus have the property: $p_j \geq 0, \ \sum p_j = 1$. Only this finite dimensional case is considered in what follows.

A general design objective is to ensure good estimation, in some form, of the parameters. To obtain a good estimation of all parameters in $\underline{\beta}$ we wish to make the covariance matrix, $\mathrm{Cov}(\hat{\underline{\beta}})$, of their maximum likelihood estimators "small". Appealing to asymptotic theory of maximum likelihood estimation we presume that $\mathrm{Cov}(\hat{\underline{\beta}}) \propto M^{-1}(p)$ where

$$M(p) = \sum p_j I_{\underline{x}_j}$$

is the per observation Fisher information matrix with

$$I_{\underline{x}} = E(\underline{v}\underline{v}^{\mathrm{T}}), \ \ \underline{v} = \partial \log p(y|\underline{x}, \underline{\beta})/\partial \underline{\beta}.$$

If $E(y|\underline{x}, \underline{\beta}) = \underline{\beta}^{\mathrm{T}} \underline{f}(\underline{x})$ and $\mathrm{Var}(y|\underline{x}, \underline{\beta}) = \sigma^2$, then $\underline{v} = \underline{f}(\underline{x})$ is independent of $\underline{\beta}$, and likewise $I_{\underline{x}}$.

We wish to make $M(p)$ large. There are various criteria which we might maximize, namely:

- D-optimality: for which $\Psi\{M\} = \log\det(M)$
- A-optimality: for which $\Psi\{M\} = -\mathrm{trace}(M^{-1})$
- c-optimality: for which $\Psi\{M\} = -\underline{c}^{\mathrm{T}} M^{-1} \underline{c}, \ \underline{c} \in \mathbb{R}^k$

We thus have an example of the following type of problem:

Problem (P) *Maximize a criterion $\phi(p)$, $p = (p_1, p_2, \ldots, p_J)$, subject to $p_i \geq 0, \ \sum p_i = 1$, where the components of p define weights or probabilities.*

There are many situations in which this problem arises in addition to optimal design problems. These include various maximum likelihood estimation problems (e.g. estimating mixing weights given data from a mixture; also latent variable models for paired comparison and ranking data), problems in survey sampling and problems in image analysis.

1.3 Derivatives and Optimality Conditions

1.3.1 Derivatives

It is necessary to be clear about the conditions of optimality. A useful tool for expressing these is the directional derivative of ϕ at p in the direction of q, in other words:

$$F_\phi(p,q) = \partial\phi[(1-\epsilon)p + \epsilon q]/\partial\epsilon \,|_{\epsilon=0+} \,.$$

If $\phi(.)$ is differentiable and $d = \partial\phi/\partial p$ then

$$F_\phi(p,q) = (q-p)^{\mathrm{T}} d = q^{\mathrm{T}} d - p^{\mathrm{T}} d.$$

Let e_j be the jth unit vector in $\mathbb{R}^J$, $d_j = \partial\phi/\partial p_j$ and

$$F_j = F_\phi(p, e_j) = d_j - p^{\mathrm{T}} d = d_j - \sum p_i d_i \,.$$

We define F_j to be the jth vertex-directional derivative of $\phi(.)$ at p. Note $\sum p_j F_j = 0$; i.e. unless all are zero some F_j are negative and some are positive.

1.3.2 Optimality Conditions for Problem (P)

We can now assert that if $\phi(.)$ is differentiable at p^*, a necessary condition for $\phi(p^*)$ to be a local maximum is

$$F_j^* = \begin{cases} F_\phi\{p^*, e_j\} = 0 \text{ for } p_j^* > 0\,, \\ F_\phi\{p^*, e_j\} \leq 0 \text{ for } p_j^* = 0\,. \end{cases}$$

Furthermore if $\phi(.)$ is concave then this is both a necessary and sufficient condition. This is known as the General Equivalence Theorem in optimal design; see Kiefer (1974) and Whittle (1973).

1.4 Algorithms

It is almost always impossible to determine optimal designs explicitly. Various classes of algorithms for construction of optimal designs have been formulated over the years. We introduce these in roughly chronological order.

1.4.1 The Wynn (Vertex-Direction) Method

Vertex-direction methods were one of the first class of methods and include Henry's earliest contributions to design theory. Others such as Andrej Pázman, Valeri Fedorov and Ivan Vuchkov have also contributed. Essentially these methods change one weight at a time. The iteration formula is

$$p^{(r+1)} = (1-\alpha_r)p^{(r)} + \alpha_r e_M \,,$$

where M is the value of j satisfying $F_M = \max_j\{F_j\}$ and α_r is a step-length. Hence, apart from a proportional change in the other weights, $p^{(r+1)}$ changes only the value of p_M; or "assigns" additional weight to $\underline{v}_M = \underline{f}(\underline{x}_M)$ through taking a step α_r from $p^{(r)}$ to e_M. This is a step towards the vertex (labelled M) which has the largest vertex-directional derivative, $F_\Phi\{p^{(r)}, q\}$ in any direction q. It is also the direction corresponding to the largest derivative in any direction. It thus has the flavour of a steepest-ascent method, but it is not precisely of that form as is noted later. A common choice of step-length is $\alpha_r \sim 1/(k+r)$; see Wynn (1972). There can however be other choices. In their paper, Wu and Wynn (1978) proved convergence under the conditions: $\alpha_r \to 0, \sum \alpha_r < \infty$. In the case of D-optimality there is an explicit solution for the optimal step-length; see Fedorov (1972). Usually $p^{(0)}$ assigns weight $(1/k)$ to k linearly independent design points.

Example 1. Wynn (1972) considered a linear model with linear, quadratic and trigonometric regression functions, namely:

$$E(y|\underline{x}, \underline{\beta}) = \underline{\beta}^{\mathrm{T}} \underline{f}(\underline{x}),\ \underline{x} = x \in X = [0,1]$$

$$\underline{f}(\underline{x}) = (x, x^2, \sin 2\pi x, \cos 2\pi x)^{\mathrm{T}},\ k = 4.$$

He discretized the design interval to:

$$x \in X_D = [0, 0.01, 0.02, .., 0.99, 1]$$

on which the D-optimal design, as found by the algorithm, is

x	0.08	0.09	0.38	0.73	0.74	1.00
$p_D^*(x)$	0.22	0.03	0.25	0.17	0.08	0.25

There are six support points, comprizing two with weight 0.25 and two pairs or clusters of neighbouring points, each with total weight 0.25.

To obtain a "closer" approximation to the optimal design on X we can follow Atwood (1976). He recommends replacing each of these pairs by a single design point formed from a convex combination of the two original points in the pair; the convex weights are to be proportional to their design weights. The point thus formed is given the total weight of the pair. This yields the design

x	0.0812	0.38	0.7332	1.00
$p^*(x)$	0.25	0.25	0.25	0.25

Note: We know that there must be an optimal design with at least 4 and at most 6 support points. In consequence there are at least 95 non-support points or zero optimal weights on X_D.

The design problem would be greatly simplified if we knew what these points were, or equivalently if we knew $\mathrm{Supp}(p^*)$ or $\mathrm{Supp}(p_D^*)$, where

$$\mathrm{Supp}(p) = \{x \in X : p(x) > 0\}.$$

A more realistic objective is to ascertain whether or not it is possible to identify most of the non-support points or at least narrow down $\mathrm{Supp}(p_D^*)$.

In an effort to shed light on this matter the following empirical investigation was reported by Torsney (1983).

The algorithm was started from 112 different initial $p^{(0)}$'s. Each $p^{(0)}$ assigned weight (1/4) to four design points in X_D. These included the four smallest and four largest values in X_D. Recorded below is the value of r (iteration count) beyond which the algorithm assigned "additional" weight only to the 13 design points in:

$$X_S = \{.07, .08, .09, .10,\ \ .37, .38, .39,\ \ .72, .73, .74, .75,\ \ .99, 1.00\}\,.$$

This set comprises the D-optimal design support points on X_D plus their immediate neighbours in X_D. The resultant frequency distribution of r is

r	1	3	4	5	6	7	8	9	10	11	12	14
$f(r)$	1	5	1	20	11	26	9	12	15	10	1	1

The average value of r was 7.4, approximately 8 which is the value of $2k$.

For the natural initial design:

x	0	0.33	0.67	1.00
$p^{(0)}(x)$	0.25	0.25	0.25	0.25

$r = 5$, approximately 4 which is the value of k. Similar results are observed for polynomial regressions. This is evidence that these algorithms can quickly "identify" (by ignoring) most non-support points; i.e. those which are not "immediate" neighbours of the members of Supp(p_D^*); see Torsney (1983); see also Harman and Pronzato (2007) and Pronzato (2003) on removing non-optimal support points in D-optimal design algorithms.

1.4.2 Multiplicative Algorithms

Now we move on to the ripples flowing from Henry's work. These update all weights "simultaneously" in some manner. In particular, the next class of algorithms updates weights in a multiplicative fashion. These algorithms have emerged in part from David Silvey's perception.

Let $f(d, \delta)$ be a function satisfying (for $\delta > 0$) the following three properties:

$$f(d,\delta) > 0;\ \ \partial f(d,\delta)/\partial d > 0 \text{ (for } \delta \geq 0);\ \ f(d,0) = \text{const}\,.$$

Examples of functions of this kind include $f(d, \delta) = \Phi(\delta d)$ and $f(d, \delta) = d^\delta$ (if $d > 0$), where $\Phi(z)$ is the c.d.f. of the standard normal distribution.

Multiplicative algorithms have the form:

$$p_j^{(r+1)} = \frac{p_j^{(r)} f(d_j^{(r)}, \delta)}{\sum p_i^{(r)} f(d_i^{(r)}, \delta)} \qquad \text{or} \qquad p_j^{(r+1)} = \frac{p_j^{(r)} f(F_j^{(r)}, \delta)}{\sum p_i^{(r)} f(F_i^{(r)}, \delta)}\,.$$

This rule clearly ensures the key constraints of $p_j^{(r)} > 0$, $\sum p_i^{(r)} = 1$. This and other properties include (see Mandal and Torsney (2000), Torsney and Alahmadi (1992)):

1. $p^{(r)}$ is always feasible.
2. $F_\phi\{p^{(r)}, p^{(r+1)}\} \geq 0$, with equality when the d_j's corresponding to non-zero p_j's have a common value $d(=\sum p_i d_i)$, in which case $p^{(r)} = p^{(r+1)}$.
3. If $\delta = 0$ there is no change in $p^{(r)}$ given $f(d, 0) =$ constant.
4. The algorithm is monotonic for sufficiently small positive δ.
5. An iterate $p^{(r)}$ is a fixed point of the iteration if derivatives $d_j^{(r)}$ corresponding to non-zero $p_j^{(r)}$ are equal; i.e. if corresponding vertex-directional derivatives $F_j^{(r)}$ are zero.

Let $f_j^{(r)} = f(d_j^{(r)}, \delta)$ or $f(F_j^{(r)}, \delta)$. Under the condition of point 5 the $f_j^{(r)}$ corresponding to non-zero $p_j^{(r)}$ share a common value; hence there is no change in weights.

A proof of the second statement follows on noting that:

$$F_\Phi(p^{(r)}, p^{(r+1)}) = \sum (p_j^{(r+1)} - p_j^{(r)}) d_j^{(r)}$$
$$= \frac{\left[\left(\sum p_j^{(r)} f_j^{(r)} d_j^{(r)}\right) - \left(\sum p_j^{(r)} f_j^{(r)}\right)\left(\sum p_j^{(r)} d_j^{(r)}\right)\right]}{\left(\sum p_j^{(r)} f_j^{(r)}\right)},$$

so

$$F_\phi\left\{p^{(r)}, p^{(r+1)}\right\} = \frac{\text{Cov}\{D, f(D, \delta)\}}{E\{f(D, \delta)\}},$$

where D is a discrete random variable with probability distribution $P(D = d_j^{(r)}) = p_j^{(r)}$. It can then be argued that $\text{Cov}\{D, f(D, \delta)\} > 0$ if $\partial f(D, \delta)/\partial D > 0$, while $E\{f(D, \delta)\} > 0$ if $f(D, \delta) > 0$.

1.4.2.1 History

This class of algorithms has evolved from a result of Fellman (1974). The result was devoted to linear criteria, not algorithms, but in effect he proved that $f(d, \delta) = d^\delta$ with $\delta = 1/2$ yields monotonicity for c-optimality. (It was David Silvey who first spotted this.) Torsney (1983) extended this result to A-optimality, while Titterington (1976) proved monotonicity of $f(d, \delta) = d^\delta$ with $\delta = 1$ for D-optimality. This latter choice is also monotonic for finding the maximum likelihood estimators of the mixing weights, given data from a mixture of distributions. It is indeed an EM algorithm; see Torsney (1977). Both choices also appear to be monotonic in determining c-optimal and D-optimal conditional designs respectively, i.e. in determining several optimizing distributions. This type of problem is discussed below. Finally the paper (Silvey et al., 1978) is an empirical study of the choice $f(d, \delta) = d^\delta$. Note that this is a feasible choice for standard design criteria and the mixture likelihood since they enjoy positive partial derivatives.

Other choices of $f(d,\delta)$ are needed if $\phi(p)$ can have negative derivatives, as in some maximum likelihood estimation problems, or if the partial derivatives d are replaced by the vertex-directional derivatives F. Torsney (1988) considers the case of $f(d,\delta) = \mathrm{e}^{\delta d}$, while objective bases for choosing $f(d,\delta)$ are to be found in Torsney and Alahmadi (1992) and Torsney and Mandal (2004, 2007).

These multiplicative algorithms could be viewed as possible sequels to vertex-direction ones; i.e. we might switch after a few iterations of the former to the latter. They too however, reveal, early on, strong indications of potential support and non-support points in plots of weights or vertex-directional derivatives against design variables. Mandal and Torsney (2006) respond to this phenomenon by proposing a clustering approach. They suggest splitting the design region into subsets centred on the peaks of these plots. They propose determining conditional designs within these subsets and marginal designs across them. This generates an optimization problem with respect to several distributions. The multiplicative algorithm readily extends to this case. Other problems requiring determination of several distributions include determining conditional designs when some variables are under control and some are not, determining maximum likelihood estimators under latent variable models for paired comparison and ranking data when treatments have a factorial structure or when the data are multivariate.

Finally Torsney and Mandal (2001), Mandal et al. (2005) and Torsney and Alahmadi (1995) use these algorithms to determine constrained optimal designs.

1.4.2.2 Implementation

Consider the problem of determining an appropriate choice of δ. We consider first the case of $d > 0$ and $f(d,\delta) = d^{\delta}$. Some simple examples are quite illustrative. Suppose that

$$\phi(p) = \{\log(p_1) + \cdots + \log(p_J)\}\ ,$$

which is a simple example of a D-optimal criterion, with $d_j = 1/p_j$. Then $p_j^* = 1/J$. The algorithm is not needed, but consider its implementation for three values of δ:

- If $\delta = 0$ there are no changes in weights.
- If $\delta = 1$ the optimum is attained in 1 step.
- If $\delta = 2$ a perfect oscillation is realized.

This suggests (monotonic) convergence for $\delta \leq 1$; slower to non-convergence for $1 \leq \delta < 2$; and divergence for $\delta \geq 2$.

Suppose now that

$$\phi(p) = -\{a_1(p_1)^{-t} + \cdots + a_J(p_J)^{-t}\},\ \ a_j > 0.$$

This is a simple example of the "trace$\{M^{-t}\}$ criterion", with

$$d_j = t\{a_j/(p_j)^{-(t+1)}\}.$$

Again there is an explicit solution, namely:

$$p_j^* = b_j/(b_1 + \cdots + b_J),\ b_j = \{a_j\}^{1/(t+1)}.$$

Consider the implementation of the algorithm for the three values of δ:

- If $\delta = 0$ there is no change in weights.
- If $\delta = 1/(t+1)$ the optimum is attained in 1 step.
- If $\delta = 2/(t+1)$ a perfect oscillation is realized.

In general, this suggests (monotonic) convergence for $\delta \le 1/(t+1)$; slower to non-convergence for $1/(t+1) \le \delta < 2/(t+1)$; and divergence for $\delta \ge 2/(t+1)$. There are implications here for homogeneous functions of degree $(-t)$. It also suggests $\delta = 1$ for D-optimality since

$$[\mathrm{trace}(M^{-t})]^{1/t} \to [\det(M)]^{-1/k} \text{ as } t \to 0.$$

Consider now the iteration

$$p_j^{(r+1)} = \frac{p_j^{(r)}\{d_j^{(r)} + \gamma\}^\delta}{\sum p_i^{(r)}\{d_i^{(r)} + \gamma\}^\delta}.$$

We have introduced a second parameter γ. One motivation for this would be to transform from negative partial derivatives to positive values in the terms $\{d_j^{(r)} + \gamma\}$. Appropriate values for γ and δ need to be determined.

Consider the criterion $\phi_\gamma(p) = \phi(p) + \gamma(\sum p_i)$. Both $\phi_\gamma(p)$ and $\phi(p)$ are optimized at the same p^*, with $d_j^{(\gamma)} = \partial\phi_\gamma/\partial p_j = (\partial\phi/\partial p_j + \gamma) = (d_j + \gamma)$. We note that for linear models with a constant term the D-optimal criterion for all parameters and the D-optimal criterion for all parameters except the constant term are related in this way for $\gamma = 1$.

Note that in models with a constant term the first diagonal element of the per observation information matrix is 1. It follows from basic results on partitioned matrices that the two criteria are equivalent. This immediately offers two options for derivatives. See Titterington (1975) for the result on derivatives. These results are extended to the case when some model terms are not subject to control in Martin-Martin et al. (2007).

What might be the value of γ? The latter observation suggests two choices, $\gamma = 0$ or 1. More generally we might choose γ such that $d_j^{(\gamma)} > 0$; for example, take $\gamma = \min\{d_j\} + \epsilon, \epsilon > 0$. If there are zero optimal weights, small ϵ should yield fast initial convergence.

For other choices of $f(d,\delta)$, such as $f(d,\delta) = \Phi(\delta d)$, further work is needed. It is sensible to take $d = F$. Then certainly the value of $|F\delta|$ must not get too large. One possible recommendation is $\delta = \min\{1, 1/\max(|F|)\}$ or $\delta = \min\{1/2, 1/\max(|F|)\}$.

1.5 A Steepest-Ascent Algorithm

We have already stated that a vertex-direction algorithm has the flavour of a steepest-ascent method, but it is not precisely of that form. It has this flavour because

$$F_M = \max_j \{F_j\} = \max_j \{F_\phi(p, e_j)\} = \max_q \{F_\phi(p, q)\}$$

where the latter maximization is over all distributions q; that is, the maximizing distribution e_M is concentrated at a single point. Note, however, that $F_\phi(p, q)$ depends on $m = (p - q)$.

Wu (1978) considered the normalized directional derivative

$$F_\phi^{(A)}(p, q) = \frac{F_\phi(p, q)}{\sqrt{m^{\mathrm{T}} A m}} \ .$$

If now q_M maximizes $F_\phi^{(A)}(p, q)$ over all distributions q, in general q_M is not necessarily concentrated at a single point. The optimizing value of $m = q - p$ is the explicit solution of a quadratic programming problem.

Wu considers the iteration

$$p^{(r+1)} = (1 - \alpha_r) p^{(r)} + \alpha_r q_M^{(r)} \ .$$

This takes a step of length α_r from $p^{(r)}$ to $q_M^{(r)}$. Wu chooses the step-length optimally. This might be described as a "Henry-high" wave ripple! It emerges that we obtain the multiplicative algorithm with $f(d, \delta) = d^\delta$ and $\delta = 1$, when

$$A = \{\mathrm{diag}(p_1, \ldots, p_J)\}^{-1} \ .$$

Wu (1978) considers the same set of examples as Silvey et al. (1978). These are successive papers in a special issue of Communications of Statistics devoted to design.

1.6 Simultaneous Approach to Optimal Weight and Support Point Determination

We now propose a new idea (a final ripple) to deal with the problem of non-support points. Consider a k-parameter linear model with one design variable x, to be selected from a design interval $X = [a, b]$. A simple case is the determination of the best k-point D-optimal design, which will often be the D-optimal design. Let $x_1, x_2, \ldots, x_k$ be its support points. Their D-optimal weights must be $p_j^* = 1/k$. We need only to determine the support points.

Let:

$$\begin{aligned} W_1 &= (x_1 - a)/(b - a) \\ W_2 &= (x_2 - x_1)/(b - a) \\ &\vdots \\ W_k &= (x_k - x_{k-1})/(b - a) \\ W_{k+1} &= (b - x_k)/(b - a). \end{aligned}$$

We have transformed from k variables to $(k+1)$ variables, but these must satisfy

$$W_h \geq 0, \quad \sum W_h = 1 \, .$$

Now consider maximizing the D-optimal criterion with respect to $W_1, W_2, \ldots,$ W_{k+1}, subject to these constraints. We have another example of Problem (P), but one in which the criterion can have positive and negative partial derivatives.

In Wynn's quadratic/trigonometric example, for which $k = 4$, we used the following choice of multiplicative algorithm : $f(F, \delta) = \Phi(\delta F), \delta = 0.1$. The four optimizing support points prove to be $\{0.0826, 0.3809, 0.7344, 1.000\}$, confirming the D-optimal design on $X = [0, 1]$. We can extend the approach to determining the best equally weighted L-point D-optimal design. For $L = 5, 6$ the two sets of support points, respectively, are

$$\{0.04825, 0.24379, 0.45004, 0.74607, 1.00000\}$$

and

$$\{0.03834, 0.20662, 0.41211, 0.70573, 0.76375, 1.00000\} \, .$$

Finally the idea extends to finding the D-optimal design through considering the problem of determining the best L-point D-optimal design. Normally we do not know the number, say L, of support points of the D-optimal design. We know only that $k \leq L \leq k(k+1)/2$. If we take $L = k(k+1)/2$ these two optimal designs must coincide.

Consider the design

$$\begin{array}{cc} x & x_1 \; x_2 \; x_3 \; \ldots \; x_L \\ p(x) & p_1 \; p_2 \; p_3 \; \ldots \; p_L . \end{array}$$

We wish to determine both the support points and the design weights optimally. Define W_h as above up to $W_{(L+1)} = (b - x_L)/(b - a)$. Our optimization problem transforms to choosing both p and W to maximize the D-optimal criterion subject to

$$p_j \geq 0, \; \sum p_j = 1; \;\; W_h \geq 0, \sum W_h = 1 \, .$$

This is yet another example of an extension of Problem (P) to that of optimizing with respect to more than one distribution. The multiplicative algorithm extends naturally to the two simultaneous multiplicative iterations:

$$p_j^{(r+1)} = \frac{p_j^{(r)} f(d_j^{(r)}, \delta)}{\sum p_i^{(r)} f(d_i^{(r)}, \delta)}, \quad f(d, \delta) = d$$

$$W_h^{(r+1)} = \frac{W_h^{(r)} f(F_h^{(r)}, \delta)}{\sum W_t^{(r)} f(F_t^{(r)}, \delta)}, \quad f(F, \delta) = \Phi(\delta F)$$

for the relevant (directional) derivatives. This is the focus of further work. These promised ripples and those which have gone before are a testament to the influence Henry has had on design theory.

References

Atwood, C.L. (1976). Convergent design sequences for sufficiently regular optimality criteria. *The Annals of Statistics,* **4**, 1124–1138.

Fedorov, V.V. (1972). *Theory of Optimal Experiments.* Academic Press, New York.

Fellman, J. (1974). On the allocation of linear observations. *Commentationes Physico-Mathematicae*, Helsinki, **44**, 27–78.

Harman, R. and Pronzato, L. (2007). Improvements on removing non-optimal support points in D-optimum design algorithms. *Statistics & Probability Letters,* **77**, 90–94.

Kiefer, J. (1974). General equivalence theory for optimum designs (approximate theory). *The Annals of Statistics,* bf 2, 849–879.

Mandal, S. and Torsney, B. (2000). Algorithms for the construction of optimizing distributions. *Communications in Statistics – Theory and Methods,* **29**, 1219–1231.

Mandal, S. and Torsney, B. (2006). Construction of optimal designs using a clustering approach. *Journal of Statistical Planning and Inference,* **136**, 1120–1134.

Mandal, S., Torsney, B., and Carriere, K.C. (2005). Constructing optimal designs with constraints. *Journal of Statistical Planning and Inference,* **128**, 609–621.

Martin Martin, R., Torsney, B., and Lopez-Fidalgo, J. (2007). Construction of marginally and conditionally restricted designs using multiplicative algorithms. *Computational Statistics and Data Analysis,* **51**, 5547–5561.

Pronzato, L. (2003). Removing non-optimal support points in D-optimum design algorithms. *Statistics & Probability Letters,* **63**, 223–228.

Silvey, S.D., Titterington, D.M., and Torsney, B. (1978). An algorithm for optimal designs on a finite design space. *Communications in Statistics A,* **7**, 1379–1389.

Titterington, D.M. (1975). Optimal design: some geometrical aspects of D-optimality. *Biometrika,* **62**, 313–320.

Titterington, D.M. (1976). Algorithms for computing D-optimal designs on a finite design space. *Conference on Information Sciences and Systems,* Department of Electrical Engineering, Johns Hopkins University, Baltimore, MD, 213–216.

Torsney, B. (1977). Contribution to discussion of a paper by Dempster, Laird and Rubin. *Journal of the Royal Statistical Society B,* **39**, 22–27.

Torsney, B. (1983). A moment inequality and monotonicity of an algorithm, *Proceedings of the International Symposium on Semi-Infinite Program and Application* (Kortanek, K.O. and Fiacco, A.V., eds.). University of Texas, Austin. Lecture Notes in Economics and Mathematical System, **215**, 249–260.

Torsney, B. (1988). Computing optimising distributions with applications in design, estimation, and image processing. *Optimal Design and Analysis of Experiments* (Dodge, Y., Fedorov, V.V. and Wynn, H.P., eds.). Elsevier Science Publishers B. V., North Holland, 361–370.

Torsney, B. and Alahmadi, A. (1992). Further development of algorithms for constructing optimizing distributions. *Model Oriented Data Analysis.* Proceedings of the second IIASA Workshop in St. Kyrik, Bulgaria (Fedorov, V.V., Muller, W.G., and Vuchkov, I.N., eds.). Physica-Verlag, Wurzburg (Wien), 121–129.

Torsney, B. and Alahmadi, A. (1995). Designing for minimally dependent observations. *Statistica Sinica,* **5**, 499–514.

Torsney, B. and Mandal, S. (2001). Construction of constrained optimal designs, *Optimum Design 2000,* (Atkinson, A., Bogacka, B., and Zhigljavsky, A.A., eds.). Kluwer, Dordecht, 141–152.

Torsney, B. and Mandal, S. (2004). Multiplicative algorithms for constructing optimizing distributions: further developments. *mODa 7 – Advances in Model Oriented Design and Analysis,* Proceedings of seventh International Workshop on Model Oriented Design and Analysis, Heeze, Netherlands, June 14–18, 2004, (Bucchiano, A.D., Lauter, H. and Wynn, H.P., eds.). Physica-Verlag, Heidelberg, 143–150.

Torsney, B. and Mandal, S. (2007). Two classes of multiplicative algorithms for constructing optimizing distributions. *Computational Statistics and Data Analysis,* **51**(3), 1592–1601.

Whittle, P. (1973). Some general points in the theory of optimal experimental design. *Journal of the Royal Statistical Society B,* **35**, 123–1390.

Wu, C.F.J. (1978). Some iterative procedures for generating non-singular optimal designs. *Communications in Statistics A,* **7**, 1399–1412.

Wu, C.F.J. and Wynn, H.P. (1978). The convergence of general step-length algorithms for regular optimum design criteria. *The Annals of Mathematical Statistics,* **6**, 1273–1285.

Wynn, H.P. (1972). Results in the theory and construction of D-optimum experimental designs (with discussion). *Journal of the Royal Statistical Society B,* **34**, 133–147, 170–186.

2

Studying Convergence of Gradient Algorithms Via Optimal Experimental Design Theory

R. Haycroft, L. Pronzato, H.P. Wynn and A. Zhigljavsky

Summary. We study the family of gradient algorithms for solving quadratic optimization problems, where the step-length γ_k is chosen according to a particular procedure. To carry out the study, we re-write the algorithms in a normalized form and make a connection with the theory of optimum experimental design. We provide the results of a numerical study which shows that some of the proposed algorithms are extremely efficient.

2.1 Introduction

In a series of papers (Pronzato et al., 2001, 2002, 2006) and the monograph (Pronzato et al., 2000), certain types of steepest-descent algorithms in $\mathbb{R}^d$ and Hilbert space have been shown to be equivalent to special algorithms for updating measures on the real line. The connection, in outline, is that when steepest-descent algorithms are applied to the minimization of the quadratic function

$$f(x) = \frac{1}{2}(Ax, x) - (x, y) \,, \tag{2.1}$$

where (x, y) is the inner product, they can be translated to the updating of measures in $[m, M]$ where

$$m = \inf_{\|x\|=1} (Ax, x), \qquad M = \sup_{\|x\|=1} (Ax, x)$$

with $0 < m < M < \infty$; m and M are the smallest and largest eigenvalues of A, respectively.

The research has developed from the well-known result, due to Akaike (1959) and revisited in Pronzato et al. (2000) Forsythe (1968) and Nocedal et al., (2002), that for standard steepest descent the renormalized iterates $x_k/\sqrt{\|x_k\|}$ converge to the two-dimensional space spanned by the eigenvectors corresponding to the eigenvalues m and M. Chapter 3 in this volume covers the generalization to the s-gradient algorithm and draws note on both the author's own work and the important paper of Forsythe (1968).

L. Pronzato, A. Zhigljavsky (eds.), *Optimal Design and Related Areas in Optimization and Statistics*, Springer Optimization and its Applications 28,
DOI 10.1007/978-0-387-79936-0_2,

The use of normalization is crucial and, in fact, it will turn out that the normalized gradient, rather than the normalized x_k will play the central role. Thus, let $g(x) = Ax - y$ be the gradient of the objective function (2.1). The steepest-descent algorithm is

$$x_{k+1} = x_k - \frac{(g_k, g_k)}{(Ag_k, g_k)} g_k \,.$$

Using the notation $\gamma_k = (g_k, g_k)/(Ag_k, g_k)$, we write the algorithm as

$$x_{k+1} = x_k - \gamma_k g_k \,.$$

This can be rewritten in terms of the gradients as

$$g_{k+1} = g_k - \gamma_k A g_k \,. \tag{2.2}$$

This is the point at which the algorithms are generalized: we now allow a varied choice of γ_k in (2.2), not necessarily that for steepest descent.

Thus, the main objective of the chapter is studying the family of algorithms (2.2) where the step-length γ_k is chosen in a general way described below. To make this study we first rewrite the algorithm (2.2) in a different (normalized) form and then make a connection with the theory of optimum experimental design.

2.2 Renormalized Version of Gradient Algorithms

Let us convert (2.2) to a "renormalized" version. First note that

$$(g_{k+1}, g_{k+1}) = (g_k, g_k) - 2\gamma_k (Ag_k, g_k) + \gamma_k^2 (A^2 g_k, g_k) \,. \tag{2.3}$$

Letting $v_k = (g_{k+1}, g_{k+1})/(g_k, g_k)$ and dividing (2.3) through by (g_k, g_k) gives

$$v_k = 1 - 2\gamma_k \frac{(Ag_k, g_k)}{(g_k, g_k)} + \gamma_k^2 \frac{(A^2 g_k, g_k)}{(g_k, g_k)} \,. \tag{2.4}$$

The value of v_k can be considered as a rate of convergence of algorithm (2.2) at iteration k. Other rates which are asymptotically equivalent to v_k can be considered as well, see Pronzato et al. (2000) for a discussion and Theorem 5 in Chapter 3 of this volume. The asymptotic rate of convergence of the gradient algorithm (2.2) can be defined as

$$R = \lim_{k \to \infty} \left(\prod_{i=1}^{k} v_i \right)^{1/k} \,. \tag{2.5}$$

Of course, this rate may depend on the initial point x_0 or, equivalently, on g_0.

To simplify the notation, we need to convert to moments and measures. Since we assume that A is a positive definite d-dimensional square matrix, we can assume, without loss of generality, that A is a diagonal matrix $\Lambda = \mathrm{diag}(\lambda_1, \ldots, \lambda_d)$; the elements $\lambda_1, \ldots, \lambda_d$ are the eigenvalues of the original matrix such that $0 < \lambda_1 \leq \cdots \leq \lambda_d$. Then for any vector $g = (g_{(1)}, \ldots, g_{(d)})^{\mathrm{T}}$ we can define

$$\mu_\alpha(g) = \frac{(A^\alpha g, g)}{(g, g)} = \frac{(\Lambda^\alpha g, g)}{(g, g)} = \frac{\sum_i g_{(i)}^2 \lambda_i^\alpha}{\sum_i g_{(i)}^2} .$$

This can be seen as the α-th moment of a distribution with mass $p_i = g_{(i)}^2 / \sum_j g_{(j)}^2$ at λ_i, $i = 1, \ldots, d$. This remark is clearly generalizable to the Hilbert space case.

Using the notation $\mu_\alpha^{(k)} = \mu_\alpha(g_k)$, where g_k are the iterates in (2.2), we can rewrite (2.4) as

$$v_k = 1 - 2\gamma_k \mu_1^{(k)} + \gamma_k^2 \mu_2^{(k)} . \tag{2.6}$$

For the steepest-descent algorithm γ_k minimizes $f(x_k - \gamma g_k)$ over γ and we have

$$\gamma_k = \frac{1}{\mu_1^{(k)}} \quad \text{and} \quad v_k = \frac{\mu_2^{(k)}}{\mu_1^{(k)2}} - 1 .$$

Write $z_k = g_k / \sqrt{(g_k, g_k)}$ for the normalized gradient and recall that $p_i = g_{(i)}^2 / \sum_j g_{(j)}^2$ is the ith probability corresponding to a vector g. The corresponding probabilities for the vectors g_k and g_{k+1} are

$$p_i^{(k)} = \frac{(g_k)_{(i)}^2}{(g_k, g_k)} \quad \text{and} \quad p_i^{(k+1)} = \frac{(g_{k+1})_{(i)}^2}{(g_{k+1}, g_{k+1})} \quad \text{for } i = 1, \ldots, d .$$

Now we are able to write down the re-normalized version of (2.2), which is the updating formula for p_i ($i = 1, \ldots, d$):

$$\begin{aligned} p_i^{(k+1)} &= \frac{(1 - \gamma_k \lambda_i)^2}{(g_k, g_k) - 2\gamma_k (A g_k, g_k) + \gamma_k^2 (A^2 g_k, g_k)} \, p_i^{(k)} \\ &= \frac{(1 - \gamma_k \lambda_i)^2}{1 - 2\gamma_k \mu_1^{(k)} + \gamma_k^2 \mu_2^{(k)}} \, p_i^{(k)} . \end{aligned} \tag{2.7}$$

When two eigenvalues of A are equal, say $\lambda_j = \lambda_{j+1}$, the updating rules for $p_j^{(k)}$ and $p_{j+1}^{(k)}$ are identical so that the analysis of the behaviour of the algorithm remains the same when $p_j^{(k)}$ and $p_{j+1}^{(k)}$ are confounded. We may thus assume that all eigenvalues of A are distinct.

2.3 A Multiplicative Algorithm for Optimal Design

Optimization in measure spaces covers a variety of areas and optimal experimental design theory is one of them. These areas often introduce algorithms which typically have two features: the measures are re-weighted in some way and the moments play an important role. Both features arise, as we have seen, in the above algorithms.

In classical optimal design theory for polynomial regression (see, e.g. Fedorov (1972)) one is interested in functionals of the moment (information) matrix $M(\xi)$ of a design measure ξ:

$$M(\xi) = \{m_{ij} : m_{ij} = \mu_{i+j}; 0 \le i, j \le K-1\} ,$$

where $\mu_\alpha = \mu_\alpha(\xi) = \int x^\alpha \mathrm{d}\xi(x)$ are the α-th moments of the measure ξ and K is an integer. For example, when $K = 2$, the case of most interest here, we have

$$M(\xi) = \begin{pmatrix} \mu_0 & \mu_1 \\ \mu_1 & \mu_2 \end{pmatrix} , \tag{2.8}$$

where $\mu_0 = 1$.

Of importance in the theory is the directional (Fréchet) derivative "towards" a discrete measure ξ_x of mass 1 at a point x. This is

$$\frac{\partial}{\partial\alpha}\Phi\left(M\left[(1-\alpha)\xi + \alpha\xi_x\right]\right)\Big|_{\alpha=0} = \mathrm{tr}\left(\overset{\circ}{\Phi}(\xi)M(\xi_x)\right) - \mathrm{tr}\left(\overset{\circ}{\Phi}(\xi)M(\xi)\right) , \tag{2.9}$$

where

$$\overset{\circ}{\Phi}(\xi) = \frac{\partial\Phi}{\partial M}\Big|_{M=M(\xi)} .$$

Here Φ is a functional on the space of $K \times K$ matrices usually considered as an optimality criterion to be maximized with respect to ξ. The first term on the right-hand side of (2.9) is

$$\varphi(x,\xi) = f^{\mathrm{T}}(x)\,\overset{\circ}{\Phi}(\xi)f(x) \tag{2.10}$$

where $f(x) = (1, x, \ldots, x^{K-1})^{\mathrm{T}}$.

A class of optimal design algorithms is based on the multiplicative updating of the weights of the current design measure $\xi^{(k)}$ with some function of $\varphi(x,\xi)$, see Chapter 1 in this volume. We show below how algorithms in this class are related to the gradient algorithms (2.2) in their re-normalized form (2.7).

Assume that our measure is discrete and concentrated on $[m, M]$. Assume also that $\partial\Phi(M)/\partial\mu_{2K-2} > 0$; that is, the (K, K)-element of the matrix $\partial\Phi(M)/\partial M$ is positive. Then $\varphi(x,\xi)$ has a well-defined minimum

$$c(\xi) = \min_{x\in\mathbb{R}} \varphi(x,\xi) > -\infty .$$

Let $\xi(x)$ be the mass at a point x and define the re-weighting at x by

$$\xi^{'}(x) = \frac{\varphi(x,\xi) - c(\xi)}{b(\xi)}\,\xi(x)\,, \tag{2.11}$$

where $b(\xi)$ is a normalizing constant

$$\begin{aligned} b(\xi) &= \int_m^M (\varphi(x,\xi) - c(\xi))\xi(\mathrm{d}x) = \int_m^M \varphi(x,\xi)\xi(\mathrm{d}x) - c(\xi) \\ &= \mathrm{tr}\left[M(\xi)\,\mathring{\Phi}(\xi)\right] - c(\xi)\,. \end{aligned}$$

We see that the first term on the left-hand side is (except for the sign) the second term in the directional derivative (2.9). We can also observe that the algorithm (2.11), considered as an algorithm for constructing Φ-optimal designs, belongs to the family of algorithms considered in Chapter 1 of this volume.

We now specialize to the case where $K = 2$. In this case,

$$f(x) = \begin{pmatrix} 1 \\ x \end{pmatrix}, \quad \mathring{\Phi}(\xi) = \begin{pmatrix} \frac{\partial\Phi}{\partial\mu_0} & \frac{1}{2}\frac{\partial\Phi}{\partial\mu_1} \\ \frac{1}{2}\frac{\partial\Phi}{\partial\mu_1} & \frac{\partial\Phi}{\partial\mu_2} \end{pmatrix}$$

and the function $\varphi(x,\xi)$ is quadratic in x:

$$\varphi(x,\xi) = (\,1\,,\,x\,)\begin{pmatrix} \frac{\partial\Phi}{\partial\mu_0} & \frac{1}{2}\frac{\partial\Phi}{\partial\mu_1} \\ \frac{1}{2}\frac{\partial\Phi}{\partial\mu_1} & \frac{\partial\Phi}{\partial\mu_2} \end{pmatrix}\begin{pmatrix} 1 \\ x \end{pmatrix} = \frac{\partial\Phi}{\partial\mu_0} + x\frac{\partial\Phi}{\partial\mu_1} + x^2\frac{\partial\Phi}{\partial\mu_2}\,. \tag{2.12}$$

Then

$$c(\xi) = \frac{\partial\Phi}{\partial\mu_0} - B(\xi)\,,$$

where

$$B(\zeta) = \frac{1}{4}\left(\frac{\partial\Phi}{\partial\mu_1}\right)^2 \Big/ \left(\frac{\partial\Phi}{\partial\mu_2}\right),$$

and the numerator on the right-hand side of (2.11) is

$$\varphi(x,\xi)-c(\xi) = \frac{\partial\Phi}{\partial\mu_2}\left(x+\frac{1}{2}\frac{\partial\Phi}{\partial\mu_1}\Big/\frac{\partial\Phi}{\partial\mu_2}\right)^2 = B(\xi)\left(1+2\frac{\partial\Phi}{\partial\mu_2}\Big/\frac{\partial\Phi}{\partial\mu_1}\,x\right)^2. \tag{2.13}$$

Let us define $\gamma = \gamma(\xi) = \gamma(\mu_1,\mu_2)$ as

$$\gamma = \gamma(\xi) = -2\frac{\partial\Phi}{\partial\mu_2}\Big/\frac{\partial\Phi}{\partial\mu_1}\,. \tag{2.14}$$

We can then write (2.13) as

$$\varphi(x,\xi) - c(\xi) = B(\xi)\,(1 - \gamma(\xi)x)^2\,. \tag{2.15}$$

Normalization is needed to ensure that the measure ξ' is a probability distribution. We obtain that the re-weighting formula (2.11) can be equivalently written as

$$\xi'(x) = \frac{(1-\gamma x)^2}{1-2\gamma\mu_1+\gamma^2\mu_2}\,\xi(x)\,. \tag{2.16}$$

The main, and we consider surprising, connection that we are seeking to make is that this is exactly the same as the general gradient algorithm in its renormalized form (2.7). To see that, we simply write the updating formula (2.16) iteratively

$$\xi^{(k+1)}(x) = \frac{(1-\gamma_k x)^2}{1-2\gamma_k\mu_1^{(k)}+\gamma_k^2\mu_2^{(k)}}\,\xi^{(k)}(x)\,. \tag{2.17}$$

2.4 Constructing Optimality Criteria which Correspond to a Given Gradient Algorithm

Consider now the reverse construction: given some $\gamma = \gamma(\mu_1, \mu_2)$, construct a criterion $\Phi = \Phi(M(\xi))$, where $M(\xi)$ is as in (2.8), which will give this γ in the algorithm (2.17). In general, this is a difficult question.

The dependence on μ_0 is not important here and we can assume that the criterion Φ is a function of the first two moments only: $\Phi(M(\xi)) = \Phi(\mu_1, \mu_2)$.

The relationship between γ and Φ is given by (2.14) and can be written in the form of the following first-order linear partial differential equation:

$$2\frac{\partial\Phi(\mu_1,\mu_2)}{\partial\mu_2} + \gamma(\mu_1,\mu_2)\,\frac{\partial\Phi(\mu_1,\mu_2)}{\partial\mu_1} = 0\,. \tag{2.18}$$

This equation does not necessarily have a solution for an arbitrary $\gamma = \gamma(\mu_1, \mu_2)$ but does for many particular forms of γ. For example, if $\gamma(\mu_1, \mu_2) = g(\mu_1)h(\mu_2)$ for some functions g and h, then a general solution to (2.18) can be written as

$$\Phi(\mu_1,\mu_2) = F\left(-\int g(\mu_1)\mathrm{d}\mu_1 + 2\int\frac{1}{h(\mu_2)}\mathrm{d}\mu_2\right),$$

where F is an arbitrary continuously differentiable function such that $\partial\Phi\,(\mu_1, \mu_2)/\partial\mu_2 > 0$ for all eligible values of (μ_1, μ_2).

Another particular case is $\gamma(\mu_1, \mu_2) = g(\mu_1)\mu_2 + h(\mu_1)\mu_2^\delta$ for some functions g and h and a constant δ. Then a general solution to (2.18) is

$$\Phi(\mu_1,\mu_2) = F\left(\mu_2^{1-\delta}\,A_1 + \frac{\delta-1}{2}\int h\,(\mu_1)\,A_1\mathrm{d}\mu_1\right)$$

with $A_1 = \exp\{\frac{1}{2}(\delta-1)\int g\,(\mu_1)\,\mathrm{d}\mu_1\}$, where F is as above (a C^1-function such that $\partial\Phi(\mu_1, \mu_2)/\partial\mu_2 > 0$).

2.5 Optimum Design Gives the Worst Rate of Convergence

Let $\Phi = \Phi(M(\xi))$ be an optimality criterion, where $M(\xi)$ is as in (2.8). Associate with it a gradient algorithm with step-length $\gamma(\mu_1, \mu_2)$ as given by (2.14).

Let ξ^* be the optimum design for Φ on $[m, M]$; that is,

$$\Phi(M(\xi^*)) = \max_{\xi} \Phi(M(\xi))$$

where the maximum is taken over all probability measures supported on $[m, M]$. Note that ξ^* is invariant for one iteration of the algorithm (2.16); that is, if $\xi = \xi^*$ in (2.16) then $\xi'(x) = \xi(x)$ for all $x \in \text{supp}(\xi)$.

In accordance with (2.6), the rate associated with the design measure ξ is defined by

$$v(\xi) = 1 - 2\gamma\mu_1 + \gamma^2\mu_2 = \frac{b(\xi)}{B(\xi)} . \tag{2.19}$$

Assume that the optimality criterion Φ is such that the optimum design ξ^* is non-degenerate (that is, ξ^* is not just supported at a single point). Note that if $\Phi(M) = -\infty$ for any singular matrix M, then this condition is satisfied.

Since the design ξ^* is optimum, all directional derivatives are non-positive:

$$\frac{\partial}{\partial\alpha} \Phi\Big[M\big((1-\alpha)\xi^* + \alpha\xi(x)\big)\Big]\Big|_{\alpha=0^+} \leq 0 ,$$

for all $x \in [m, M]$. Using (2.9), this implies

$$\max_{x\in[m,M]} \varphi(x, \xi^*) \leq t^* = \text{tr}\left[M(\xi^*) \overset{\circ}{\Phi} (\xi^*)\right] .$$

Since $\varphi(x, \xi^*)$ is a quadratic convex function of x, this is equivalent to $\varphi(m, \xi^*) < t^*$ and $\varphi(M, \xi^*) \leq t^*$. As

$$\int_m^M \varphi(x, \xi^*)\xi^*(\mathrm{d}x) - t^* ,$$

this implies that ξ^* is supported at m and M. Since ξ^* is non-degenerate, ξ^* has positive masses at both points m and M and

$$\varphi(m, \xi^*) = \varphi(M, \xi^*) = t^* .$$

As $\varphi(x, \xi^*)$ is quadratic in x with its minimum at $1/\gamma$, see (2.15), it implies that

$$\gamma^* = \gamma(\mu_1(\xi^*), \mu_2(\xi^*)) = \frac{2}{m + M} .$$

The rate $v(\xi^*)$ is therefore

$$v(\xi^*) = \frac{b(\xi^*)}{B(\xi^*)} = \frac{t^* - c(\xi^*)}{B(\xi^*)} = (1 - m\gamma^*)^2 = (1 - M\gamma^*)^2 = R_{\max}\,,$$

where

$$R_{\max} = \frac{(M-m)^2}{(M+m)^2}\,. \tag{2.20}$$

Assume now that the optimum design ξ^* is degenerate and is supported at a single point x^*. Note that since $\varphi(x, \xi^*)$ is both quadratic and convex, x^* is either m or M. Since the optimum design is invariant in one iteration of the algorithm (2.16), γ^* is constant and

$$\max_{\xi} v(\xi) = \max\left[(1 - m\gamma^*)^2, (1 - M\gamma^*)^2\right] \geq R_{\max}$$

with the inequality replaced by an equality if and only if $\gamma^* = 2/(M+m)$.

2.6 Some Special Cases

In Table 2.1, we provide a few examples of gradient algorithms (2.2) and indicate the corresponding functions of the probability measures ξ. We only restrict ourselves to the optimality criteria $\Phi(\xi)$ that have the form $\Phi(\xi)=\Phi(M(\xi))$, where $M(\xi)$ is the moment matrix (2.8). Neither other moments of ξ nor information about the support of ξ are used for constructing the algorithms below.

For a number of algorithms (most of them have not previously been considered in literature), the table provides the following functions.

- The optimality criterion $\Phi(\xi)$.
- The step-length γ_k in the algorithm (2.2); here γ_k is expressed in the form of $\gamma(\xi)$ as defined in (2.14).
- The rate function $v(\xi)$ as defined in (2.19); this is equivalent to the rate $v_k = (g_{k+1}, g_{k+1})/(g_k, g_k)$ at iteration k for the original algorithm (2.2).
- The φ-function $\varphi(x, \xi)$ as defined in (2.10), see also (2.12).
- The expression for $\mathrm{tr}[M(\xi)\overset{\circ}{\Phi}(\xi)]$; this is the quantity that often appears in the right-hand side in the conditions for the optimality of designs.
- The minimum of the φ-function: $c(\xi) = \min_x \varphi(x, \xi)$.

The steepest-descent algorithm corresponds to the case when $\Phi(\xi)$ is the D-optimality criterion. Two forms of this criterion are given in the table; of course, they correspond to the same optimization algorithm. It is well known that the asymptotic rate of the steepest-descent algorithm is always close to the value $R_{\max}$ defined in (2.20). The asymptotic behaviour of the steepest-descent algorithm has already been extensively studied, see, e.g. Pronzato et al. (2000), Akaike (1959) and Nocedal et al. (2002).

Table 2.1. Examples of gradient algorithms

Algorithm	$\Phi(M(\xi))$	$\gamma(\xi)$	$v(\xi)$	$\varphi(x,\xi) = f^{\mathrm{T}}(x)\,\overset{\circ}{\Phi}(\xi)f(x)$	$\mathrm{tr}[M(\xi)\,\overset{\circ}{\Phi}(\xi)]$	$c(\xi) = \min \varphi(x,\xi)$
Steepest descent	$\log(\mu_2 - \mu_1^2)$	$\frac{1}{\mu_1}$	$\frac{\mu_2}{\mu_1^2} - 1$	$\frac{\mu_2 - 2x\mu_1 + x^2}{\mu_2 - \mu_1^2}$	2	1
(D-optimality)	$\mu_2 - \mu_1^2$			$\mu_2 - 2x\mu_1 + x^2$	$2\Phi(\xi)$	$\Phi(\xi)$
Constant $\gamma_k = \gamma$	$\gamma\mu_2 - 2\mu_1$	γ	$1 - 2\gamma\mu_1 + \gamma^2\mu_2$	$\gamma\mu_2 - 2x + \gamma x^2$	$2(\gamma\mu_2 - \mu_1)$	$\gamma\mu_2 - \frac{1}{\gamma}$
Steepest descent with relaxation	$\varepsilon\mu_2 - \mu_1^2$	$\frac{\varepsilon}{\mu_1}$	$1 - 2\varepsilon + \varepsilon^2\frac{\mu_2}{\mu_1^2}$	$\mu_2 - 2x\mu_1 + \varepsilon x^2$	$\mu_2(1+\varepsilon) - 2\mu_1^2$	$\Phi(\xi)/\varepsilon$
Square-root	$\sqrt{\mu_2} - \mu_1$	$\frac{1}{\sqrt{\mu_2}}$	$2\left(1 - \frac{\mu_1}{\sqrt{\mu_2}}\right)$	$\frac{\sqrt{\mu_2}}{2}\left(1 - \frac{x}{\sqrt{\mu_2}}\right)^2$	$\Phi(\xi)$	0
α-root	$\mu_2^\alpha - \mu_1^{2\alpha}$	$\frac{\mu_2^{\alpha-1}}{\mu_1^{2\alpha-1}}$	$\left(\frac{\mu_2}{\mu_1^2}\right)^{2\alpha-1} - 2\left(\frac{\mu_2}{\mu_1^2}\right)^{\alpha-1} + 1$	$\alpha(\mu_2^\alpha - 2\mu_1^{2\alpha-1}x + \mu_2^{\alpha-1}x^2)$	$2\alpha\Phi(\xi)$	$\alpha\frac{\mu_2^{2\alpha-1} - \mu_1^{4\alpha-2}}{\mu_2^{\alpha-1}}$
α-root with relaxation	$\varepsilon\mu_2^\alpha - \mu_1^{2\alpha}$	$\varepsilon\frac{\mu_2^{\alpha-1}}{\mu_1^{2\alpha-1}}$	$\varepsilon^2\left(\frac{\mu_2}{\mu_1^2}\right)^{2\alpha-1} - 2\varepsilon\left(\frac{\mu_2}{\mu_1^2}\right)^{\alpha-1} + 1$	$\alpha(\varepsilon^2\mu_2^\alpha - 2\mu_1^{2\alpha-1}x + \varepsilon\mu_2^{\alpha-1}x^2)$	$2\alpha\Phi(\xi)$	$\alpha\frac{\varepsilon^2\mu_2^{2\alpha-1} - \mu_1^{4\alpha-2}}{\varepsilon\mu_2^{\alpha-1}}$
Minimum residues (c-optimality)	$1 - \frac{\mu_1^2}{\mu_2}$	$\frac{\mu_1}{\mu_2}$	$1 - \frac{\mu_1^2}{\mu_2}$	$\frac{(x\mu_1 - \mu_2)^2}{\mu_2^2}$	$\Phi(\xi)$	0
A-optimality	$1/\mathrm{tr}D(\xi)$ $= \frac{\mu_2 - \mu_1^2}{1+\mu_2}$	$\frac{(1+\mu_1^2)}{\mu_1(1+\mu_2)}$	$\frac{(1+2\mu_1^2+\mu_1^2\mu_2)(\mu_2-\mu_1^2)}{\mu_1^2(1+\mu_2)^2}$	$\frac{(x-\mu_1)^2 + (x\mu_1 - \mu_2)^2}{(\mu_2+1)^2}$	$\Phi(\xi)$	$\frac{(\mu_2-\mu_1^2)^2}{(\mu_1^2+1)(\mu_2+1)^2}$

The gradient algorithm with constant step-length $\gamma_k = \gamma$ is well known in literature. It converges slowly; its rate of convergence can easily be analysed without using the technique of the present chapter. We do not study this algorithm and provide its characteristics in the table below only for the sake of completeness.

The steepest-descent algorithm with relaxation is also well known in literature on optimization. It is known that for suitable values of the relaxation parameter ε, this algorithm has a faster convergence rate than the ordinary steepest-descent algorithm. However, the reasons why this occurs were not known. In Sect. 2.7, we try to explain this phenomenon. In addition, we prove that if the relaxation parameter is either too small ($\varepsilon < 2m/(m+M)$) or too large ($\varepsilon > 2M/(m+M)$) then the rate of the steepest descent with relaxation becomes worse than $R_{\max}$, the worst-case rate of the standard steepest-descent algorithm.

The square-root algorithm can be considered as a modification of the steepest-descent algorithm. The asymptotic behaviour of this algorithm is now well understood, see Theorem 2 below.

The α-root algorithm is a natural extension of the steepest-descent and the square-root algorithms. The optimality criterion used to construct the α-root algorithm can be considered as the D-optimality criterion applied to the matrix which is obtained from the moment matrix (2.8) by the transformation

$$M(\xi) = \begin{pmatrix} 1 & \mu_1 \\ \mu_1 & \mu_2 \end{pmatrix} \to M^{(\alpha)}(\xi) = \begin{pmatrix} 1 & \mu_1^\alpha \\ \mu_1^\alpha & \mu_2^\alpha \end{pmatrix} . \tag{2.21}$$

(Of course, other optimality criteria can be applied to the matrix $M^{(\alpha)}(\xi)$, not only the D-criterion.) The asymptotic rate of the α-root algorithm is studied numerically, see Figs. 2.7–2.9. The conclusion is that this algorithm has an extremely fast rate when α is slightly larger than 1.

The α-root algorithm with relaxation (this class of algorithms includes the steepest-descent and square-root algorithms with relaxation) is an obvious generalization of the algorithm of steepest descent with relaxation. Its asymptotic behaviour is also similar: for a fixed α, for very small and very large values of the relaxation parameter ε, the algorithm either diverges or converges with the rate $\geq R_{\max}$. Unless α itself is either too small or too large, there is always a range of values of the relaxation parameter ε for which the rates are much better than $R_{\max}$ and where the algorithm behaves chaotically. The algorithm corresponds to the D-optimality criterion applied to the matrix which is obtained from the original moment matrix (2.8) by the transformation

$$\begin{pmatrix} 1 & \mu_1 \\ \mu_1 & \mu_2 \end{pmatrix} \to \begin{pmatrix} \varepsilon & \mu_1^\alpha \\ \mu_1^\alpha & \mu_2^\alpha \end{pmatrix} .$$

Of course, this transformation generalizes (2.21).

One can use many different optimality criteria Φ for constructing new gradient algorithms. In Table 2.1, we provide the characteristics of the algorithms associated with two other criteria, both celebrated in the theory of optimal design, namely, c-optimality and A-optimality. The algorithm corresponding to c-optimality, with $\Phi_c(M(\xi)) = (1\,,\ \mu_1)M^{-1}(\xi)(1\,,\ \mu_1)^{\mathrm{T}}$, is the so-called minimum residues optimization algorithm, see Kozjakin and Krasnosel'skii (1982). As shown in Pronzato et al. (2006), its asymptotic behaviour is equivalent to that of the steepest-descent algorithm (and it is therefore not considered below).

The A-optimality criterion gives rise to a new optimization algorithm for which the expression for the step-length would otherwise have been difficult to develop. This algorithm is very easy to implement and its asymptotic rate is reasonably fast, see Fig. 2.10.

Many other optimality criteria (and their mixtures) generate gradient algorithms with fast asymptotic rates. This is true, in particular, for the Φ_p-criteria

$$\Phi_p(M(\xi)) = \left(\mathrm{tr}M^{-p}(\xi)\right)^{\frac{1}{p}}$$

for some values of p. For example, a very fast asymptotic rate (see Fig. 2.10) is obtained in the case $p = 2$, where we have

$$\begin{aligned}
\Phi_2(M(\xi)) = 1/\mathrm{tr}\, M^{-2}(\xi) &= \frac{(\mu_2 - {\mu_1}^2)^2}{1{+}{\mu_2}^2{+}2{\mu_1}^2}\,, \qquad (2.22)\\
\gamma(\xi) &= \frac{1 + 2{\mu_1}^2 + {\mu_1}^2\mu_2}{\mu_1(1{+}\mu_2{+}{\mu_1}^2{+}{\mu_2}^2)}\,,\\
\varphi(x,\xi) - c(\xi) &= \frac{2(\mu_2{-}{\mu_1}^2)(x{-}\mu_1{-}\mu_1\mu_2{-}{\mu_2}^2\mu_1{+}2x{\mu_1}^2{+}x{\mu_1}^2\mu_2{-}{\mu_1}^3)^2}{({\mu_1}^2\mu_2 + 2{\mu_1}^2 + 1)({\mu_2}^2 + 2{\mu_1}^2 + 1)^2}\,,
\end{aligned}$$

and

$$v(\xi) = \frac{(\mu_2 - {\mu_1}^2)({\mu_1}^2{\mu_2}^3 + 2{\mu_1}^2{\mu_2}^2 + 3\mu_2{\mu_1}^2 + 2{\mu_1}^4\mu_2 + 4{\mu_1}^2 + 1 + 3{\mu_1}^4)}{{\mu_1}^2(\mu_2 + {\mu_1}^2 + 1 + {\mu_2}^2)^2}\,.$$

2.7 The Steepest-Descent Algorithm with Relaxation

Steepest descent with relaxation is defined as the algorithm (2.2) with

$$\gamma_k = \frac{\varepsilon}{\mu_1}\,, \qquad (2.23)$$

where ε is some fixed positive number. The main updating formula (2.7) has the form

$$p_i^{(k+1)} = \frac{(1 - \frac{\varepsilon}{\mu_1}\lambda_i)^2}{1 - 2\varepsilon + \varepsilon^2\frac{\mu_2}{\mu_1^2}}\, p_i^{(k)}\,. \qquad (2.24)$$

Table 2.2. Values of $\xi^*(m)$, $\Phi(M(\xi^*))$ and $v(\xi^*)$

	$0 < \varepsilon \le \frac{2m}{m+M}$	$\frac{2m}{m+M} \le \varepsilon \le \frac{2M}{m+M}$	$\varepsilon \ge \frac{2M}{m+M}$
$\xi^*(m)$	1	$\frac{2M-\varepsilon(M+m)}{2(M-m)}$	0
$\Phi(M(\xi^*))$	$m^2(\varepsilon-1)$	$\frac{1}{4}\varepsilon^2(m+M)^2-\varepsilon mM$	$M^2(\varepsilon-1)$
$v(\xi^*)$	$(1-\varepsilon)^2$	$R_{\max}$	$(\varepsilon-1)^2$

This can also be represented in the form (2.11) for the design optimality criterion

$$\Phi(M(\xi)) = \varepsilon\mu_2(\xi) - \mu_1^2(\xi)\,. \tag{2.25}$$

The rate $v(\xi)$ associated with a design ξ was defined in (2.19) and is simplified to

$$v(\xi) = 1 - 2\,\varepsilon + \varepsilon^2\mu_2/\mu_1^2\,.$$

The following lemma reveals properties of the optimum designs for the optimality criteria (2.25).

Lemma 1. *Let* $0 < m < M$, $\varepsilon > 0$ *and let* ξ^* *be the optimum design corresponding to the optimality criterion* (2.25). *Then* ξ^* *is supported at two points* m *and* M *with the weight* $\xi^*(m)$ *as in Table 2.2 and* $\xi^*(M) = 1 - \xi^*(m)$.

The proof is straightforward.

In addition to the values of the weights $\xi^*(m)$, Table 2.2 contains the values of the optimality criterion $\Phi(M(\xi))$ and the rate function $v(\xi)$ for the optimum design $\xi = \xi^*$.

Theorem 1 below shows that if the relaxation coefficient ε is either small ($\varepsilon < 4Mm/(M+m)^2$) or large ($\varepsilon > 1$), then for almost all starting points the algorithm asymptotically behaves as if it has started at the worst possible initial point. However, for some values of ε the rate does not attract to a constant value and often exhibits chaotic behaviour. Typical behaviour of the asymptotic rate (2.5) is shown in Fig. 2.1 where we display the asymptotic rates in the case $M/m = 10$.

In this figure and all other figures in this chapter we assume that $d = 100$ and all the eigenvalues are equally spaced. We have established numerically that the dependence on the dimension d is insignificant as long as $d \ge 10$. In particular, we found that there is virtually no difference in the values of the asymptotic rates corresponding to the cases $d = 100$ and $d = 10^6$, for all the algorithms studied. In addition, choosing equally spaced eigenvalues is effectively the same as choosing eigenvalues uniformly distributed on $[m, M]$ and taking expected values of the asymptotic rates.

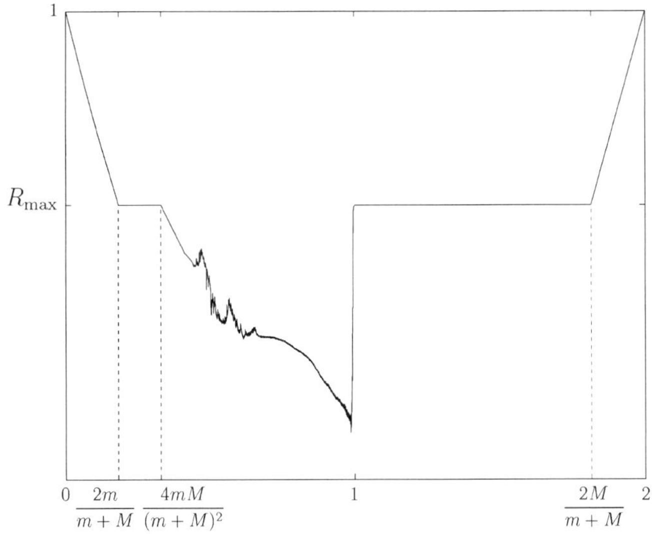

Fig. 2.1. Asymptotic rate of convergence as a function of ε for the steepest-descent algorithm with relaxation ε

Theorem 1. *Assume that ε is such that either $0 < \varepsilon < 4Mm/(m+M)^2$ or $\varepsilon > 1$. Let ξ_0 be any non-degenerate probability measure with support $\{\lambda_1, \ldots, \lambda_d\}$ and let the sequence of probability measures $\{\xi^{(k)}\}$ be defined via the updating formula (2.24) where $p_i^{(k)}$ are the masses $\xi^{(k)}(\lambda_i)$. Then the following statements hold:*

- *For any starting point x_0, the sequence $\Phi_k = \Phi(M(\xi^{(k)}))$ monotonously increases ($\Phi_0 \leq \Phi_1 \leq \cdots \leq \Phi_k \leq \cdots$) and converges to a limit $\lim_{k\to\infty} \Phi_k$.*
- *For almost all starting points x_0 (with respect to the uniform distribution of $g_0/\sqrt{||g_0||}$ on the unit sphere in $\mathbb{R}^d$),*
  – the limit $\lim_{k\to\infty} \Phi_k$ does not depend on the initial measure $\xi^{(0)}$ and is equal to $\Phi(M(\xi^*))$ as defined in Table 2.2;
  – the sequence of probability measures $\{\xi^{(k)}\}$ converges (as $k \to \infty$) to the optimum design ξ^* defined in Lemma 1 and
  – the asymptotic rate R of the steepest-descent algorithm with relaxation, as defined in (2.2) and (2.23), is equal to $v(\xi^*)$ defined in Table 2.2.

Proof. Note that $\Phi_k \leq 0$ for all $\xi^{(k)}$ if

$$0 < \varepsilon < \min_{\xi} \frac{\mu_1^2(\xi)}{\mu_2(\xi)} = \frac{4Mm}{(m+M)^2} .$$

If $\varepsilon \geq 1$ then $\Phi_k \geq 0$ for any $\xi^{(k)}$ in view of the Cauchy–Schwarz inequality. Additionally, $|\Phi_k| \leq (1+\varepsilon)M^2$ for all $\xi^{(k)}$ so that $\{\Phi_k\}$ is always a bounded sequence.

The updating formulae for the moments are

$$\mu_1^{'} = (\mu_1^3 - 2\varepsilon\mu_1\mu_2 + \varepsilon^2\mu_3)/W \quad \text{and} \quad \mu_2^{'} = (\mu_1^2\mu_2 - 2\varepsilon\mu_1\mu_3 + \varepsilon^2\mu_4)/W$$

where $W = \varepsilon^2\mu_2 + \mu_1^2(1-2\varepsilon)$ and

$$\mu_\alpha = \mu_\alpha(\xi^{(k)}), \;\; \mu_\alpha^{'} = \mu_\alpha(\xi^{(k+1)}), \; \forall\alpha, \; k = 0, 1, \ldots \tag{2.26}$$

The sequence Φ_k is non-decreasing if $\Phi_{k+1} - \Phi_k \geq 0$ holds for any probability measure $\xi = \xi^{(k)}$. If for some k the measure $\xi^{(k)}$ is degenerate (that is, has mass 1 at one point) then $\xi^{(k+1)} = \xi^{(k)}$ and the first statement of the theorem holds. Let us suppose that $\xi = \xi^{(k)}$ is non-degenerate for all k. In particular, this implies $\mu_2 > \mu_1^2$ and

$$W = \varepsilon^2\mu_2 + \mu_1^2(1-2\varepsilon) = \mu_1^2(1-\varepsilon)^2 + \varepsilon^2(\mu_2 - \mu_1^2) > 0\,.$$

We have

$$\Phi_{k+1} - \Phi_k \geq 0 \Longleftrightarrow \left(\varepsilon\mu_2^{'} - (\mu_1^{'})^2\right) - \left(\varepsilon\mu_2 - \mu_1^2\right) \geq 0\,. \tag{2.27}$$

We can represent the left-hand side of the second inequality in (2.27) as

$$\left(\varepsilon\mu_2^{'} - (\mu_1^{'})^2\right) - \left(\varepsilon\mu_2 - \mu_1^2\right) = \varepsilon\frac{U}{W^2}\,,$$

where

$$\begin{aligned} U &= W(\mu_1^2\mu_2 - 2\varepsilon\mu_1\mu_3 + \varepsilon^2\mu_4 - W\mu_2) \\ &+ (W\mu_1 + \mu_1^3 - 2\varepsilon\mu_1\mu_2 + \varepsilon^2\mu_3)(\varepsilon\mu_1\mu_2 - \varepsilon\mu_3 - 2\,{\mu_1}^3 + 2\,\mu_1\mu_2)\,. \end{aligned}$$

As W^2 will always remain strictly positive, the problem is reduced to determining whether or not $U \geq 0$. To establish the inequality $U \geq 0$, we show that $U = V(a,b)$ for some a and b, where

$$V(a,b) = \mathrm{var}(aX + bX^2) = a^2\mu_2 + 2ab\mu_3 + b^2\mu_4 - (a\mu_1 + b\mu_2)^2 \geq 0 \tag{2.28}$$

and X is the random variable with distribution ξ.

Consider $U - V(a,b) = 0$ as an equation with respect to a and b and prove that there is a solution to this equation. First, choose b to eliminate the μ_4 term: $b = b_0 = \varepsilon\sqrt{W}$. The value of b_0 is correctly defined as $W > 0$.

The next step is to prove that there is a solution to the equation $U - V(a,b_0) = 0$ with respect to a. Note that $U - V(a,b_0)$ is a quadratic function in a. Let D be the discriminant of this quadratic function; it can be simplified to

$$D = (\varepsilon - 1)\left(\varepsilon\mu_2 - {\mu_1}^2\right)\left(\mu_3\varepsilon + 2\,{\mu_1}^3 - \varepsilon\mu_1\mu_2 - 2\,\mu_2\mu_1\right)^2\,.$$

This is clearly non-negative for $\varepsilon > 1$ and $0 < \varepsilon < 4Mm/(m+M)^2$. Therefore, there exist some a and b such that $U = V(a,b)$. This implies $\Phi_{k+1} - \Phi_k \geq 0$

and therefore $\{\Phi_k\}$ is a monotonously increasing bounded sequence converging to some limit $\Phi_* = \lim_{k\to\infty}\Phi_k$.

Consider now the second part of the theorem. Assume that the initial measure ξ_0 is such that $\xi_0(m) > 0$ and $\xi_0(M) > 0$.

From the sequence of measures $\{\xi^{(k)}\}$ choose a subsequence weakly converging to some measure ξ_* (we can always find such a sequence as all the measures are supported on an interval). Let $X = X_*$ be the random variable defined by the probability measure ξ_*. For this measure, the value of V defined in (2.28) is zero as $\Phi(M(\xi^{(k+1)})) = \Phi(M(\xi^{(k)}))$ if $\xi^{(k)} = \xi_*$. Therefore, the random variable $aX_* + bX_*^2$ is degenerated (here a and b are some coefficients) and therefore X_* is concentrated at either a single point or at two distinct points.

Assume that $\varepsilon < 1$ and therefore $0 < \varepsilon < 4Mm/(m+M)^2$ (similar arguments work in the case $\varepsilon > 1$). Then, similar to the proof for the steepest-descent algorithm (see Pronzato et al. (2000), p. 175) one can see that the masses $\xi^{(k)}(m)$ are bounded away from 0; that is, $\xi^{(k)}(m) \geq c$ for some $c > 0$ and all k, as long as $\xi_0(m) > 0$. Therefore, the point m always belongs to the support of the measure ξ_*. If $0 < \varepsilon < 2m/(m+M)$ then one can easily check that ξ_* must be concentrated at a single point (otherwise $U = U(\xi_*) \neq 0$); this point is necessarily m and therefore the limiting design ξ_* coincides with the optimal design ξ^*. If $2m/(m+M) < \varepsilon < 4Mm/(m+M)^2$, then using similar arguments one can find that the second support point of ξ_* is M for almost all starting points.[1] As long as the support of ξ_* is established, one can easily see that the only two-point design that has $U(\xi_*) = 0$ is the optimal design ξ^* with respect to the criterion (2.25). □

One of the implications of the theorem is that if the relaxation parameter is either too small ($\varepsilon < 2m/(m+M)$) or too large ($\varepsilon > 2M/(m+M)$) then the rate of the steepest-descent algorithm with relaxation becomes worse than $R_{\max}$, the worst-case rate of the standard steepest-descent algorithm. As a consequence, we also obtain a well-known result that if the value of the relaxation coefficient is either $\varepsilon < 0$ or $\varepsilon > 2$, then the steepest descent with relaxation diverges. When $4Mm/(m+M)^2 < \varepsilon \leq 1$ the relaxed steepest-descent algorithm does not necessarily converge to the optimum design. It is within this range of ε that improved asymptotic rates of convergence are demonstrated, see Fig. 2.1.

The convergence rate of all gradient-type algorithms depends on, amongst other things, the condition number $\varrho = M/m$. As one would expect, an increase in ϱ gives rise to a worsened rate of convergence. The improvement yielded by the addition of a suitable relaxation coefficient to the steepest-descent algorithm however, produces significantly better asymptotic rates of

[1] The only possibility for M to vanish from the support of the limiting design ξ_* is to obtain $\mu_1(\xi^{(k)})/\varepsilon = M$ at some iteration k, but this obviously almost never (with respect to the distribution of $g_0/\sqrt{||g_0||}$) happens. Note that in this case M will be replaced with λ_{d-i} for some $i \geq 1$ which can only improve the asymptotic rate of convergence of the optimization algorithm.

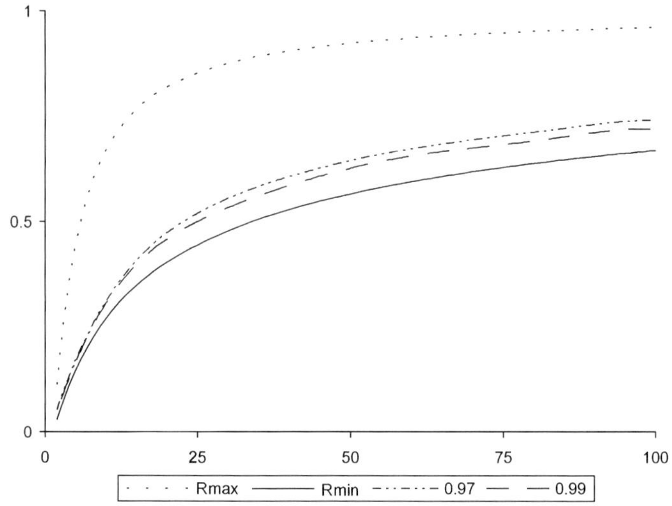

Fig. 2.2. Asymptotic rate of convergence as a function of ϱ for steepest descent with relaxation coefficients $\varepsilon = 0.97$ and $\varepsilon = 0.99$

convergence than standard steepest descent. Figure 2.2 shows the effect of increasing the value of ϱ on the rates of convergence for the steepest-descent algorithm with relaxation coefficients $\varepsilon = 0.97$ and $\varepsilon = 0.99$ (see also Fig. 2.10). For large d and ϱ, we expect that the asymptotic convergence rates for this family of algorithms will be bounded above by $R_{\max}$ and below by $R_{\min}$ where

$$R_{\min} = \left(\frac{\sqrt{M} - \sqrt{m}}{\sqrt{M} + \sqrt{m}}\right)^2 = \left(\frac{\sqrt{\varrho} - 1}{\sqrt{\varrho} + 1}\right)^2 ; \tag{2.29}$$

the rate $R_{\min}$ is exactly the same as the rate N_∞^* defined in (3.22), Chap. 3. The relaxation coefficient $\varepsilon = 0.99$ produces asymptotic rates approaching $R_{\min}$ (see also Fig. 2.10).

In Fig. 2.3, we display 250 rates $v(\xi_k)$, $750 < k \leq 1{,}000$ which the steepest-descent algorithms with relaxation coefficients $\varepsilon \in [0.4, 1]$ attained starting at the random initial designs ξ_0. In this figure, we observe bifurcations and transitions to chaotic regimes occurring for certain values of ε.

Figure 2.4 shows the same results but in the form of the log-rates, $-\log(v(\xi_k))$. Using the log-rates rather than the original rates helps to see the variety of small values of the rates, which is very important as small rates $v(\xi_k)$ force the final asymptotic rate to be small.

Figure 2.5 shows the log-rates, $-\log(v(\xi_k))$, occurring for $\varepsilon \in [0.99, 1.0]$. This figure illustrates the effect of bifurcation to chaos when we decrease the values of ε starting at 1.

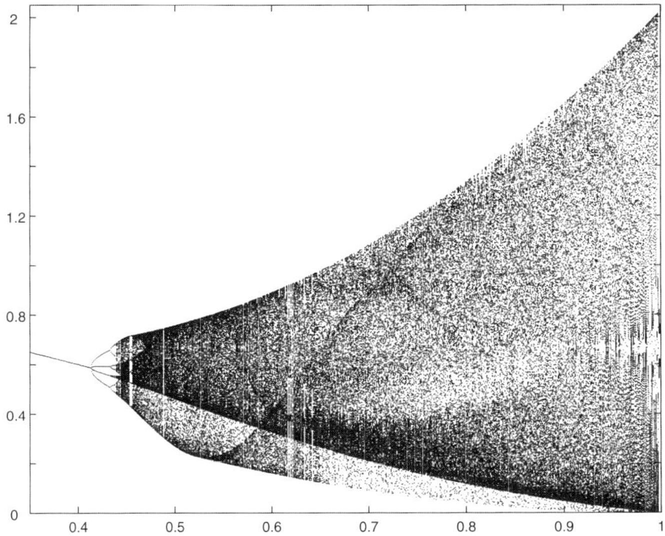

Fig. 2.3. Rates $v(\xi_k)$ $(750 < k \leq 1{,}000)$ for steepest descent with relaxation; varying ε, $\varrho = 10$

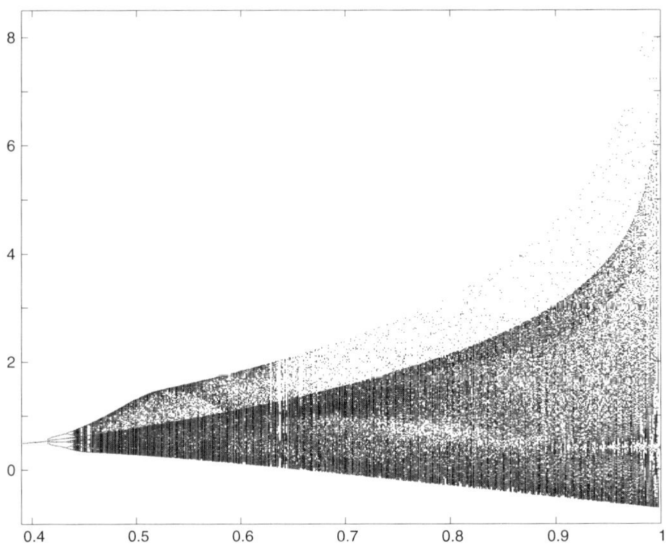

Fig. 2.4. Log-rates, $-\log(v(\xi_k))$ $(750 < k \leq 1{,}000)$ for steepest descent with relaxation; varying ε, $\varrho = 10$

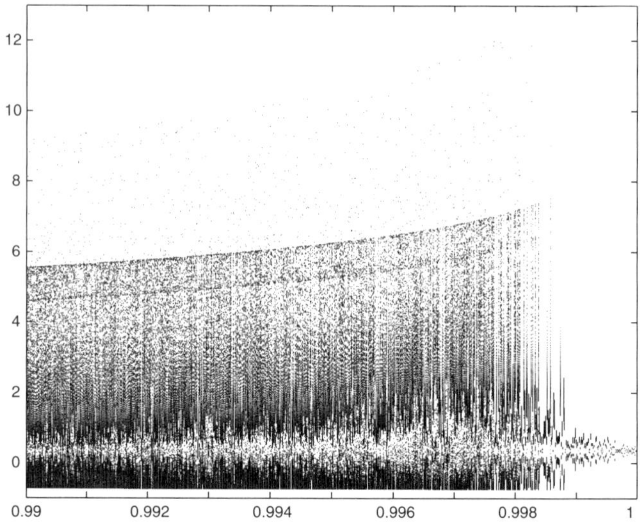

Fig. 2.5. Log-rates, $-\log(v(\xi_k))$ $(750 < k \le 1{,}000)$ for steepest descent with relaxation; $\varrho = 10$, $\varepsilon \in [0.99, 1.0]$

2.8 Square-Root Algorithm

For the square-root algorithm, we have:

$$\Phi(M(\xi)) = \sqrt{\mu_2} - \mu_1, \quad v = v(\xi) = 2\left(1 - \frac{\mu_1}{\sqrt{\mu_2}}\right).$$

In the present case, the optimum design ξ^* is concentrated at the points m and M with weights

$$\xi^*(M) = \frac{3m + M}{4(m + M)}, \quad \xi^*(m) = \frac{m + 3M}{4(m + M)}. \tag{2.30}$$

For the optimum design, we have

$$\Phi(M(\xi^*)) = \frac{1}{4}\frac{(M - m)^2}{m + M} \quad \text{and} \quad v(\xi^*) = R_{\max}.$$

The main updating formula (2.7) has the form

$$p_i^{(k+1)} = \frac{(1 - \frac{1}{\sqrt{\mu_2}}\lambda_i)^2}{2(1 - \frac{\mu_1}{\sqrt{\mu_2}})}\, p_i^{(k)}. \tag{2.31}$$

Theorem 2. *Let ξ_0 be any non-degenerate probability measure with support $\{\lambda_1, \ldots, \lambda_d\}$ and let the sequence of probability measures $\{\xi^{(k)}\}$ be defined via the updating formula (2.31) where $p_i^{(k)}$ are the masses $\xi^{(k)}(\lambda_i)$. Then the following statements hold:*

- *For any starting point x_0, the sequence $\Phi_k = \Phi(M(\xi^{(k)}))$ monotonously increases ($\Phi_0 \leq \Phi_1 \leq \cdots \leq \Phi_k \leq \cdots$) and converges to a limit $\lim_{k\to\infty} \Phi_k$.*
- *If the starting point x_0 of the optimization algorithm is such that $\xi^{(0)}(\lambda_1) > 0$ and $\xi^{(0)}(\lambda_d) > 0$, then*
  – the limit $\lim_{k\to\infty} \Phi_k$ does not depend on the initial measure $\xi^{(0)}$ and is equal to $\Phi(M(\xi^*)) = (M-m)^2/[4(m+M)]$;
  – the sequence of probability measures $\{\xi^{(k)}\}$ converges (as $k \to \infty$) to the optimum design ξ^* defined in (2.30) and
  – the asymptotic rate R of the square-root optimization algorithm is equal to $R_{\max}$.

Proof. The proof is similar to (and simpler than) the proof of Theorem 1.

For the square-root algorithm, we have

$$\mu_1' = \left(\mu_1 - 2\sqrt{\mu_2} + \frac{\mu_3}{\mu_2}\right)/v\,, \quad \mu_2' = \left(\mu_2 - 2\frac{\mu_3}{\sqrt{\mu_2}} + \frac{\mu_4}{\mu_2}\right)/v\,,$$

where we use the notation for the moments introduced in (2.26).

Furthermore,

$$\Phi_{k+1} - \Phi_k = \sqrt{\mu_2'} - \mu_1' - \sqrt{\mu_2} + \mu_1\,,$$

$$\Phi_{k+1} \geq \Phi_k \Longleftrightarrow \sqrt{\mu_2'} \geq \mu_1' + \sqrt{\mu_2} - \mu_1 = \frac{\mu_3 - \mu_1\mu_2}{2\sqrt{\mu_2}\left(\sqrt{\mu_2} - \mu_1\right)} - \mu_1\,.$$

The inequality

$$\mu_2' \geq (\mu_1' + \sqrt{\mu_2} - \mu_1)^2 \tag{2.32}$$

would therefore imply $\Phi_{k+1} \geq \Phi_k$ for any design $\xi = \xi^{(k)}$.

Let X be a random variable with probability distribution $\xi = \xi_k$. Then we can see that the difference $\mu_2' - (\mu_1' + \sqrt{\mu_2} - \mu_1)^2$ can be represented as

$$\mu_2' - (\mu_1' + \sqrt{\mu_2} - \mu_1)^2 = \frac{1}{2\sqrt{\mu_2}(\sqrt{\mu_2} - \mu_1)} \mathrm{var}(aX + X^2) \geq 0$$

where

$$a = \kappa + \frac{\kappa + 2\sqrt{\mu_2}}{\sqrt{2}}\sqrt{1 - \frac{\mu_1}{\sqrt{\mu_2}}} \quad \text{and} \quad \kappa = \frac{\mu_1\mu_2 - \mu_3}{\mu_2 - \mu_1^2}\,.$$

This implies (2.32) and therefore the monotonic convergence of the sequence $\{\Phi_k\}$.

As a consequence, any limiting design for the sequence $\{\xi_k\}$ is concentrated at two points. If $\xi^{(0)}(\lambda_1) > 0$ and $\xi^{(0)}(\lambda_d) > 0$, then there is a constant c_0 such that $\xi^{(k)}(\lambda_1) > c_0$ and $\xi^{(k)}(\lambda_d) > c_0$ for all k implying that the limiting design is concentrated at m and M. The only design with support $\{m, M\}$ that leaves the value of $\Phi(M(\xi))$ unchanged is the optimal design ξ^* with weights (2.30). The rate $v(\xi^*)$ for this algorithm is $R_{\max}$. □

2.9 *A*-Optimality

Consider the behaviour of the gradient algorithm generated by the A-optimality criterion in the two-dimensional case; that is, when $d = 2$, $\lambda_1 = m$ and $\lambda_2 = M$. Assume that the initial point x_0 is such that $0 < \xi^{(0)}(m) < 1$ (otherwise the initial design $\xi^{(0)}$ is degenerated and so are all other designs $\xi^{(k)}$, $k \geq 1$).

Denote $p_k = \xi^{(k)}(m)$ for $k = 0, 1, \ldots$. As $d = 2$, all the designs $\xi^{(k)}$ are fully described by the corresponding values of p_k. In the case of A-optimality, the updating formula for the p_k's is $p_{k+1} = f(p_k)$ where

$$f(p) = \left(1 - \frac{(1+{\mu_1}^2)}{\mu_1\,(1+\mu_2)}m\right)^2 \frac{{\mu_1}^2\,(1+\mu_2)^2}{(1+2\,{\mu_1}^2+{\mu_1}^2\mu_2)\,(\mu_2-{\mu_1}^2)}\,p\,,$$

$\mu_1 = pm + (1-p)M$ and $\mu_2 = pm^2 + (1-p)M^2$. The fixed point of the transformation $p_{k+1} = f(p_k)$ is

$$p_* = \frac{M^2+1-\sqrt{(M^2+1)(m^2+1)}}{M^2-m^2}\,.$$

For this point we have $p_* = f(p_*)$ and the design with the mass p_* at m and mass $1-p_*$ at M is the A-optimum design for the linear regression model $y_j = \theta_0 + \theta_1 x_j + \varepsilon_j$ on the interval $[m, M]$ (and any subset of this interval that includes m and M). This fixed point p_* is unstable for the mapping $p \to f(p)$ as $|f'(p_*)| > 1$.

For the transformation $f^2(\cdot) = f(f(\cdot))$, see Fig. 2.6 for an illustration of this map, there are two stable fixed points which are 0 and 1. The fact that the points 0 and 1 are stable for the mapping $p \to f^2(p)$ follows from

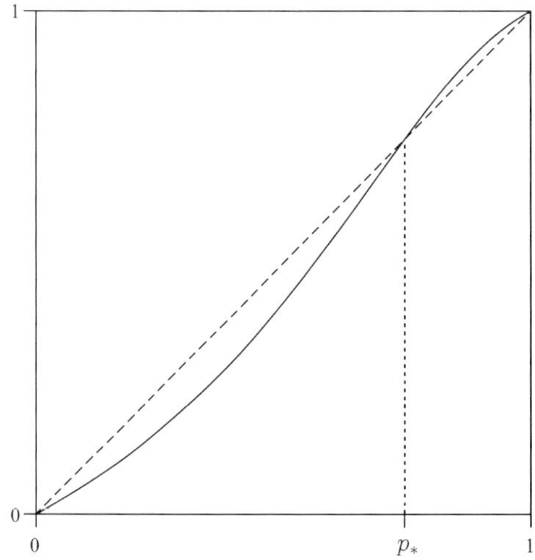

Fig. 2.6. Graph depicting the transformation $f^2(\cdot)$ for $m = 1$, $M = 4$

$$\left.(f(f(p)))'\right|_{p=0} = \left.(f(f(p)))'\right|_{p=1} = f'(0)f'(1) = \frac{(Mm+1)^4}{(m^2+1)^2(M^2+1)^2};$$

the right-hand side of this formula is always positive and less than 1. There is a third fixed point for the mapping $p \to f^2(p)$; this is of course p_* which is clearly unstable.

This implies that in the two-dimensional case the sequence of measures $\xi^{(k)}$ attracts (as $k \to \infty$) to a cycle of oscillations between two degenerate measures, one is concentrated at m and the other one is concentrated at M. The superlinear convergence of the corresponding gradient algorithm follows from the fact that the rates $v(\xi)$ at these two degenerate measures are 0 (implying $v_k \to 0$ as $k \to \infty$ for the sequence of rates v_k).

To summarize the result, we formulate it as a theorem.

Theorem 3. *For almost all starting points x_0, the gradient algorithm corresponding to the A-optimality criterion for $d = 2$ has a superlinear convergence in the sense that the sequence of rates v_k tends to 0 as $k \to \infty$.*

If the dimension d is larger than 2, then the convergence of the optimization algorithm generated by the A-criterion is no longer superlinear. In fact, the algorithm tries to attract to the two-dimensional plane with the basis e_1, e_d by reducing the weights of the designs $\xi^{(k)}$ at the other eigenvalues. However, when it gets close to the plane, its convergence rate accelerates and the updating rule quickly recovers the weights of the other components. Then the process restarts basically at random, which creates chaos. (This phenomenon is observed in many other gradient algorithms with fast asymptotic convergence rate).

The asymptotic rate (in the form of efficiency with respect to $R_{\min}$) of the gradient algorithm generated by the A-criterion, is shown in Fig. 2.10.

2.10 α-Root Algorithm and Comparisons

In Fig. 2.7, we display the numerically computed asymptotic rates for the α- root gradient algorithm with $\alpha \in [0.75, 2]$, $\varrho = 50$ and $\varrho = 100$. This figure illustrates that for $\alpha < 1$ the asymptotic rate of the algorithm is $R_{\max}$. The asymptotic rate becomes much better for values of α slightly larger than 1. Numerical simulations show that the optimal value of α depends on ϱ and on the intermediate eigenvalues. For $\varrho \le 85$, the optimal value of α tends to be around 1.015; for $\varrho = 90 \pm 5$, the optimal value of α switches to a value of around 1.05, where it stays for larger values of ϱ.

In Fig. 2.8, we display 250 log-rates, $-\log(v(\xi_k))$, $750 < k \le 1{,}000$, which the α-root gradient algorithm attained starting at the random initial design ξ_0, for different α. This figure is similar to Fig. 2.4 for the steepest-descent algorithm with relaxation.

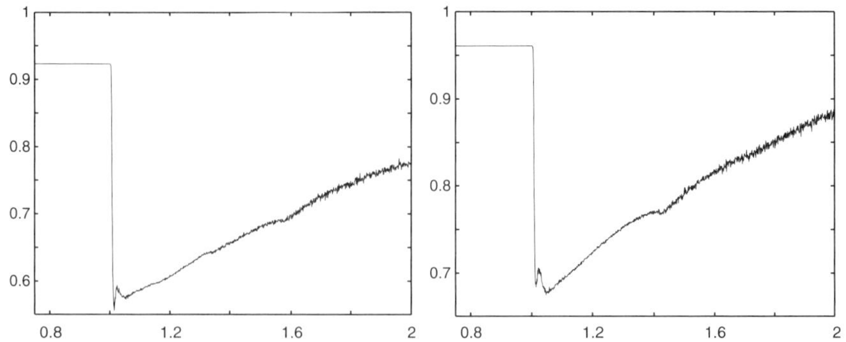

Fig. 2.7. Asymptotic rates for the α-root gradient algorithm; *left*: $\varrho = 50$, *right*: $\varrho = 100$

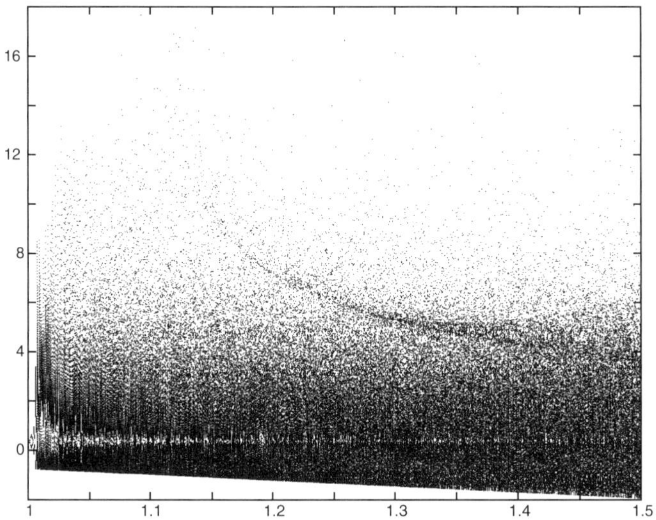

Fig. 2.8. Log-rates, $-\log(v(\xi_k))$ $(750 < k \leq 1{,}000)$ for the α-root gradient algorithm; varying α, $\varrho = 10$

Figure 2.9 is similar to Fig. 2.5 and shows the log-rates, $-\log(v(\xi_k))$, for the α-root gradient algorithm occurring for $\alpha \in [1.0, 1.01]$.

In Fig. 2.10, we compare the asymptotic rates of the following gradient algorithms: (a) α-root algorithm with $\alpha = 1.05$; (b) the algorithm based on the Φ_2-optimality criterion, see (2.22); (c) steepest descent with relaxation $\varepsilon = 0.99$, see Sect. 2.7, and (d) the algorithm based on A-optimality criterion, see Sect. 2.9. The asymptotic rates are displayed in the form of efficiencies with respect to $R_{\min}$ as defined in (2.29); that is, as the ratios $R_{\min}/R$, where R is the asymptotic rate of the respective algorithm.

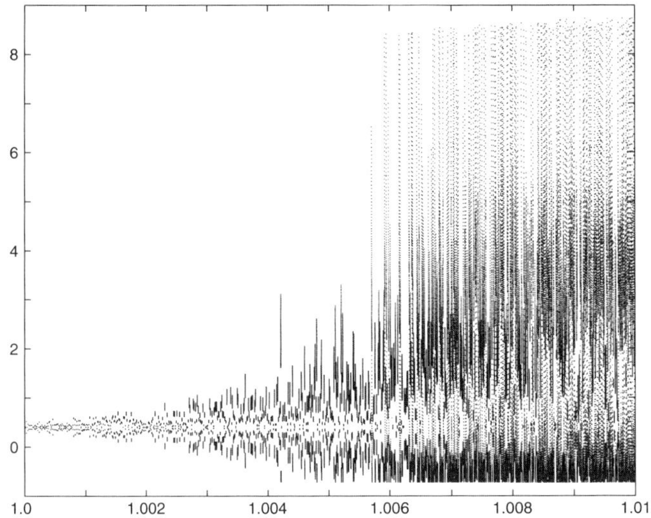

Fig. 2.9. Log-rates, $-\log(v(\xi_k))$ $(750 < k \leq 1{,}000)$ the α-root gradient algorithm; $\varrho = 10$, $\alpha \in [1.0, 1.01]$

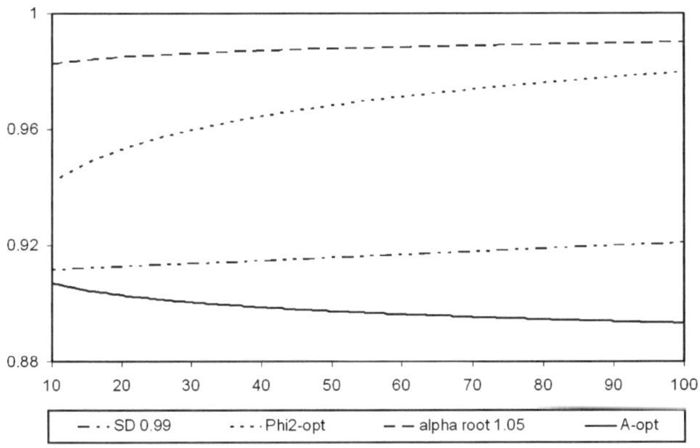

Fig. 2.10. Efficiency relative to $R_{\min}$ for various algorithms, varying ϱ

In Fig. 2.11, we compare the asymptotic rates (in the form of efficiencies with respect to $R_{\min}$) of the following gradient algorithms: (a) α-root algorithm with optimal value of α; (b) steepest descent with optimal value of the relaxation coefficient ε; (c) Cauchy–Barzilai–Borwein method (CBB) as defined in (Raydan and Svaiter, 2002); (d) Barzilai–Borwein method (BB) as defined in (Barzilai and Borwein, 1988).

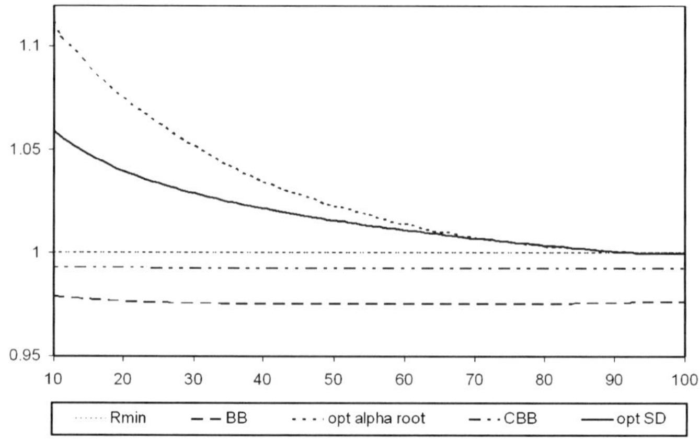

Fig. 2.11. Efficiency relative to $R_{\min}$ for various algorithms, varying ϱ (other algorithms)

References

Akaike, H. (1959). On a successive transformation of probability distribution and its application to the analysis of the optimum gradient method. *Annals of the Institute of Statistical Mathematics*, Tokyo, **11**, 1–16.

Barzilai, J. and Borwein, J. (1988). Two-point step size gradient methods. *IMA Journal of Numerical Analysis*, **8**, 141–148.

Fedorov, V. (1972). *Theory of Optimal Experiments.* Academic Press, New York.

Forsythe, G. (1968). On the asymptotic directions of the s-dimensional optimum gradient method. *Numerische Mathematik*, **11**, 57–76.

Kozjakin, V. and Krasnosel'skii, M. (1982). Some remarks on the method of minimal residues. *Numerical Functional Analysis and Optimization*, **4**(3), 211–239.

Nocedal, J., Sartenaer, A., and Zhu, C. (2002). On the behavior of the gradient norm in the steepest descent method. *Computational Optimization and Applications*, **22**, 5–35.

Pronzato, L., Wynn, H., and Zhigljavsky, A. (2000). *Dynamical Search.* Chapman & Hall/CRC, Boca Raton.

Pronzato, L., Wynn, H., and Zhigljavsky, A. (2001). Renormalised steepest descent in Hilbert space converges to a two-point attractor. *Acta Applicandae Mathematicae*, **67**, 1–18.

Pronzato, L., Wynn, H., and Zhigljavsky, A. (2002). An introduction to dynamical search. In P. Pardalos and H. Romeijn, editors, *Handbook of Global Optimization*, volume 2, Chap. 4, pages 115–150. Kluwer, Dordrecht.

Pronzato, L., Wynn, H., and Zhigljavsky, A. (2006). Asymptotic behaviour of a family of gradient algorithms in $\mathbb{R}^d$ and Hilbert spaces. *Mathematical Programming*, **A107**, 409–438.

Raydan, M. and Svaiter, B. (2002). Relaxed steepest descent and Cauchy-Barzilai-Borwein method. *Computational Optimization and Applications*, **21**, 155–167.

3

A Dynamical-System Analysis of the Optimum s-Gradient Algorithm

L. Pronzato, H.P. Wynn and A. Zhigljavsky

Summary. We study the asymptotic behaviour of Forsythe's s-optimum gradient algorithm for the minimization of a quadratic function in $\mathbb{R}^d$ using a renormalization that converts the algorithm into iterations applied to a probability measure. Bounds on the performance of the algorithm (rate of convergence) are obtained through optimum design theory and the limiting behaviour of the algorithm for $s = 2$ is investigated into details. Algorithms that switch periodically between $s = 1$ and $s = 2$ are shown to converge much faster than when s is fixed at 2.

3.1 Introduction

The asymptotic behaviour of the steepest-descent algorithm (that is, the optimum 1-gradient method) for the minimization of a quadratic function in $\mathbb{R}^d$ is well known, see Akaike (1959), Nocedal et al. (1998, 2002) and Chap. 7 of Pronzato et al. (2000). Any vector y of norm one with two non-zero components only is a fixed point for two iterations of the algorithm after a suitable renormalization. The main result is that, in the renormalized space, one typically observes convergence to a two-point limit set which lies in the space spanned by the eigenvectors corresponding to the smallest and largest eigenvalues of the matrix A of the quadratic function. The proof for bounded quadratic operators in Hilbert space is similar to the proof for $\mathbb{R}^d$ although more technical, see Pronzato et al. (2001, 2006). In both cases, the method consists of converting the renormalized algorithm into iterations applied to a measure ν_k supported on the spectrum of A. The additional technicalities arise from the fact that in the Hilbert space case the measure may be continuous. For $s = 1$, the well-known inequality of Kantorovich gives a bound on the rate of convergence of the algorithm, see Kantorovich and Akilov (1982) and Luenberger (1973, p. 151). However, the actual asymptotic rate of convergence, although satisfying the Kantorovich bound, depends on the starting point and is difficult to predict; a lower bound can be obtained Pronzato et al. (2001, 2006) from considerations on the stability of the fixed points for the attractor.

L. Pronzato, A. Zhigljavsky (eds.), *Optimal Design and Related Areas in Optimization and Statistics*, Springer Optimization and its Applications 28, DOI 10.1007/978-0-387-79936-0_3,

The situation is much more complicated for the optimum s-gradient algorithm with $s \geq 2$ and the chapter extends the results presented in Forsythe (1968) in several directions. First, two different sequences are shown to be monotonically increasing along the trajectory followed by the algorithm (after a suitable renormalization) and a link with optimum design theory is established for the construction of upper bounds for these sequences. Second, the case $s = 2$ is investigated into details and a precise characterization of the limiting behaviour of the renormalized algorithm is given. Finally, we show how switching periodically between the algorithms with, respectively, $s = 1$ and $s = 2$ drastically improves the rate of convergence. The resulting algorithm is shown to have superlinear convergence in $\mathbb{R}^3$ and we give some explanations for the fast convergence observed in simulations in $\mathbb{R}^d$ with d large: by switching periodically between algorithms one destroys the stability of the limiting behaviour obtained when s is fixed (which is always associated with slow convergence).

The chapter is organized as follows. Section 3.2 presents the optimum s-gradient algorithm for the minimization of a quadratic function in $\mathbb{R}^d$, first in the original space and then, after a suitable renormalization, as a transformation applied to a probability measure. Rates of convergence are defined in the same section. The asymptotic behaviour of the optimum s-gradient algorithm in $\mathbb{R}^d$ is considered in Sect. 3.3 where some of the properties established in Forsythe (1968) are recalled. The analysis for the case $s = 2$ is detailed in Sect. 3.4. Switching strategies that periodically alternate between $s = 1$ and $s = 2$ are considered in Sect. 3.5.

3.2 The Optimum s-Gradient Algorithm for the Minimization of a Quadratic Function

Let A be a real bounded self-adjoint (symmetric) operator in a real Hilbert space $\mathcal{H}$ with inner product (x, y) and norm given by $\|x\| = (x, x)^{1/2}$. We shall assume that A is positive, bounded below, and its spectral boundaries will be denoted by m and M:

$$m = \inf_{\|x\|=1} (Ax, x), \quad M = \sup_{\|x\|=1} (Ax, x),$$

with $0 < m < M < \infty$. The function f_0 to be minimized with respect to $t \in \mathcal{H}$ is the quadratic form

$$f_0(t) = \frac{1}{2}(At, t) - (t, y)$$

for some $y \in \mathcal{H}$, the minimum of which is located at $t^* = A^{-1}y$. By a translation of the origin, which corresponds to the definition of $x = t - t^*$ as the variable of interest, the minimization of f_0 becomes equivalent to that of f defined by

$$f(x) = \frac{1}{2}(Ax, x), \tag{3.1}$$

which is minimum at $x^* = 0$. The directional derivative of f at x in the direction u is

$$\nabla_u f(x) = (Ax, u).$$

The direction of steepest descent at x is $-g$, with $g = g(x) = Ax$ the gradient of f at x. The minimum of f along the line $\mathcal{L}_1(x) = \{x + \gamma Ax\,,\ \gamma \in \mathbb{R}\}$ is obtained for the optimum step-length

$$\gamma^* = -\frac{(g, g)}{(Ag, g)},$$

which corresponds to the usual steepest-descent algorithm. One iteration of the steepest-descent algorithm, or optimum 1-gradient method, is thus

$$x_{k+1} = x_k - \frac{(g_k, g_k)}{(Ag_k, g_k)} g_k, \tag{3.2}$$

with $g_k = Ax_k$ and x_0 some initial element in $\mathcal{H}$. For any integer $s \geq 1$, define the s-dimensional plane of steepest descent by

$$\mathcal{L}_s(x) = \{x + \sum_{i=1}^{s} \gamma_i A^i x\,,\ \gamma_i \in \mathbb{R} \text{ for all } i\}.$$

In the optimum s-gradient method, x_{k+1} is chosen as the point in $\mathcal{L}_s(x_k)$ that minimizes f. When $\mathcal{H} = \mathbb{R}^d$, A is $d \times d$ symmetric positive-definite matrix with minimum and maximum eigenvalues, m and M, respectively, and x_{k+1} is uniquely defined provided that the d eigenvalues of A are all distinct. Also, in that case $\mathcal{L}_d(x_k) = \mathbb{R}^d$ and only the case $s \leq d$ is of interest. We shall give special attention to the case $s = 2$.

3.2.1 Updating Rules

Similarly to Pronzato et al. (2001, 2006) and Chap. 2 of this volume, consider the renormalized gradient

$$z(x) = \frac{g(x)}{(g(x), g(x))^{1/2}},$$

so that $(z(x), z(x)) = 1$ and denote $z_k = z(x_k)$ for all k. Also define

$$\mu_j^k = (A^j z_k, z_k)\,,\ j \in \mathbb{Z}\,, \tag{3.3}$$

so that $\mu_0^k = 1$ for any k and the optimum step-length of the optimum 1-gradient at step k is $-1/\mu_1^k$, see (3.2). The optimum choice of the s γ_i's in the optimum s-gradient can be obtained by direct minimization of f over

$\mathcal{L}_s(x_k)$. A simpler construction follows from the observation that g_{k+1}, and thus z_{k+1}, must be orthogonal to $\mathcal{L}_s(x_k)$, and thus to $z_k, Az_k, \ldots, A^{s-1}z_k$. The vector of optimum step-lengthes at step k, $\overrightarrow{\gamma}_k = (\gamma_1^k, \ldots, \gamma_s^k)^\top$, is thus solution of the following system of s linear equations

$$\mathbf{M}_{s,1}^k \overrightarrow{\gamma}_k = -(1, \mu_1^k, \ldots, \mu_{s-1}^k)^\top , \qquad (3.4)$$

where $\mathbf{M}_{s,1}^k$ is the $s \times s$ (symmetric) matrix with element (i,j) given by $\{\mathbf{M}_{s,1}^k\}_{i,j} = \mu_{i+j-1}^k$.

The following remark will be important later on, when we shall compare the rates of convergence of different algorithms.

Remark 1. One may notice that one step of the optimum s-gradient method starting from some x in $\mathcal{H}$ corresponds to s successive steps of the conjugate gradient algorithm starting from the same x, see Luenberger (1973, p. 179).

The next remark shows the connection with optimum design of experiments, which will be further considered in Sect. 3.2.3 (see also Pronzato et al. (2005) where the connection is developed around the case of the steepest-descent algorithm).

Remark 2. Consider least-squares (LS) estimation in a regression model

$$\sum_{i=1}^{s} \gamma_i \, t^i = -1 + \varepsilon_i$$

with (ε_i) a sequence of i.i.d. errors with zero mean. Assume that the t_i's are generated according to a probability (design) measure ξ. Then, the LS estimator of the parameters γ_i, $i = 1, \ldots, s$ is

$$\hat{\gamma} = -\left[\int (t, t^2, \ldots, t^s)^\top (t, t^2, \ldots, t^s) \, \xi(\mathrm{d}t)\right]^{-1} \int (t, t^2, \ldots, t^s)^\top \, \xi(\mathrm{d}t)$$

and coincides with $\overrightarrow{\gamma}_k$ when ξ is such that

$$\int t^{j+1} \, \xi(\mathrm{d}t) = \mu_j^k \, , \; j = 0, 1, 2 \ldots$$

The information matrix $\mathbf{M}(\xi)$ for this LS estimation problem then coincides with $\mathbf{M}_{s,1}^k$.

Using (3.4), one iteration of the optimum s-gradient method thus gives

$$x_{k+1} = Q_s^k(A) x_k \, , \quad g_{k+1} = Q_s^k(A) g_k \qquad (3.5)$$

where $Q_s^k(t)$ is the polynomial $Q_s^k(t) = 1 + \sum_{i=1}^{s} \gamma_i^k t^i$ with the γ_i^k solutions of (3.4). Note that the use of any other polynomial $P(t)$ of degree s or less, and such that $P(0) = 1$, yields a larger value for $f(x_{k+1})$. Using (3.4), we obtain

$$Q_s^k(t) = 1 - (1, \mu_1^k, \ldots, \mu_{s-1}^k)\,[\mathbf{M}_{s,1}^k]^{-1} \begin{pmatrix} t \\ \vdots \\ t^s \end{pmatrix}$$

and direct calculations give

$$Q_s^k(t) = \frac{\begin{vmatrix} 1 & \mu_1^k & \cdots & \mu_{s-1}^k & 1 \\ \mu_1^k & \mu_2^k & \cdots & \mu_s^k & t \\ \vdots & \vdots & \cdots & \vdots & \vdots \\ \mu_s^k & \mu_{s+1}^k & \cdots & \mu_{2s-1}^k & t^s \end{vmatrix}}{|\mathbf{M}_{s,1}^k|} \tag{3.6}$$

where, for any square matrix $\mathbf{M}$, $|\mathbf{M}|$ denotes its determinant. The derivation of the updating rule for the normalized gradient z_k relies on the computation of the inner product (g_{k+1}, g_{k+1}). From the orthogonality property of g_{k+1} to $g_k, Ag_k, \ldots, A^{s-1}g_k$ we get

$$\begin{aligned}(g_{k+1}, g_{k+1}) &= (g_{k+1}, \gamma_s^k A^s g_k) = \gamma_s^k\,(Q_s^k(A)A^s g_k, g_k) \\ &= \frac{\begin{vmatrix} 1 & \mu_1^k & \cdots & \mu_{s-1}^k & \mu_s^k \\ \mu_1^k & \mu_2^k & \cdots & \mu_s^k & \mu_{s+1}^k \\ \vdots & \vdots & \cdots & \vdots & \vdots \\ \mu_s^k & \mu_{s+1}^k & \cdots & \mu_{2s-1}^k & \mu_{2s}^k \end{vmatrix}}{|\mathbf{M}_{s,1}^k|}\,\gamma_s^k\,(g_k, g_k)\,,\end{aligned} \tag{3.7}$$

where γ_s^k, the coefficient of t^s in $Q_s^k(t)$, is given by

$$\gamma_s^k = \frac{\begin{vmatrix} 1 & \mu_1^k & \cdots & \mu_{s-1}^k \\ \mu_1^k & \mu_2^k & \cdots & \mu_s^k \\ \vdots & \vdots & \cdots & \vdots \\ \mu_{s-1}^k & \mu_s^k & \cdots & \mu_{2s-2}^k \end{vmatrix}}{|\mathbf{M}_{s,1}^k|}\,. \tag{3.8}$$

3.2.2 The Optimum s-Gradient Algorithm as a Sequence of Transformations of a Probability Measure

When $\mathcal{H} = \mathbb{R}^d$, we can assume that A is already diagonalized, with eigenvalues $0 < m = \lambda_1 \le \lambda_2 \le \cdots \le \lambda_d = M$, and consider $[z_k]_i^2$, with $[z_k]_i$ the ith component of z_k, as a mass on the eigenvalue λ_i (note that $\sum_{i=1}^d [z_k]_i^2 = \mu_0^k = 1$). Define the discrete probability measure ν_k supported on $(\lambda_1, \ldots, \lambda_d)$ by $\nu_k(\lambda_i) = [z_k]_i^2$, so that its jth moment is μ_j^k, $j \in \mathbb{Z}$, see (3.3).

Remark 3. When two eigenvalues λ_i and λ_j of A are equal, their masses $[z_k]_i^2$ and $[z_k]_j^2$ can be added since the updating rule is the same for the two components $[z_k]_i$ and $[z_k]_j$. Concerning the analysis of the rate of convergence of the optimum s-gradient algorithm, confounding masses associated with equal eigenvalues simply amounts to reducing the dimension of $\mathcal{H}$ and we shall therefore assume that all eigenvalues of A are different when studying the evolution of ν_k.

In the general case where $\mathcal{H}$ is a Hilbert space, let E_λ denote the spectral family associated with the operator A; we then define the measure ν_k by $\nu_k(\mathrm{d}\lambda) = \mathrm{d}(E_\lambda z_k, z_k)$, $m \leq \lambda \leq M$. In both cases $\mathcal{H} = \mathbb{R}^d$ and $\mathcal{H}$ a Hilbert space, we consider ν_k as the spectral measure of A at the iteration k of the algorithm, and write

$$\mu_j^k = \int t^i \, \nu_k(\mathrm{d}t) \, .$$

For any measure ν on the interval $[m, M]$, any $\alpha \in \mathbb{R}$ and any positive integer m define

$$\mathbf{M}_{m,\alpha}(\nu) = \int t^\alpha \, (1, t, t^2, \ldots, t^m)^\top (1, t, t^2, \ldots, t^m) \, \nu(\mathrm{d}t) \, . \tag{3.9}$$

For both $\mathcal{H} = \mathbb{R}^d$ and $\mathcal{H}$ a Hilbert space, the iteration on z_k can be written as

$$\begin{aligned} z_k \to z_{k+1} = T_z(z_k) &= \frac{g_{k+1}}{(g_{k+1}, g_{k+1})^{1/2}} \\ &= \frac{(g_k, g_k)^{1/2}}{(g_{k+1}, g_{k+1})^{1/2}} \, Q_s^k(A) z_k \\ &= \frac{|\mathbf{M}_{s,1}^k|}{|\mathbf{M}_{s,0}^k|^{1/2} |\mathbf{M}_{s-1,0}^k|^{1/2}} \, Q_s^k(A) z_k \, , \end{aligned} \tag{3.10}$$

with $\mathbf{M}_{s,1}^k = \mathbf{M}_{s-1,1}(\nu_k)$ and $\mathbf{M}_{m,0}^k = \mathbf{M}_{m,0}(\nu_k)$, i.e. the $(m+1) \times (m+1)$ matrix with element (i, j) given by $\{\mathbf{M}_{m,0}^k\}_{i,j} = \mu_{i+j-2}^k$. The iteration on z_k can be interpreted as a transformation of the measure ν_k

$$\nu_k \to \nu_{k+1} = T_\nu(\nu_k) \quad \text{with} \quad \nu_{k+1}(\mathrm{d}x) = H_k(x) \nu_k(\mathrm{d}x) \, , \tag{3.11}$$

where, using (3.5–3.8), we have

$$\begin{aligned} H_k(x) &= \frac{[Q_s^k(x)]^2 \, |\mathbf{M}_{s,1}^k|^2}{|\mathbf{M}_{s,0}^k| |\mathbf{M}_{s-1,0}^k|} \\ &= (1, x, \ldots, x^s)[\mathbf{M}_{s,0}^k]^{-1} \begin{pmatrix} 1 \\ x \\ \vdots \\ x^s \end{pmatrix} - (1, x, \ldots, x^{s-1})[\mathbf{M}_{s-1,0}^k]^{-1} \begin{pmatrix} 1 \\ x \\ \vdots \\ x^{s-1} \end{pmatrix} . \end{aligned} \tag{3.12}$$

As moment matrices, $\mathbf{M}_{s,1}^k$ and $\mathbf{M}_{m,0}^k$ are positive semi-definite and $|\mathbf{M}_{s,1}^k| \geq 0$ for any $s \geq 1$ (respectively $|\mathbf{M}_{m,0}^k| \geq 0$ for any $m \geq 0$), with equality if and only if ν_k is supported on strictly less that s points (respectively, on m points or less). Also note that from the construction of the polynomial $Q_s^k(t)$, see (3.5), we have

$$\int Q_s^k(t)\, t^i\, \nu_k(\mathrm{d}t) = (Q_s^k(A)g_k, A^i g_k) = 0\,,\ i = 0, \ldots, s-1\,, \tag{3.13}$$

which can be interpreted as an orthogonality property between the polynomials $Q_s^k(t)$ and t^i for $i = 0, \ldots, s-1$. From this we can easily deduce the following.

Theorem 1. *Assume that ν_k is supported on $s+1$ points at least. Then the polynomial $Q_s^k(t)$ defined by (3.5) has s roots in the open interval (m, M).*

Proof. Let ζ_i, $i = 1, \ldots, q-1$, denote the roots of $Q_s^k(t)$ in (m, M). Suppose that $q - 1 < s$. Consider the polynomial $T(t) = (-1)^{q-1}\, Q_s^k(m) \prod_{i=1}^{q-1}(t - \zeta_i)$, it satisfies $T(t)Q_s^k(t) > 0$ for all $t \in (m, M)$, $t \neq \zeta_i$. Therefore, $\int T(t)Q_s^k(t)\nu_k(\mathrm{d}t) > 0$, which contradicts (3.13) since $T(t)$ has degree $q - 1 \leq s - 1$. □

Remark 4. Theorem 1 implies that $Q_s^k(m)$ has the same sign as $Q_s^k(0) = (-1)^s|\mathbf{M}_{s-1,0}^k|/|\mathbf{M}_{s,1}^k|$, that is, $(-1)^s$. Similarly, $Q_s^k(M)$ has the same sign as $\lim_{t\to\infty} Q_s^k(t)$ which is positive.

Using the orthogonality property (3.13), we can also prove the next two theorems concerning the support of ν_k, see Forsythe (1968).

Theorem 2. *Assume that the measure ν_k is supported on $s+1$ points at least. Then this is also true for the measure ν_{k+1} obtained through (3.11).*

Proof. We only need to consider the case when the support $\mathcal{S}_k$ of ν_k is finite, that is, when $\mathcal{H} = \mathbb{R}^d$. Suppose that ν_k is supported on n points, $n \geq s+1$. The determinants $|\mathbf{M}_{s,1}^k|$, $|\mathbf{M}_{s,0}^k|$ and $|\mathbf{M}_{s-1,1}^k|$ in (3.12) are thus strictly positive. Let q be the largest integer such that there exist $\lambda_{i_1} < \lambda_{i_2} < \cdots < \lambda_{i_q}$ in $\mathcal{S}_k$ with $Q_s^k(\lambda_{i_j})Q_s^k(\lambda_{i_{j+1}}) < 0$, $j = 1, \ldots, q-1$. We shall prove that $q \geq s+1$. From (3.11, 3.12) it implies that ν_k is supported on $s+1$ points at least.

From (3.13), $\int Q_s^k(t)\, \nu_k(\mathrm{d}t) = 0$, so that there exist λ_{i_1} and λ_{i_2} in $\mathcal{S}_k$ with $Q_s^k(\lambda_{i_1})Q_s^k(\lambda_{i_2}) < 0$, therefore $q \geq 2$. Suppose that $q \leq s$. By construction $Q_s^k(\lambda_{i_1})$ is of the same sign as $Q_s^k(m)$ and we can construct q disjoint open intervals Λ_j, $j = 1, \ldots, q$ such that $\lambda_{i_j} \in \Lambda_j$ and $Q_s^k(\lambda_i)Q_s^k(\lambda_{i_j}) \geq 0$ for all $\lambda_i \in \mathcal{S}_k \cap \Lambda_j$ with $\cup_{j=1}^q \bar{\Lambda}_j = [m, M]$, where $\bar{\Lambda}_j$ is the closure of Λ_j (notice that $Q_s^k(\lambda)$ may change sign in Λ_j but all the $Q_s^k(\lambda_i)$'s are of the same sign for $\lambda_i \in \mathcal{S}_k \cap \Lambda_j$). Consider the $q-1$ scalars ζ_i, $i = 1, \ldots, q-1$, defined by the endpoints of the Λ_j's, m and M excluded; they satisfy $\lambda_{i_1} < \zeta_1 < \lambda_{i_2} < \cdots < \zeta_{q-1} < \lambda_{i_q}$. Form now the polynomial $T(t) = (-1)^{s+q-1}(t - \zeta_1) \times \cdots \times (t - \zeta_{q-1})$, one can check that $Q_s^k(\lambda_i)T(\lambda_i) \geq 0$ for all λ_i in $\mathcal{S}_k$. Also, $Q_s^k(\lambda_{i_j})T(\lambda_{i_j}) > 0$ for $j = 1, \ldots, q$. This implies $\sum_{i=1}^d T(\lambda_i)Q_s^k(\lambda_i)[z_k]_i^2 > 0$, which contradicts (3.13) since $T(t)$ has degree $q - 1 \leq s - 1$. □

Corollary 1. *If x_0 is such that ν_0 is supported on $n_0 = s+1$ points at least, then $(g_k, g_k) > 0$ for all $k \geq 1$. Also, the determinants of all $(m+1)\times(m+1)$ moment matrices $\mathbf{M}_{m,\alpha}(\nu_k)$, see (3.9), are strictly positive for all $k \geq 1$.*

Theorem 3. *Assume that ν_0 is supported on $n_0 = s+1$ points. Then $\nu_{2k} = \nu_0$ for all k.*

Proof. It is enough to prove that g_{2k} is parallel to g_0, and thus that g_2 is parallel to g_0. Since the updating rule only concerns nonzero components, we may assume that $d = s+1$. We have $g_1 = Q_s^0(A)g_0$, $g_2 = Q_s^1(A)g_1$, g_1 is orthogonal to $g_0, Ag_0, \ldots, A^{s-1}g_0$, which are independent, and g_2 is orthogonal to g_1. We can thus decompose g_2 with respect to the basis $g_0, Ag_0, \ldots, A^{s-1}g_0$ as

$$g_2 = \sum_{i=0}^{s-1} \alpha_i\, A^i g_0 \,.$$

Now, g_2 is orthogonal to Ag_1, and thus

$$g_1^\top A g_2 = 0 = \sum_{i=0}^{s-1} \alpha_i\, g_1^\top A^{i+1} g_0 = \alpha_{s-1}\, g_1^\top A^s g_0$$

with $g_1^\top A^s g_0 \neq 0$ since otherwise g_1 would be zero. Therefore, $\alpha_{s-1} = 0$. Similarly, g_2 is orthogonal to $A^2 g_1$, which gives

$$0 = \sum_{i=0}^{s-2} \alpha_i\, g_1^\top A^{i+2} g_0 = \alpha_{s-2}\, g_1^\top A^s g_0$$

so that $\alpha_{s-2} = 0$. Continuing like that up to $g_1^\top A^{s-1} g_2$ we obtain $\alpha_1 = \alpha_2 = \cdots = \alpha_{s-1} = 0$ and $g_2 = \alpha_0 g_0$. Notice that $\alpha_0 > 0$ since $(A^{-1}g_2, g_0) = (A^{-1}g_1, g_1)$, see (3.18). □

The transformation $z_k \to z_{k+1} = T_z(z_k)$ (respectively $\nu_k \to \nu_{k+1} = T_\nu(\nu_k)$) can be considered as defining a dynamical system with state $z_k \in \mathcal{H}$ at iteration k (respectively, $\nu_k \in \Pi$, the set of probability measures defined on the spectrum of A). One purpose of the chapter is to investigate the limit set of the orbit of the system starting at z_0 or ν_0. As it is classical in the study of stability of dynamical systems where Lyapunov functions often play a key role (through the Lyapunov Stability Theorem or Lasalle's Invariance Principle, see, e.g. Elaydi (2005, Chap. 4), the presence of monotone sequences in the dynamics of the renormalized algorithm will be an important ingredient of the analysis. Theorem 3 shows that the behaviour of the renormalized algorithm may be periodic with period 2. We shall see that this type of behaviour is typical, although the structure of the attractor may be rather complicated.

3.2.3 Rates of Convergence and Monotone Sequences

3.2.3.1 Rates of Convergence

Consider the following rate of convergence of the algorithm at iteration k,

$$r_k = \frac{f(x_{k+1})}{f(x_k)} = \frac{(Ax_{k+1}, x_{k+1})}{(Ax_k, x_k)} = \frac{(A^{-1}g_{k+1}, g_{k+1})}{(A^{-1}g_k, g_k)} . \tag{3.14}$$

From the orthogonality property of g_{k+1} we have

$$(A^{-1}g_{k+1}, g_{k+1}) = (A^{-1}Q_s^k(A)g_k, g_{k+1}) = (A^{-1}g_k, g_{k+1})$$

and thus, using (3.6),

$$r_k = \frac{|\mathbf{M}_{s,-1}^k|}{|\mathbf{M}_{s,1}^k|\,\mu_{-1}^k}$$

with $\mathbf{M}_{s,-1}^k = \mathbf{M}_{s,-1}(\nu_k)$, see (3.9), that is, the $(s+1)\times(s+1)$ moment matrix with element (i,j) given by $\{\mathbf{M}_{s,-1}^k\}_{i,j} = \mu_{i+j-3}^k$. Using the orthogonality property of g_{k+1} again, we can easily prove that the sequence of rates (r_k) is non-decreasing along the trajectory followed by the algorithm.

Theorem 4. *When x_0 is such that ν_0 is supported on $s+1$ points at least, the rate of convergence r_k defined by (3.14) is non-decreasing along the path followed by the optimum s-gradient algorithm. It also satisfies*

$$r_k \le R_s^* = T_s^{-2}\left(\frac{\varrho+1}{\varrho-1}\right) \tag{3.15}$$

where $\varrho = M/m$ is the condition number of A and $T_s(t)$ is the sth Chebyshev polynomial (normalized so that $\max_{t\in[-1,1]}|T_s(t)| = 1$)

$$T_s(t) = \cos[s\arccos(t)] = \frac{(t+\sqrt{t^2-1})^s + (t-\sqrt{t^2-1})^s}{2} . \tag{3.16}$$

Moreover, the equality in (3.15) is obtained when ν_k is the measure ν_s^* defined by

$$\nu_s^*(y_0) = \nu_s^*(y_s) = 1/(2s)\,,\ \nu_s^*(y_j) = 1/s\,,\ 1 \le j \le s-1\,, \tag{3.17}$$

where $y_j = (M+m)/2 + [\cos(j\pi/s)](M-m)/2$.

Proof. From Corollary 1, $(g_0, g_0) > 0$ implies $(g_k, g_k) > 0$ for all k and r_k is thus well defined. Straightforward manipulations give

$$(A^{-1}g_{k+1}, g_{k+1}) - (A^{-1}g_{k+2}, g_k) = \left(A^{-1}[Q_s^k(A) \quad Q_s^{k+1}(A)]g_k, Q_s^k(A)g_k\right)$$
$$= \left(A^{-1}\sum_{i=1}^{s}(\gamma_i^k - \gamma_i^{k+1})A^i g_k, Q_s^k(A)g_k\right) = 0 \tag{3.18}$$

where equality to zero follows from (3.13). Therefore, from the Cauchy–Schwarz inequality,

$$(A^{-1}g_{k+1}, g_{k+1})^2 = (A^{-1}g_{k+2}, g_k)^2 \leq (A^{-1}g_{k+2}, g_{k+2})(A^{-1}g_k, g_k)$$

and $r_k \leq r_{k+1}$, with equality if and only if $g_{k+2} = \alpha g_k$ for some $\alpha \in \mathbb{R}^+$ ($\alpha > 0$ since $(A^{-1}g_{k+2}, g_k) = (A^{-1}g_{k+1}, g_{k+1})$). This shows that r_k is non-decreasing.

The rate (3.14) can also be written as

$$r_k = \left[\mu_{-1}^k \{(\mathbf{M}_{s,-1}^k)^{-1}\}_{1,1}\right]^{-1} .$$

Define the measure $\bar{\nu}_k$ by $\bar{\nu}_k(\mathrm{d}t) = \nu_k(\mathrm{d}t)/(t\mu_{-1}^k)$ (so that $\int \bar{\nu}_k(\mathrm{d}t) = 1$) and denote $\bar{\mathbf{M}}_{m,n}^k$ the matrix obtained by substituting $\bar{\nu}_k$ for ν_k in $\mathbf{M}_{m,n}^k$ for any n, m. Then, $\bar{\mathbf{M}}_{s,0}^k = \mathbf{M}_{s,-1}^k/\mu_{-1}^k$ and $r_k = \left[\{(\bar{\mathbf{M}}_{s,0}^k)^{-1}\}_{1,1}\right]^{-1}$. The maximum value for r_k is thus obtained for the D_s-optimal measure $\bar{\nu}_s^*$ on $[m, M]$ for the estimation θ_0 in the linear regression model $\eta(\theta, x) = \sum_{i=0}^{s} \theta_i x^i$ with i.i.d. errors, see, e.g. Fedorov (1972, p. 144) and Silvey (1980, p. 10) ($\bar{\nu}_s^*$ is also c-optimal for $c = (1, 0, \ldots, 0)^\top$). This measure is uniquely defined, see Hoel and Levine (1964) and Sahm (1998, p. 52): it is supported at the $s+1$ points $y_j = (M+m)/2 + [\cos(j\pi/s)](M-m)/2$, $j = 0, \ldots, s$, and each y_j receives a weight proportional to α_j/y_j, with $\alpha_0 = \alpha_s = 1/2$ and $\alpha_j = 1$ for $j = 1, \ldots, s-1$. Applying the transformation $\nu(\mathrm{d}t) = t\bar{\nu}(\mathrm{d}t)\mu_{-1}$ we obtain the measure ν_s^* given by (3.17). ■

Remark 5. Meinardus (1963) and Forsythe (1968) arrive at the result (3.15) by a different route. They write

$$r_k = \frac{(A^{-1}g_{k+1}, g_{k+1})}{(A^{-1}g_k, g_k)} = \frac{\int [Q_s^k(t)]^2\, t^{-1}\, \nu_k(\mathrm{d}t)}{\mu_{-1}^k} .$$

Since $Q_s^k(t)$ minimizes $f(x_{k+1})$, $r_k \leq (1/\mu_{-1}^k) \int P^2(t)\, t^{-1}\, \nu_k(\mathrm{d}t)$ for any s-degree polynomial $P(t)$ such that $P(0) = 1$. Equivalently,

$$r_k \leq \frac{\int S^2(t)\, t^{-1}\, \nu_k(\mathrm{d}t)}{S^2(0)\mu_{-1}^k}$$

for any s-degree polynomial $S(t)$. Take $S(t) = S^*(t) = T_s[(M+m-2t)/(M-m)]$, so that $S^2(t) \leq 1$ for $t \in [m, M]$, then r_k satisfies $r_k \leq [S^*(0)]^{-2} = R_s^*$ with R_s^* given by (3.15). □

Notice that $T_s[(\varrho+1)/(\varrho-1)] > 1$ in (3.15), so that we have the following.

Corollary 2. *If x_0 is such that ν_0 is supported on $s+1$ points at least, then the optimum s-gradient algorithm converges linearly to the optimum, that is,*

$$0 < c_1 = \frac{f(x_1)}{f(x_0)} \leq \frac{f(x_{k+1})}{f(x_k)} \leq R_s^* < 1\,, \quad \textit{for all } k\,.$$

Moreover, the convergence slows down monotonically on the route to the optimum and the rate r_k given by (3.14) tends to a limit r_∞.

The monotonicity of the sequence (r_k), together with Theorem 3, has the following consequence.

Corollary 3. *Assume that ν_0 is supported on $n_0 = s+1$ points. Then $r_{k+1} = r_k$ for all k.*

Other rates of convergence can be defined as

$$R_k(W) = \frac{(Wg_{k+1}, g_{k+1})}{(Wg_k, g_k)} \tag{3.19}$$

with W be a bounded positive self-adjoint operator in $\mathcal{H}$. However, the following theorem shows that all such rates are asymptotically equivalent, see Pronzato et al. (2006).

Theorem 5. *Let W be a bounded positive self-adjoint operator in $\mathcal{H}$, with bounds c and C such that $0 < c < C < \infty$ (when $\mathcal{H} = \mathbb{R}^d$, W is a $d \times d$ positive-definite matrix with minimum and maximum eigenvalues respectively, c and C). Consider the rate of convergence defined by (3.19) if $\|g_k\| \neq 0$ and $R_k(W) = 1$ otherwise. Apply the optimum s-gradient algorithm (3.5), initialized at $g_0 = g(x_0)$, for the minimization of $f(x)$ given by (3.1). Then the limit*

$$R(W, x_0) = \lim_{n\to\infty} \left[\prod_{k=0}^{n-1} R_k(W)\right]^{1/n}$$

exists for all x_0 in $\mathcal{H}$ and $R(W, x_0) = R(x_0)$ does not depend on W. In particular,

$$R(W, x_0) = r_\infty = \lim_{n\to\infty} \left(\prod_{k=0}^{n-1} r_k\right)^{1/n} \tag{3.20}$$

with r_k defined by (3.14).

Proof. Assume that x_0 is such that for some $k \geq 0$, $\|g_{k+1}\| = 0$ with $\|g_i\| > 0$ for all $i \leq k$ (that is, $x_{k+1} = x^*$ and $x_i \neq x^*$ for $i \leq k$). This implies $R_k(W) = 0$ for any W, and therefore $R(W, x_0) = R(x_0) = 0$.

Assume now that $\|g_k\| > 0$ for all k. Consider

$$V_n = \left[\prod_{k=0}^{n-1} R_k(W)\right]^{1/n} = \left[\prod_{k=0}^{n-1} \frac{(Wg_{k+1}, g_{k+1})}{(Wg_k, g_k)}\right]^{1/n} = \left[\frac{(Wg_n, g_n)}{(Wg_0, g_0)}\right]^{1/n}.$$

We have,

$$\forall z \in \mathcal{H},\ c\|z\|^2 \leq (Wz, z) \leq C\|z\|^2,$$

and thus

$$(c/C)^{1/n} \left[\frac{(g_n, g_n)}{(g_0, g_0)}\right]^{1/n} \leq V_n \leq (C/c)^{1/n} \left[\frac{(g_n, g_n)}{(g_0, g_0)}\right]^{1/n}.$$

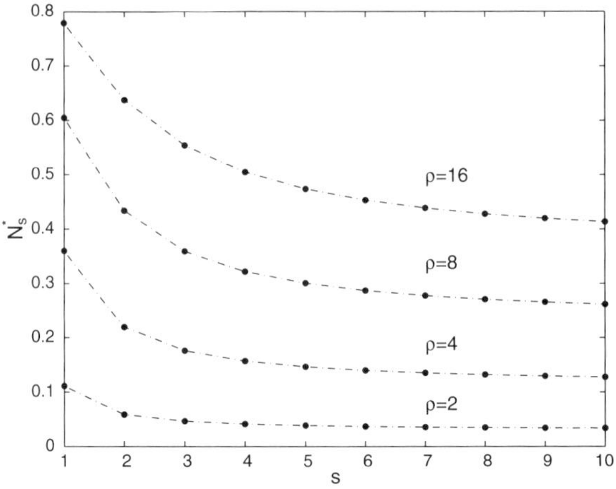

Fig. 3.1. Upper bounds N_s^*, see (3.21), as functions of s for different values of the condition number ϱ

Since $(c/C)^{1/n} \to 1$ and $(C/c)^{1/n} \to 1$ as $n \to \infty$, $\liminf_{n\to\infty} V_n$ and $\limsup_{n\to\infty} V_n$ do not depend on W. Taking $W = A^{-1}$ we get $R_k(W) = r_k$. The sequence (r_k) is not decreasing, and thus $\lim_{n\to\infty} V_n = r_\infty$ for any W. □

For any fixed $\varrho = M/m$, the bound R_s^* given by (3.15) tends to zero as s tends to infinity whatever the dimension d when $\mathcal{H} = \mathbb{R}^d$, and also when $\mathcal{H}$ is a Hilbert space. However, since one step of the optimum s-gradient method corresponds to s successive steps of the conjugate gradient algorithm, see Remark 1, a normalized version of the convergence rate allowing comparison with classical steepest descent is $r_k^{1/s}$, which is bounded by

$$N_s^* = (R_s^*)^{1/s} = T_s^{-2/s}\left(\frac{\varrho+1}{\varrho-1}\right) \tag{3.21}$$

where $T_s(t)$ is the s-th Chebyshev polynomial, see (3.16). The quantity N_s^* is a decreasing function of s, see Fig. 3.1, but has a positive limit when s tends to infinity,

$$\lim_{s\to\infty} N_s^* = N_\infty^* = \frac{(\sqrt{\varrho}-1)^2}{(\sqrt{\varrho}+1)^2}\,. \tag{3.22}$$

Figure 3.2 shows the evolution of N_∞^* as a function of the condition number ϱ.

3.2.3.2 A Second Monotone-Bounded Sequence

Another quantity q_k also turns out to be non-decreasing along the trajectory followed by the algorithm, as shown in the following theorem.

Theorem 6. *When x_0 is such that ν_0 is supported on $s+1$ points at least, the quantity q_k defined by*

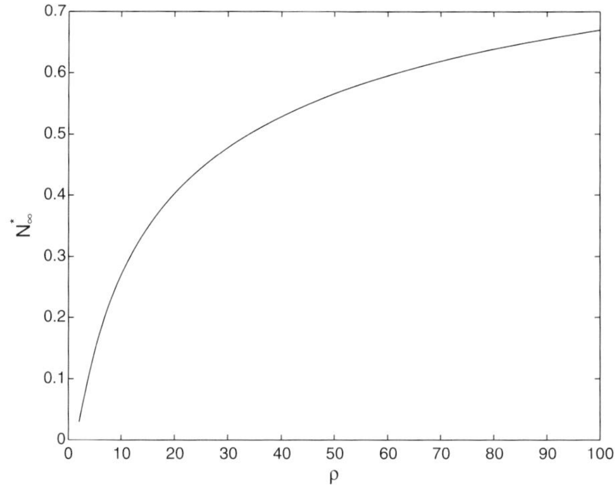

Fig. 3.2. Limiting value N_∞^* as a function of ϱ

$$q_k = \frac{(g_{k+1}, g_{k+1})}{(\gamma_s^k)^2\,(g_k, g_k)}\,, \tag{3.23}$$

with γ_s^k given by (3.8), is non-decreasing along the path followed by the optimum s-gradient algorithm. Moreover, it satisfies

$$q_k \leq q_s^* = \frac{(M-m)^{2s}}{2^{4s-2}}\,, \quad \textit{for all } k\,, \tag{3.24}$$

where the equality is obtained when ν_k is the measure (3.17) of Theorem 4.

Proof. From Corollary 1, $(g_0, g_0) > 0$ implies $(g_k, g_k) > 0$ and $\gamma_s^k > 0$ for all k so that q_k is well defined. Using the same approach as for r_k in Theorem 4, we write

$$\begin{aligned}
(g_{k+1}, g_{k+1})/\gamma_s^k - (g_{k+2}, g_k)/\gamma_s^{k+1} &= \left(Q_s^k(A)g_k, g_{k+1}\right)/\gamma_s^k \\
&\qquad \left(Q_s^{k+1}(A)g_k, g_{k+1}\right)/\gamma_s^{k+1} \\
&= \left([Q_s^k(A)/\gamma_s^k - Q_s^{k+1}(A)/\gamma_s^{k+1}]g_k, g_{k+1}\right) \\
&= \left(\left[(1/\gamma_s^k - 1/\gamma_s^{k+1})I + \sum_{i=1}^{s-1}(\gamma_i^k/\gamma_s^k - \gamma_i^{k+1}/\gamma_s^{k+1})A^i\right] g_k, g_{k+1}\right) \\
&= 0\,,
\end{aligned}$$

where equality to zero follows from (3.13). The Cauchy–Schwarz inequality then implies

$$(g_{k+1}, g_{k+1})^2/(\gamma_s^k)^2 = (g_{k+2}, g_k)^2/(\gamma_s^{k+1})^2 \leq (g_{k+2}, g_{k+2})(g_k, g_k)/(\gamma_s^{k+1})^2$$

and $q_k \leq q_{k+1}$ with equality if and only if $g_{k+2} = \alpha g_k$ for some $\alpha \in \mathbb{R}^+$ ($\alpha > 0$ since $(g_{k+2}, g_k) = (g_{k+1}, g_{k+1})\gamma_s^{k+1}/\gamma_s^k$ and $\gamma_s^k > 0$ for all k, see Corollary 1). This shows that q_k is non-decreasing.

The determination of the probability measure that maximizes q_k is again related to optimal design theory. Using (3.7) and (3.8), we obtain

$$q_k = \frac{|\mathbf{M}_{s,0}^k|}{|\mathbf{M}_{s-1,0}^k|}\,.$$

Hence, using the inversion of a partitioned matrix, we can write

$$q_k = \mu_{2s}^k - (1, \mu_1^k, \ldots, \mu_{s-1}^k)(\mathbf{M}_{s-1,0}^k)^{-1}\begin{pmatrix} 1 \\ \mu_1^k \\ \vdots \\ \mu_{s-1}^k \end{pmatrix} = \left(\left\{(\mathbf{M}_{s,0}^k)^{-1}\right\}_{s+1,s+1}\right)^{-1},$$

so that the maximization of q_k with respect to ν_k is equivalent to the determination of a D_s-optimum measure on $[m, M]$ for the estimation of θ_s in the linear regression model $\eta(\theta, x) = \sum_{i=0}^{s} \theta_i x^i$ with i.i.d. errors (or to the determination of a c-optimal measure on $[m, M]$ with $c = (0, \ldots, 0, 1)^\top$). This measure is uniquely defined, see (Kiefer and Wolfowitz, 1959, p. 283): when the design interval is normalized to $[-1, 1]$, the optimum measure ξ^* is supported on $s+1$ points given by ± 1 and the $s-1$ zeros of the derivative of s-th Chebyshev polynomial $T_s(t)$ given by (3.16) and the weights are

$$\xi^*(-1) = \xi^*(1) = 1/(2s)\,,\ \xi^*(\cos[j\pi/s]) = 1/s\,,\ 1 \leq j \leq s-1\,.$$

The transformation $t \in [-1, 1] \mapsto z = (M+m)/2 + t(M-m)/2 \in [m, M]$ gives the measure ν_s^* on $[m, M]$. The associated maximum value for q_k is q_s^* given by (3.24), see Kiefer and Wolfowitz (1959, p. 283). □

The monotonicity of the sequence (q_k), together with Theorem 3, implies the following analogue to Corollary 3.

Corollary 4. *Assume that ν_0 is supported on $n_0 = s+1$ points. Then $q_{k+1} = q_k$ for all k.*

As a non-decreasing and bounded sequence, (q_k) tends to a limit q_∞. The existence of limiting values r_∞ and q_∞ will be essential for studying the limit points of the orbits (z_{2k}) or (ν_{2k}) in the next sections. In the rest of the chapter we only consider the case where $\mathcal{H} = \mathbb{R}^d$ and assume that A is diagonalized with d distinct eigenvalues $0 < m = \lambda_1 < \lambda_2 < \cdots < \lambda_d = M$.

3.3 Asymptotic Behaviour of the Optimum s-Gradient Algorithm in $\mathbb{R}^d$

The situation is much more complex when $s \geq 2$ than for $s = 1$, and we shall recall some of the properties established in Forsythe (1968) for $\mathcal{H} = \mathbb{R}^d$.

If x_0 is such that the initial measure ν_0 is supported on s points or less, the algorithm terminates in one step. In the rest of the section we thus suppose that ν_0 is supported on $n_0 \geq s+1$ points. The algorithm then converges linearly to the optimum, see Corollary 2. The set $\mathcal{Z}(x_0)$ of limit points of the sequence of renormalized gradients z_{2k} satisfies the following.

Theorem 7. *If x_0 is such that ν_0 has $s+1$ support points at least, the set $\mathcal{Z}(x_0)$ of limit points of the sequence of renormalized gradients z_{2k}, $k = 0, 1, 2, \ldots$, is a closed connected subset of the d-dimensional unit sphere $\mathcal{S}_d$. Any y in $\mathcal{Z}(x_0)$ satisfies $T_z^2(y) = y$, where $T_z^2(y) = T_z[T_z(y)]$ with T_z defined by (3.10).*

Proof. Using (3.18) we get

$$r_{k+1} - r_k = \frac{(A^{-1}g_{k+2}, g_{k+2})}{(A^{-1}g_{k+1}, g_{k+1})} \times \left[1 - \frac{(A^{-1}g_{k+2}, g_k)^2}{(A^{-1}g_{k+2}, g_{k+2})(A^{-1}g_k, g_k)}\right]$$
$$= r_{k+1} \times \left[1 - \frac{(A^{-1}z_{k+2}, z_k)^2}{(A^{-1}z_{k+2}, z_{k+2})(A^{-1}z_k, z_k)}\right].$$

Since r_k is non-decreasing and bounded, see Theorem 4, r_k tends to a limit $r_\infty(x_0)$ and, using Cauchy–Schwarz inequality $\|z_{k+2} - z_k\| \to 0$. The set $\mathcal{Z}(x_0)$ of limit points of the sequence z_{2k} is thus a continuum on $\mathcal{S}_d$.

Take any $y \in \mathcal{Z}(x_0)$. There exists a subsequence (k_i) such that $z_{2k_i} \to y$ as $i \to \infty$, and $z_{2k_i+2} = T_z^2(z_{2k_i}) \to y$ since $\|z_{2k_i+2} - z_{2k_i}\| \to 0$. The continuity of T_z then implies $z_{2k_i+2} \to T_z^2(y)$ and thus $T_z^2(y) = y$. □

Obviously, the sequence (z_{2k+1}) satisfies a similar property (consider ν_1 as a new initial measure ν_0), so that we only need to consider the sequence of even iterates.

Remark 6. Forsythe (1968) conjectures that the continuum for $\mathcal{Z}(x_0)$ is in fact always a single point. Although it is confirmed by numerical simulations, we are not aware of any proof of attraction to a single point. One may however think of $\mathcal{Z}(x_0)$ as the set of possible limit points for the sequence (z_{2k}), leaving open the possibility for attraction to a particular point y^* in $\mathcal{Z}(x_0)$. Examples of sets $\mathcal{Z}(x_0)$ will be presented in Sect. 3.4. Note that we shall speak of attraction and attractors although the terms are somewhat inaccurate: starting from x'_0 such that $z'_0 = g(x'_0)/\|g(x'_0)\|$ is arbitrarily close to some y in $\mathcal{Z}(x_0)$ yields a limit set $\mathcal{Z}(x'_0)$ for the iterates z'_{2k} close to $\mathcal{Z}(x_0)$ but $\mathcal{Z}(x'_0) \neq \mathcal{Z}(x_0)$. □

Some coordinates $[y]_i$ of a given y in $\mathcal{Z}(x_0)$ may equal zero, $i \in \{1, \ldots, d\}$. Define the asymptotic spectrum $\mathcal{S}(x_0, y)$ at $y \in \mathcal{Z}(x_0)$ as the set of eigenvalues λ_i such that $[y]_i \neq 0$ and let $n = n(x_0, y)$ be the number of points in $\mathcal{S}(x_0, y)$. We shall then say that $\mathcal{S}(x_0, y)$ is an n-point asymptotic spectrum. We know from Theorem 3 that if ν_0 is supported on exactly $n_0 = s+1$ points then

$n(x_0, y) = s+1$ (and $\nu_{2k} = \nu_0$ for all k so that $\mathcal{Z}(x_0)$ is the singleton $\{z_0\}$). In the more general situation where ν_0 is supported on $n_0 \geq s+1$ points, $n(x_0, y)$ satisfies the following, see Forsythe (1968).

Theorem 8. *Assume that ν_0 is supported on $n_0 > s+1$ points. Then the number of points $n(x_0, y)$ in the asymptotic spectrum $\mathcal{S}(x_0, y)$ of any $y \in \mathcal{Z}(x_0)$ satisfies*

$$s + 1 \leq n(x_0, y) \leq 2s\,.$$

Proof. Take any y in $\mathcal{Z}(x_0)$, let n be the number of its non-zero components. To this y we associate a measure ν through $\nu(\lambda_i) = [y]_i^2$, $i = 1, \ldots, d$ and we construct a polynomial $Q_s(t)$ from the moments of ν, see (3.6). Applying the transformation (3.11) to ν we get the measure ν' from which we construct the polynomial $Q'_s(t)$. The invariance property $T_z^2(y) = T_z[T_z(y)] = y$, with T_z defined by (3.10), implies $Q_s(\lambda_i)Q'_s(\lambda_i)[y]_i = c[y]_i$, $c > 0$, where $[y]_i$ is any non-zero component of y, $i = 1, \ldots, n$. The equation $Q_s(t)Q'_s(t) = c > 0$ can have between 1 and $2s$ solutions in (m, M). We know already from Theorem 2 that $n(x_0, y) \geq s+1$. □

The following theorem shows that when m and M are support points of ν_0, then the asymptotic spectrum of any $y \in \mathcal{Z}(x_0)$ also contains m and M.

Theorem 9. *Assume that ν_0 is supported on $n_0 \geq s+1$ points and that $\nu_0(m) > 0$, $\nu_0(M) > 0$. Then $\liminf_{k\to\infty} \nu_k(m) > 0$ and $\liminf_{k\to\infty} \nu_k(M) > 0$.*

Proof. We only consider the case for M, the proof being similar for m.

First notice that from Theorem 1, all roots of the polynomials $Q_s^k(t)$ lie in the open interval (m, M), so that $\nu_k(M) > 0$ for any k.

Suppose that $\liminf_{k\to\infty} \nu_k(M) = 0$. Then there exists a subsequence (k_i) such that z_{2k_i} tends to some y in $\mathcal{Z}(x_0)$ and $\lim_{i\to\infty} \nu_{2k_i}(M) = 0$. To this y we associate a measure ν as in the proof of Theorem 8 and construct a polynomial $Q_s(t)$ from the moments of ν, see (3.6). Since $\nu_{2k_i}(M) \to 0$, $\nu(M) = 0$. Let λ_j be the largest eigenvalue of A such that $\nu(\lambda_j) > 0$. Then, all zeros of $Q_s(t)$ lie in (m, λ_j) and the same is true for the polynomial $Q'_s(t)$ constructed from the measure $\nu' = T_\nu(\nu)$ obtained by the transformation (3.11). Hence, $Q_s(t)$ and $Q'_s(t)$ are increasing (and positive, see Remark 4) for t between λ_j and M, so that $Q_s(M)Q'_s(M) > Q_s(\lambda_j)Q'_s(\lambda_j)$. This implies by continuity

$$Q_s^{2k_i}(M)Q_s^{2k_i+1}(M) \geq c\,Q_s^{2k_i}(\lambda_j)Q_s^{2k_i+1}(\lambda_j)$$

for some $c > 1$ and all i larger than some i_0. Therefore,

$$\frac{[g_{2k_i+2}]_d^2}{[g_{2k_i}]_d^2} \geq c^2 \frac{[g_{2k_i+2}]_j^2}{[g_{2k_i}]_j^2}\,, \quad i > i_0\,,$$

see (3.6), and thus

$$\frac{[z_{2k_i+2}]_d^2}{[z_{2k_i+2}]_j^2} \geq c^2 \frac{[z_{2k_i}]_d^2}{[z_{2k_i}]_j^2}, \quad i > i_0 .$$

Since $[z_k]_d^2 > 0$ for all k and $[z_{2k_i}]_j^2 \to \nu(\lambda_j) > 0$ this implies $[z_{2k_i}]_d^2 \to \infty$ as $i \to \infty$, which is impossible. Therefore, $\liminf_{k\to\infty} \nu_k(M) > 0$. □

The properties above explain the asymptotic behaviour of the steepest-descent algorithm in $\mathbb{R}^d$: when $s = 1$ and ν_0 is supported on two points at least, including m and M, then $n(x_0, y) = 2$ for any y in $\mathcal{Z}(x_0)$ and m and M are in the asymptotic spectrum $S(x_0, y)$ of any $y \in \mathcal{Z}(x_0)$. Therefore, $S(x_0, y) = \{m, M\}$ for all $y \in \mathcal{Z}(x_0)$. Since $\mathcal{Z}(x_0)$ is a part of the unit sphere $\mathcal{S}_d$, $\|y\| = 1$ and there is only one degree of freedom. The limiting value r_∞ of r_k then defines the attractor uniquely and $\mathcal{Z}(x_0)$ is a singleton.

In the case where s is even, Forsythe (1968) gives examples of invariant measures ν_0 satisfying $\nu_{k+2} = \nu_k$ and supported on $2q$ points with $s + 1 < 2q \leq 2s$, or supported on $2q + 1$ points with $s + 1 \leq 2q + 1 < 2s$. The nature of the sets $\mathcal{Z}(x_0)$ and $\mathcal{S}(x_0, y)$ is investigated more deeply in the next section for the case $s = 2$.

3.4 The Optimum 2-Gradient Algorithm in $\mathbb{R}^d$

In all the sections, we omit the index k in the moments μ_j^k and matrices $\mathbf{M}_{m,n}^k$. The polynomial $Q_2^k(t)$ defined by (3.6) is then

$$Q_2^k(t) = \frac{\begin{vmatrix} 1 & \mu_1 & 1 \\ \mu_1 & \mu_2 & t \\ \mu_2 & \mu_3 & t^2 \end{vmatrix}}{|\mathbf{M}_{2,1}|} = \frac{\begin{vmatrix} 1 & \mu_1 & 1 \\ \mu_1 & \mu_2 & t \\ \mu_2 & \mu_3 & t^2 \end{vmatrix}}{\begin{vmatrix} \mu_1 & \mu_2 \\ \mu_2 & \mu_3 \end{vmatrix}}$$

and the function $H_k(x)$, see (3.12), is given by

$$H_k(x) = \frac{\begin{vmatrix} 1 & \mu_1 & 1 \\ \mu_1 & \mu_2 & x \\ \mu_2 & \mu_3 & x^2 \end{vmatrix}^2}{\begin{vmatrix} 1 & \mu_1 \\ \mu_1 & \mu_2 \end{vmatrix} \begin{vmatrix} 1 & \mu_1 & \mu_2 \\ \mu_1 & \mu_2 & \mu_3 \\ \mu_2 & \mu_3 & \mu_4 \end{vmatrix}} \tag{3.25}$$

$$= \begin{pmatrix} 1 & x & x^2 \end{pmatrix} \begin{pmatrix} 1 & \mu_1 & \mu_2 \\ \mu_1 & \mu_2 & \mu_3 \\ \mu_2 & \mu_3 & \mu_4 \end{pmatrix}^{-1} \begin{pmatrix} 1 \\ x \\ x^2 \end{pmatrix} - \begin{pmatrix} 1 & x \end{pmatrix} \begin{pmatrix} 1 & \mu_1 \\ \mu_1 & \mu_2 \end{pmatrix}^{-1} \begin{pmatrix} 1 \\ x \end{pmatrix} .$$

The monotone sequences (r_k) and (q_k) of Sect. 3.2.3, see (3.14, 3.23), are given by

$$r_k = \frac{\begin{vmatrix} \mu_{-1} & 1 & \mu_1 \\ 1 & \mu_1 & \mu_2 \\ \mu_1 & \mu_2 & \mu_3 \end{vmatrix}}{\mu_{-1}\begin{vmatrix} \mu_1 & \mu_2 \\ \mu_2 & \mu_3 \end{vmatrix}}, \quad q_k = \frac{\begin{vmatrix} 1 & \mu_1 & \mu_2 \\ \mu_1 & \mu_2 & \mu_3 \\ \mu_2 & \mu_3 & \mu_4 \end{vmatrix}}{\begin{vmatrix} 1 & \mu_1 \\ \mu_1 & \mu_2 \end{vmatrix}}. \tag{3.26}$$

When ν_0 is supported on three points, $\nu_{k+2} = \nu_0$ for all k from Theorem 3 and, when ν_0 is supported on less than three points, the algorithm converges in one iteration. In the rest of the section, we thus assume that ν_0 is supported on more than three points. Without any loss of generality, we may take d as the number of components in the support of ν_0 and m and M, respectively, as the minimum and maximum values of these components.

3.4.1 A Characterization of Limit Points through the Transformation $\nu_k \to \nu_{k+1}$

From Theorem 8, the number of components $n(x_0, y)$ of the asymptotic spectrum $\mathcal{S}(x_0, y)$ of any $y \in \mathcal{Z}(x_0)$ satisfies $3 \leq n(x_0, y) \leq 4$ and, from Theorem 9, $\mathcal{S}(x_0, y)$ always contains m and M.

Consider the function

$$\bar{Q}_2^k(t) = \frac{Q_2^k(t)|\mathbf{M}_{2,1}|}{|\mathbf{M}_{1,0}|^{1/2}|\mathbf{M}_{2,0}|^{1/2}} = \frac{\begin{vmatrix} 1 & \mu_1 & 1 \\ \mu_1 & \mu_2 & t \\ \mu_2 & \mu_3 & t^2 \end{vmatrix}}{\begin{vmatrix} 1 & \mu_1 \\ \mu_1 & \mu_2 \end{vmatrix}^{1/2} \begin{vmatrix} 1 & \mu_1 & \mu_2 \\ \mu_1 & \mu_2 & \mu_3 \\ \mu_2 & \mu_3 & \mu_4 \end{vmatrix}^{1/2}}. \tag{3.27}$$

It satisfies $[\bar{Q}_2^k(t)]^2 = H_k(t)$, $z_{k+1} = \bar{Q}_2^k(A)z_k$, see (3.10), and can be considered as a normalized version of $Q_2^k(t)$; $z_{k+2} = z_k$ is equivalent to $\bar{Q}_2^{k+1}(A)\bar{Q}_2^k(A)z_k = z_k$, that is $\bar{Q}_2^k(\lambda_i)\bar{Q}_2^{k+1}(\lambda_i) = 1$ for all i's such that $[z_k]_i \neq 0$. $\bar{Q}_2^k(t)$ and $\bar{Q}_2^{k+1}(t)$ are second-order polynomials in t with two zeros in (m, M), see Theorem 1, and we write

$$\bar{Q}_2^k(t) = \alpha_k t^2 + \beta_k t + \omega_k\,, \quad \bar{Q}_2^{k+1}(t) = \alpha_{k+1} t^2 + \beta_{k+1} t + \omega_{k+1}\,.$$

From the expressions (3.26, 3.27), $\alpha_k = 1/\sqrt{q_k}$ for any k, so that both α_k and α_{k+1} tend to some limit $1/\sqrt{q_\infty}$, see Theorem 6. From (3.27) and (3.7, 3.8),

$$\omega_k^2 = \frac{(g_k, g_k)}{(g_{k+1}, g_{k+1})}, \quad \omega_{k+1}^2 = \frac{(g_{k+1}, g_{k+1})}{(g_{k+2}, g_{k+2})}.$$

Since $\|z_{2k+2} - z_{2k}\|$ tends to zero, see Theorem 7, Theorem 5 implies that $\omega_k\omega_{k+1}$ tend to $1/r_\infty$ as k tends to infinity.

3.4.1.1 Three-Point Asymptotic Spectra

Assume that x_0 is such that ν_0 has more than three support points. To any y in $\mathcal{Z}(x_0)$ we associate the measure ν defined by $\nu(\lambda_i) = [y]_i^2$, $i = 1, \ldots, d$ and denote $Q_2(t)$ the polynomial obtained through (3.6) from the moments of ν. Denote ν' the iterate of ν through T_ν; to ν' we associate $Q_2'(t)$ and write

$$Q_2(t) = \alpha t^2 + \beta t + \omega \,, \quad Q_2'(t) = \alpha' t^2 + \beta' t + \omega' \,, \tag{3.28}$$

where the coefficients satisfy $\alpha = \alpha' = 1/\sqrt{q_\infty}$ and $\omega\omega' = 1/r_\infty$.

Suppose that $n(x_0, y) = 3$, with $\mathcal{S}(x_0, y) = \{m, \lambda_j, M\}$, where λ_j is some eigenvalue of A in (m, M). We thus have

$$Q_2(m)Q_2'(m) = Q_2(\lambda_j)Q_2'(\lambda_j) = Q_2(M)Q_2'(M) = 1 \,,$$

so that $Q_2(t)$ and $Q_2'(t)$ are uniquely defined, in the sense that the number of solutions in (β, β', ω) is finite. (ω is a root of a 6th degree polynomial equation, with one value for β and β' associated with each root. There is always one solution at least: any measure supported on m, λ_j, M is invariant, so that at least two roots exist for ω. The numerical solution of a series of examples shows that only two roots exist, which renders the product $Q_2(t)Q_2'(t)$ unique, due to the possible permutation between (β, ω) and (β', ω').) Figure 3.3 presents a plot of the function $Q_2(t)Q_2'(t)$ when ν gives respectively the weights $1/4$, $1/4$ and $1/2$ to the points $m = 1$, $\lambda_j = 4/3$ and $M = 2$ (which gives $Q_2(t)Q_2'(t) = (81t^2 - 249t + 176)(12t^2 - 33t + 22)/8$).

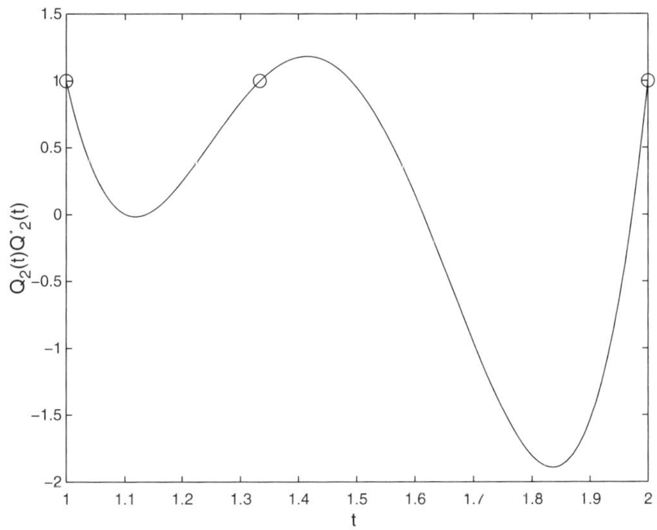

Fig. 3.3. $Q_2(t)Q_2'(t)$ when $\nu(1) = 1/4$, $\nu(4/3) = 1/4$ and $\nu(2) = 1/2$

The orthogonality property (3.13) for $i = 0$ gives $\int Q_2(t)\, \nu(\mathrm{d}t) = 0$, that is,

$$\alpha\mu_2 + \beta\mu_1 + \omega = 0 \tag{3.29}$$

where $\mu_i = p_1 m^i + p_2 \lambda_j^i + (1 - p_1 - p_2)M^i$, $i = 1, 2 \ldots$, with $p_1 = [y]_1^2$ and $p_2 = [y]_j^2$. Since ν has three support points, its moments can be expressed as linear combinations of μ_1 and μ_2 through the equations

$$\int t^i (t - m)(t - \lambda_j)(t - M)\, \nu(\mathrm{d}t) = 0\,, \quad i \in \mathbb{Z}\,.$$

Using (3.27) and (3.29), μ_1 and μ_2 can thus be determined from two coefficients of $Q_2(t)$ only. After calculation, the Jacobians J_1, J_2 of the transformations $(\mu_1, \mu_2) \to (\alpha, \beta)$ and $(\mu_1, \mu_2) \to (\alpha, \omega)$ are found to be equal to

$$J_1 = \frac{Mm + m\lambda_j + \lambda_j M - 2\mu_1(m + \lambda_j + M) + \mu_1^2 + 2\mu_2}{2|\mathbf{M}_{2,0}||\mathbf{M}_{1,0}|}\,,$$
$$J_2 = \frac{\mu_1^2(m + \lambda_j + M) - 2\mu_1\mu_2 - m\lambda_j M}{2|\mathbf{M}_{2,0}||\mathbf{M}_{1,0}|}\,.$$

The only measure $\tilde{\nu}$ supported on $\{m, \lambda_j, M\}$ for which $J_1 = J_2 = 0$ is given by $\mu_1 = \lambda_j$, $\mu_2 = [\lambda_j(m + \lambda_j + M) - mM]/2$, or equivalently $\tilde{\nu}(m) = (M - \lambda_j)/[2(M - m)]$, $\tilde{\nu}(\lambda_j) = 1/2$. The solution for μ_1, μ_2 (and thus for ν supported at m, λ_j, M) associated with a given polynomial $Q_2(t)$ through the pair r_∞, q_∞ is thus locally unique and there is no continuum for three-point asymptotic spectra, $\mathcal{Z}(x_0)$ is a singleton. As Fig. 3.3 illustrates, the existence of a continuum would require the presence of an eigenvalue λ^* in the spectrum of A to which some weight could be transferred from ν. This is only possible if $Q_2(\lambda^*)Q_2'(\lambda^*) = 1$, so that λ^* is uniquely defined ($\lambda^* = 161/108$ in Fig. 3.3). When this happens, it corresponds to a four-point asymptotic spectrum, a situation considered next.

3.4.1.2 Four-Point Asymptotic Spectra

We consider the same setup as above with now $n(x_0, y) = 4$ and $\mathcal{S}(x_0, y) = \{m, \lambda_j, \lambda_k, M\}$, where $\lambda_j < \lambda_k$ are two eigenvalues of A in (m, M). We thus have

$$Q_2(m)Q_2'(m) = Q_2(\lambda_j)Q_2'(\lambda_j) = Q_2(\lambda_k)Q_2'(\lambda_k) = Q_2(M)Q_2'(M) = 1\,,$$

where $Q_2(t), Q_2'(t)$ are given by (3.28) and satisfy $\alpha = \alpha' = 1/\sqrt{q_\infty}$, $\omega\omega' = 1/r_\infty$. The system of equations in $(\alpha, \beta, \omega, \alpha', \beta', \omega')$ is over-determined, which implies the existence of a relation between r_∞ and q_∞. As it is the case for three-point asymptotic spectra, to a given value for q_∞ corresponds a unique polynomial $Q_2(t)$ (up to the permutation with $Q_2'(t)$). The measures ν associated with $Q_2(t)$ can be characterized by their three moments μ_1, μ_2, μ_3, which can be obtained from the values of α, β, ω once μ_4 has been expressed as a function of μ_1, μ_2 and μ_3 using

$$\int (t-m)(t-\lambda_j)(t-\lambda_k)(t-M)\,\nu(\mathrm{d}t) = 0\,.$$

Consider the Jacobian J of the transformation $(\mu_1, \mu_2, \mu_3) \to (\alpha, \beta, \omega)$. Setting J to zero defines a two-dimensional manifold in the space of moments μ_1, μ_2, μ_3, or equivalently in the space of weights p_1, p_2, p_3 with $p_1 = \nu(m)$, $p_2 = \nu(\lambda_j)$, $p_3 = \nu(\lambda_k)$ (and $\nu(M) = 1 - p_1 - p_2 - p_3$). By setting some value to the limit r_∞ or q_∞ one removes one degree of freedom and the manifold becomes one dimensional. Since $[y]_1^2 = p_1$, $[y]_j^2 = p_2$, $[y]_k^2 = p_3$, $[y]_d^2 = 1 - p_1 - p_2 - p_3$, the other components being zero, this also characterizes the limit set $\mathcal{Z}(x_0)$.

Let $x_1 < x_2$ (respectively, $x_1' < x_2'$) denote the two zeros of $\bar{Q}_2(t)$ (respectively, $\bar{Q}_2'(t)$). Suppose that $\lambda_j < x_1$. Then $\bar{Q}_2(\lambda_j) > 0$ and $\bar{Q}_2(\lambda_j)\bar{Q}_2'(\lambda_j) = 1$ implies $\bar{Q}_2'(\lambda_j) > 0$ and thus $\lambda_j < x_1'$. But then, $\bar{Q}_2(\lambda_j) < \bar{Q}_2(m)$ and $\bar{Q}_2'(\lambda_j) < \bar{Q}_2'(m)$ which contradicts $\bar{Q}_2(\lambda_j)\bar{Q}_2'(\lambda_j) = \bar{Q}_2(m)\bar{Q}_2'(m) = 1$. Therefore, $\lambda_j > x_1$, and similarly $x_2 > \lambda_k$, that is

$$m < x_1 < \lambda_j < \lambda_k < x_2 < M\,. \tag{3.30}$$

Denote

$$\begin{aligned}
&S = x_1 + x_2 = -\frac{\beta}{\alpha}\,, \quad P = x_1\,x_2 = \frac{\omega}{\alpha}\,,\\
&S_\lambda = \lambda_j + \lambda_k\,, \quad P_\lambda = \lambda_j\,\lambda_k\,,\\
&S_m = m + M\,, \quad P_m = m\,M\,,
\end{aligned}$$

and

$$E = (S_\lambda - S)[S(S - S_m) + (P_m - P)] - (P_\lambda - P)(S - S_m)\,. \tag{3.31}$$

One can easily check by direct calculation that

$$J = \frac{|\mathbf{M}_{1,0}|^2}{2\alpha|\mathbf{M}_{2,0}|^3}\,E$$

so that the set of limit points y in $\mathcal{Z}(x_0)$ with $n(x_0, y) = 4$ is characterized by $E = 0$. In the next section, we investigate the form of the corresponding manifold into more details in the case where the spectrum $\mathcal{S}(x_0, y)$ is symmetric with respect to $c = (m + M)/2$.

3.4.1.3 Four-Point Symmetric Asymptotic Spectra

When the spectrum is symmetric with respect to $c = (m + M)/2$, $S_\lambda = S_m$ so that the equation $E = 0$, with E given by (3.31), becomes

$$(S - S_m)[S(S - S_m) + (P_m - P) + (P_\lambda - P)] = 0\,.$$

This defines a two-dimensional manifold with two branches: $\mathcal{M}_1$ defined by $S = S_m$ and $\mathcal{M}_2$ defined by $S(S - S_m) + (P_m - P) + (P_\lambda - P) = 0$. The manifolds $\mathcal{M}_1$ and $\mathcal{M}_2$ only depend on the spectrum $\{m, \lambda_j, \lambda_k, M\}$. Note that on the branch $\mathcal{M}_1$ we have $(x_1 + x_2)/2 = (m + M)/2 = c$ so that $\bar{Q}_2(t)$ is symmetric with respect to c. One may also notice that $\bar{Q}_2(t)$ symmetric with respect to c implies that the spectrum $\mathcal{S}(x_0, y)$ is symmetric with respect to c when $E = 0$. Indeed, $\bar{Q}_2(t)$ symmetric implies $S = S_m$, (3.30) then implies $P \neq P_m$, so that $E = 0$ implies $S_\lambda = S = S_m$.

The branch $\mathcal{M}_1$ can be parameterized in P, and the values of r_∞, q_∞ satisfy

$$\begin{aligned} r_\infty &= \frac{(P_m - P)(P - P_\lambda)}{P(P_m + P_\lambda - P)}, \\ q_\infty &= (P_m - P)(P - P_\lambda). \end{aligned}$$

Both r_∞ and q_∞ are maximum for $P = (P_\lambda + P_m)/2$. On $\mathcal{M}_1$, p_1, p_2, p_3 satisfy the following

$$p_1 = \frac{[p_2 (M - \lambda_j)(\lambda_j - \lambda_k) - (P - P_m)](P - P_\lambda)}{(P - P_m)(M - m)(M - \lambda_j)}, \tag{3.32}$$

$$p_3 = \frac{P - P_m}{(M - \lambda_j)(\lambda_j - m)} - p_2, \tag{3.33}$$

where the value of P is fixed by r_∞ or q_∞, and the one-dimensional manifold for p_1, p_2, p_3 is a linear segment in $\mathbb{R}^3$.

We parameterize the branch $\mathcal{M}_2$ in S, and obtain

$$\begin{aligned} r_\infty &= \frac{(m + \lambda_j - S)(M + \lambda_j - S)(m + \lambda_k - S)(M + \lambda_k - S)}{[(\lambda_k - S)(\lambda_j - S) + P_m][(\lambda_k - S)(\lambda_j - S) + P_m + 2S_m(S_m - S)]}, \\ q_\infty &= \frac{(m + \lambda_j - S)(M + \lambda_j - S)(m + \lambda_k - S)(M + \lambda_k - S)}{4}. \end{aligned}$$

Both r_∞ and q_∞ are maximum for $S = S_m = m + M$. Hence, for each branch the maximum value for r_∞ and q_∞ is obtained on the intersection $\mathcal{M}_1 \cap \mathcal{M}_2$ where $S = S_m$ and $P = (P_m + P_\lambda)/2$. On $\mathcal{M}_2$, p_1, p_2, p_3 satisfy the following

$$\begin{aligned} p_2 &= \frac{\lambda_k + m - S}{\lambda_k - \lambda_j} \left[\frac{\lambda_k + M - S}{2(M - \lambda_j)} - p_1 \frac{M - m}{\lambda_j + M - S} \right], \\ p_3 &= \frac{\lambda_j + m - S}{\lambda_k - \lambda_j} \left[- \frac{\lambda_j + M - S}{2(\lambda_j - m)} + p_1 \frac{M - m}{\lambda_k + M - S} \right], \end{aligned}$$

where now the value of S is fixed by r_∞ or q_∞; the one-dimensional manifold for p_1, p_2, p_3 is again a linear segment in $\mathbb{R}^3$.

Figures 3.4 and 3.5 present the two manifolds $\mathcal{M}_1$ and $\mathcal{M}_2$ in the space (p_1, p_2, p_3) when $m = 1$, $\lambda_j = 4/3$, $\lambda_k = 5/3$ and $M = 2$. The line segment C, C' on Fig. 3.4 corresponds to symmetric distributions for which $p_2 = p_3$.

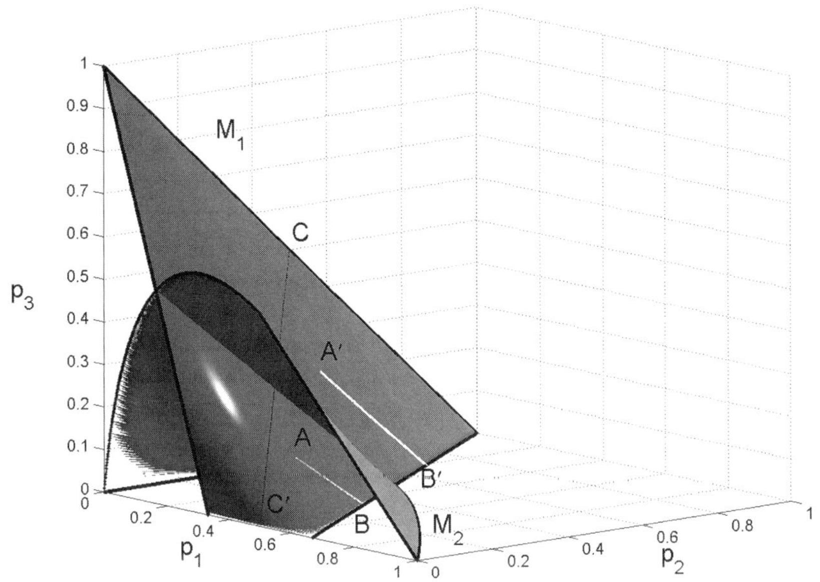

Fig. 3.4. Sequence of iterates close to the manifold $\mathcal{M}_1$: A, B for z_{2k}, A', B' for z_{2k+1}; the line segment C, C' corresponds to symmetric distributions on $\mathcal{M}_1$ ($p_2 = p_3$)

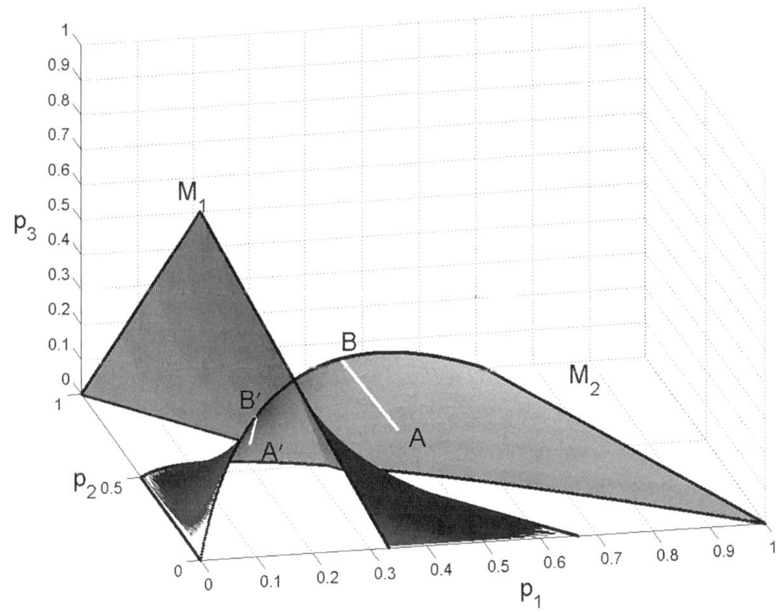

Fig. 3.5. Sequence of iterates close to the manifold $\mathcal{M}_2$: A, B for z_{2k}, A', B' for z_{2k+1}

On both figures, when starting the algorithm at x_0 such that the point with coordinates $([z_0]_1^2, [z_0]_2^2, [z_0]_3^2)$ is in A, to the even iterates z_{2k} correspond points that evolve along the line segment A, B, and to the odd iterates z_{2k+1} corresponds the line A', B'. The initial z_0 is chosen such that A is close to the manifold $\mathcal{M}_1$ in Fig. 3.4 and to the manifold $\mathcal{M}_2$ in Fig. 3.5. In both cases the limit set $\mathcal{Z}(x_0)$ is a singleton $\{y\}$ with $n(x_0, y) = 3$: $[y]_3 = 0$ in Fig. 3.4 and $[y]_2 = 0$ in Fig. 3.5.

3.4.2 A Characterization of Limit Points Through Monotone Sequences

Assume that x_0 is such that ν_0 has more than three support points. The limit point for the orbit (z_k) are such that $r_{k+1} = r_k$. To any y in $\mathcal{Z}(x_0)$ we associate the measure ν defined by $\nu(\lambda_i) = [y]_i^2$, $i = 1, \ldots, d$ and then $\nu' = T_\nu(\nu)$ with T_ν given by (3.11); with ν and ν' we associate respectively, $r(\nu)$ and $r(\nu')$, which are defined from their moments by (3.26). Then, $y \in \mathcal{Z}(x_0)$ implies $T_\nu[T_\nu(\nu)] = \nu$ and

$$\Delta(\nu) = r(\nu') - r(\nu) = 0\,.$$

We thus investigate the nature of the sets of measures satisfying $\Delta(\nu) = 0$.

3.4.2.1 Distributions That are Symmetric with Respect to μ_1

When ν is *symmetric* with respect to μ_1 direct calculation gives

$$\Delta(\nu) = \frac{|\mathbf{M}_{4,-1}||\mathbf{M}_{2,1}|}{\mu_1|\mathbf{M}_{3,0}||\mathbf{M}_{2,0}|}\,,$$

which is zero for any four-point distribution. Any four-point distribution ν that is symmetric with respect to μ_1 is thus invariant in two iterations, that is $T_\nu[T_\nu(\nu)] = \nu$. The expression above for $\Delta(\nu)$ is not valid when ν is not symmetric with respect to μ_1, a situation considered below.

3.4.2.2 General Situation

Direct (but lengthy) calculations give

$$\Delta(\nu) = \frac{N}{D}$$

with

$$\begin{aligned} D = \mu_{-1}|\mathbf{M}_{2,1}||\mathbf{M}_{2,-1}| \left[|\mathbf{M}_{1,0}|^2(|\mathbf{M}_{1,-1}||\mathbf{M}_{4,1}| - \mu_1|\mathbf{M}_{4,-1}|)\right. \\ \left. + |\mathbf{M}_{2,0}|^2(\mu_{-1}|\mathbf{M}_{3,1}| - |\mathbf{M}_{3,-1}|) - 2|\mathbf{M}_{1,0}||\mathbf{M}_{2,0}||\mathbf{M}_{3,0}|\right] \end{aligned} \quad (3.34)$$

$$\begin{aligned} N = |\mathbf{M}_{4,-1}||\mathbf{M}_{1,0}|^2 \left[\mu_1(\mu_{-1}|\mathbf{M}_{2,1}| - |\mathbf{M}_{2,-1}|)^2 - \mu_{-1}|\mathbf{M}_{2,1}|^2\right] \\ +a^2|\mathbf{M}_{4,1}| + b^2|\mathbf{M}_{3,-1}| - 2ab|\mathbf{M}_{3,0}| \end{aligned} \quad (3.35)$$

with $a = |\mathbf{M}_{1,0}||\mathbf{M}_{2,-1}|$ and $b = (\mu_{-1}|\mathbf{M}_{2,1}| - |\mathbf{M}_{2,-1}|)|\mathbf{M}_{2,0}|$. The determinants that are involved satisfy some special identities

$$\begin{aligned} |\mathbf{M}_{1,0}|^2 &= |\mathbf{M}_{1,-1}||\mathbf{M}_{2,1}| - \mu_1|\mathbf{M}_{2,-1}| \\ |\mathbf{M}_{2,0}|^2 &= |\mathbf{M}_{2,-1}||\mathbf{M}_{3,1}| - |\mathbf{M}_{2,1}||\mathbf{M}_{3,-1}| \\ |\mathbf{M}_{3,0}|^2 &= |\mathbf{M}_{3,-1}||\mathbf{M}_{4,1}| - |\mathbf{M}_{3,1}||\mathbf{M}_{4,-1}| \end{aligned} \tag{3.36}$$

and

$$\begin{aligned} &\mu_{-1}|\mathbf{M}_{1,1}| - |\mathbf{M}_{1,-1}| = 1\,, \ |\mathbf{M}_{1,-1}||\mathbf{M}_{1,1}| - \mu_1|\mathbf{M}_{1,-1}| = 0 \\ &\quad (|\mathbf{M}_{1,-1}||\mathbf{M}_{2,1}| - \mu_1|\mathbf{M}_{2,-1}|)(\mu_{-1}|\mathbf{M}_{1,1}| - |\mathbf{M}_{1,-1}|) = |\mathbf{M}_{1,0}|^2 \\ &\quad (|\mathbf{M}_{1,-1}||\mathbf{M}_{3,1}| - \mu_1|\mathbf{M}_{3,-1}|)(\mu_{-1}|\mathbf{M}_{2,1}| - |\mathbf{M}_{2,-1}|) = |\mathbf{M}_{2,0}|^2 \\ &\qquad\qquad + (|\mathbf{M}_{1,-1}||\mathbf{M}_{2,1}| - \mu_1|\mathbf{M}_{2,-1}|)(\mu_{-1}|\mathbf{M}_{3,1}| - |\mathbf{M}_{3,-1}|) \\ &\quad (|\mathbf{M}_{1,-1}||\mathbf{M}_{4,1}| - \mu_1|\mathbf{M}_{4,-1}|)(\mu_{-1}|\mathbf{M}_{3,1}| - |\mathbf{M}_{3,-1}|) = |\mathbf{M}_{3,0}|^2 \\ &\qquad\qquad + (|\mathbf{M}_{1,-1}||\mathbf{M}_{3,1}| - \mu_1|\mathbf{M}_{3,-1}|)(\mu_{-1}|\mathbf{M}_{4,1}| - |\mathbf{M}_{4,-1}|) \end{aligned}$$

where all the terms inside the brackets are non-negative. Using these identities we obtain

$$\begin{aligned} D = \mu_{-1}|\mathbf{M}_{2,1}||\mathbf{M}_{2,-1}| \Bigg\{ &\Big[|\mathbf{M}_{1,0}|(|\mathbf{M}_{1,-1}||\mathbf{M}_{4,1}| - \mu_1|\mathbf{M}_{4,-1}|)^{1/2} \\ &\qquad + |\mathbf{M}_{2,0}|(\mu_{-1}|\mathbf{M}_{3,1}| - |\mathbf{M}_{3,-1}|)^{1/2} \Big]^2 \\ + 2|\mathbf{M}_{1,0}||\mathbf{M}_{2,0}| & \\ \times \, & \frac{(|\mathbf{M}_{1,-1}||\mathbf{M}_{3,1}| - \mu_1|\mathbf{M}_{3,-1}|)(\mu_{-1}|\mathbf{M}_{4,1}| - |\mathbf{M}_{4,-1}|)}{(|\mathbf{M}_{1,-1}||\mathbf{M}_{4,1}| - \mu_1|\mathbf{M}_{4,-1}|)^{1/2}(\mu_{-1}|\mathbf{M}_{3,1}| - |\mathbf{M}_{3,-1}|)^{1/2} + |\mathbf{M}_{3,0}|} \Bigg\} \end{aligned}$$

and thus $D > 0$ when ν has three support points or more.

We also get

$$a^2|\mathbf{M}_{4,1}| + b^2|\mathbf{M}_{3,-1}| - 2ab|\mathbf{M}_{3,0}| \geq |\mathbf{M}_{1,0}|^2|\mathbf{M}_{2,1}|^2 \frac{|\mathbf{M}_{3,1}||\mathbf{M}_{4,-1}|}{|\mathbf{M}_{3,-1}|}$$

which gives

$$\begin{aligned} N &\geq \frac{|\mathbf{M}_{1,0}|^2|\mathbf{M}_{4,-1}|}{|\mathbf{M}_{3,-1}|} \left[|\mathbf{M}_{1,0}|^2(\mu_{-1}|\mathbf{M}_{2,1}| - |\mathbf{M}_{2,-1}|)|\mathbf{M}_{3,-1}| + |\mathbf{M}_{2,0}|^3|\mathbf{M}_{2,-1}| \right] \\ &\geq 0\,. \end{aligned}$$

Now, $y \in \mathcal{Z}(x_0)$ implies $N = 0$. Since $|\mathbf{M}_{2,-1}| > 0$ when ν has three support points or more, $N = 0$ implies $|\mathbf{M}_{4,-1}| = 0$, that is, ν has three or four support points only, and we recover the result of Theorem 8.

Setting $|\mathbf{M}_{4,-1}| = 0$ in (3.35) we obtain that $y \in \mathcal{Z}(x_0)$ implies

$$a^2|\mathbf{M}_{4,1}| - 2ab|\mathbf{M}_{3,0}| + b^2|\mathbf{M}_{3,-1}| = 0\,. \tag{3.37}$$

The condition (3.37) is satisfied for any three-point distribution. For a four point-distribution it is equivalent to b/a being the (double) root of the following quadratic equation in t, $|\mathbf{M}_{4,1}| - 2t|\mathbf{M}_{3,0}| + |\mathbf{M}_{3,-1}|t^2 = 0$. This condition can be written as

$$\delta(\nu) = (\mu_{-1}|\mathbf{M}_{2,1}| - |\mathbf{M}_{2,-1}|)|\mathbf{M}_{2,0}||\mathbf{M}_{3,-1}| - |\mathbf{M}_{1,0}||\mathbf{M}_{3,0}||\mathbf{M}_{2,-1}| = 0\,. \tag{3.38}$$

(One may notice that when ν is symmetric with respect to μ_1, $\delta(\nu)$ becomes $\delta(\nu) = [|\mathbf{M}_{1,0}||\mathbf{M}_{2,0}|^2/|\mathbf{M}_{3,0}|]\ |\mathbf{M}_{4,-1}|$ which is equal to zero for a four-point distribution. We thus recover the result of Sect. 3.4.2.)

To summarize, $y \in \mathcal{Z}(x_0)$ implies that ν is supported on three or four points and satisfies (3.38). In the case of a four-point distribution supported on $m, \lambda_j, \lambda_k, M$, after expressing μ_{-1}, μ_4, μ_5 and μ_6 as functions of μ_1, μ_2, μ_3 though

$$\int t^i(t-m)(t-\lambda_j)(t-\lambda_k)(t-M)\,\nu(\mathrm{d}t) = 0\,, \quad i \in \mathbb{Z}\,,$$

we obtain $\delta(\nu) = KJ$ with

$$\begin{aligned} K = {} & 2\frac{\alpha|\mathbf{M}_{2,0}|^3|\mathbf{M}_{1,0}|}{m\lambda_j\lambda_k M}\,p_1p_2p_3(1-p_1-p_2-p_3) \\ & \times(\lambda_k-\lambda_j)^2(\lambda_k-m)^2(M-\lambda_k)^2(\lambda_j-m)^2(M-\lambda_j)^2(M-m)^2 > 0\,, \end{aligned}$$

where p_1, p_2, p_3, α and J are defined as in Sect. 3.4.1. Since $K > 0$, the attractor is equivalently defined by $J = 0$, which is precisely the situation considered in Sect. 3.4.1.

3.4.3 Stability

Not all three or four-point asymptotic spectra considered in Sect. 3.4.1 correspond to stable attractors. Although ν associated with some vector y on the unit sphere $\mathcal{S}_d$ may be invariant by two iterations of (3.11), a measure ν_k arbitrarily close to ν (that is, associated with a renormalized gradient z_k close to y) may lead to an iterate ν_{k+2} far from ν_k. The situation can be explained from the example of a three-point distribution considered in Fig. 3.3 of Sect. 3.4.1. The measure ν is invariant in two iteration of T_ν given by (3.11). Take a measure $\nu_k = (1-\kappa)\nu + \kappa\nu'$ where $0 < \kappa < 1$ and ν' is a measure on (m, M) that puts some positive weight to some point λ^* in the intervals $(4/3, 161/108)$ or $(1.7039, 1.9337)$. Then, for κ small enough the function $Q_2(t)Q_2'(t)$ obtained for ν' is similar to that plotted for ν on Fig. 3.3, and the weight of λ^* will increase in two iterations since $|Q_2(\lambda^*)Q_2'(\lambda^*)| > 1$. The invariant measure ν is thus an unstable fixed point for T_ν^2 when the spectrum of A contains some eigenvalues in $(4/3, 161/108) \cup (1.7039, 1.9337)$. The analysis is thus similar to that in Pronzato et al. (2001, 2006) for the steepest-descent algorithm ($s = 1$), even if the precise derivation of stability regions (in terms of the weights that two-step invariant measures give to their three or four support points), for a given spectrum for A, is much more difficult for $s = 2$.

3.4.4 Open Questions and Extension to $s > 2$

For $s = 2$, as shown in Sect. 3.4.1, if x_0 is such that z_0 is exactly on one of the manifolds $\mathcal{M}_1$ or $\mathcal{M}_2$, then by construction $z_{2k} = z_0$ for all k. Although it might be possible that choosing z_0 close enough to $\mathcal{M}_1$ (respectively, $\mathcal{M}_2$) would force z_{2k} to converge to a limit point before reaching the plane $p_3 = 0$ (respectively, $p_2 = 0$), the monotonicity of the trajectory along the line segment A, B indicates that there is no continuum and the limit set $\mathcal{Z}(x_0)$ is a single point. This was conjectured by Forsythe (1968) and is still an open question. We conjecture that additionally for almost all initial points the trajectory always attracts to a three-point spectrum (though the attraction make take a large number of iterations when z_k is very close to $\mathcal{M}_1 \cup \mathcal{M}_2$ for some k). One might think of using the asymptotic equivalence of rates of convergence, as stated in Theorem 5, to prove this conjecture. However, numerical calculations show that for any point on the manifold $\mathcal{M}_1$ defined in Sect. 3.4.1, the product of the rates at two consecutive iterations satisfies $R_k(W)R_{k+1}(W) = r_\infty^2$ for any positive-definite matrix W (so that $R_k(W)R_{k+1}(W)$ does not depend on p_2 on the linear segment defined by r_∞ on $\mathcal{M}_1$, see (3.32 and 3.33), and all points on this segment can thus be considered as asymptotically equivalent).

Extending the approach of Sect. 3.4.2 for the characterization of limit points to the case $s > 2$ seems rather difficult, and the method used in Sect. 3.4.1 is more promising. A function $\bar{Q}_s^k(t)$ can be defined similarly to (3.27), leading to two s-degree polynomials $Q_s(t)$, $Q_s'(t)$, see (3.28), each of them having s roots in (m, M). Let α and α' denote the coefficients of terms of highest degree in $Q_s(t)$ and $Q_s'(t)$, respectively, then $\alpha = \alpha' = 1/\sqrt{q_\infty}$. Also, let ω and ω' denote the constant terms in $Q_s(t)$ and $Q_s'(t)$; Theorem 5 implies that $\omega\omega' = 1/r_\infty$. Let n be the number of components in the asymptotic spectrum $\mathcal{S}(x_0, y)$, with $s+1 \leq n \leq 2s$ from Theorem 8; we thus have $2(s + 1)$ coefficients to determine, with $n + 3$ equations:

$$\alpha = \alpha' = 1/\sqrt{q_\infty}\,, \ \omega\omega' = 1/r_\infty \ \text{ and } \ Q_s(\lambda_i)Q_s'(\lambda_i) = 1\,, \ i = 1, \ldots, n$$

where the λ_i's are the eigenvalues of A in $\mathcal{S}(x_0, y)$ (including m and M, see Theorem 9). We can then demonstrate that the functions $Q_s(t)$ and $Q_s'(t)$ are uniquely defined when $n = 2s$ and $n = 2s-1$ (which are the only possible cases when $s = 2$). When $n = 2s$, the system is over-determined, as it is the case in Sect. 3.4.1 for four-point asymptotic spectra when $s = 2$. The limit set $\mathcal{Z}(x_0)$ corresponds to measures ν for which the weights p_i, $i = 1, \ldots, n-1$ belong to a $n-2$-dimensional manifold. By setting some value to r_∞ or q_∞ the manifold becomes $(n - 3)$-dimensional. When $n = 2s - 1$, ν can be characterized by its $2s - 2$ first moments, which cannot be determined uniquely when $s > 2$.

Therefore, although $Q_s(t)$ and $Q'_s(t)$ are always uniquely defined for $n = 2s$ and $n = 2s - 1$, the possibility of a continuum for $\mathcal{Z}(x_0)$ still exists. The situation is even more complex when $s + 1 \leq n \leq 2s - 2$ and $s > 2$.

3.5 Switching Algorithms

The bound $N_2^* = (R_2^*)^{1/2}$ on the convergence rate of the optimum 2-gradient algorithm, see (3.21), is smaller than the bound $N_1^* = R_1^*$, which indicates a slower convergence for the latter. Let ϵ denote a required precision on the squared norm of the gradient g_k, then the number of gradient evaluations needed to reach the precision ϵ is bounded by $\log(\epsilon)/\log(N_2^*)$ for the optimum 2-gradient algorithm and by $\log(\epsilon)/\log(N_1^*)$ for the steepest-descent algorithm. To compare the number of gradient evaluations for the two algorithms we thus compute the ratio $L_{1/2}^* = \log(N_1^*)/\log(N_2^*)$. The evolution of $L_{1/2}^*$ as a function of ϱ is presented in Fig. 3.6 in solid line. The improvement of the optimum 2-gradient over steepest descent is small for small ϱ but $L_{1/2}^*$ tends to $1/2$ as ϱ tends to infinity. (More generally, the ratio $L_{1/s}^* = \log(N_1^*)/\log(N_s^*)$ tends to $1/s$ as $\varrho \to \infty$.) The ratio $L_{1/\infty}^* = \log(N_1^*)/\log(N_\infty^*)$, where N_∞^* is defined in (3.22), is also presented in Fig. 3.6. It tends to zero as $1/\sqrt{\varrho}$ when ϱ tends to ∞.

The slow convergence of steepest descent, or, more generally, of the optimum s-gradient algorithm, is partly due to the existence of a measure ν_s^* associated with a large value for the rate of convergence R_s^*, see Theorem 4,

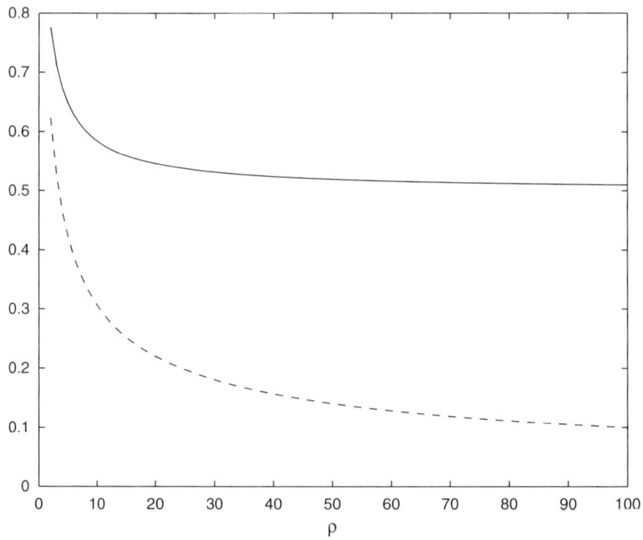

Fig. 3.6. Ratios $L_{1/2}^*$ (*solid line*) and $L_{1/\infty}^*$ (*dashed line*) as functions of ϱ

but mainly to the fact that ν_s^* is supported on $s+1$ points and is thus invariant in two steps of the algorithm, that is, $T_\nu[T_\nu(\nu_s^*)] = \nu_s^*$, see Theorem 3 (in fact, ν_s^* is even invariant in one step, i.e. $T_\nu(\nu_s^*) = \nu_s^*$).

Switching between algorithms may then be seen as an attractive option to destroy the stability of this worst-case behaviour. The rest of the chapter is devoted to the analysis of the performance obtained when the two algorithms for $s = 1$ and $s = 2$ are combined, in the sense that the resulting algorithm switches between steepest descent and optimum 2-gradient. Notice that no measure exists that is invariant in two iterations for both the steepest descent and the optimum 2-gradient algorithms (this is no longer true for larger values of s, for instance, a symmetric four-point distribution is invariant in two iterations for $s = 2$, see Sect. 3.4.2, and $s = 3$, see Theorem 3).

3.5.1 Superlinear Convergence in $\mathbb{R}^3$

We suppose that $d = 3$, A is already diagonalized with eigenvalues $m < \lambda < M = \varrho m$, and that x_0 is such that ν_0 puts a positive weight on each of them. We denote $p_k = \nu_k(m)$, $t_k = \nu_k(\lambda)$ (so that $\nu_k(M) = 1 - p_k - t_k$). Consider the following algorithm.

Algorithm A

Step 0: Fix n, the total number iterations allowed, choose ϵ, a small positive number; go to Step 1.
Step 1 ($s = 1$): Use steepest descent from $k = 0$ to k^*, the first value of k such that $q_{k+1} - q_k < \epsilon$, where q_k is given by (3.23) with $s = 1$; go to Step 2.
Step 2 ($s = 2$): Use the optimum 2-gradient algorithm for iterations k^* to n.

Notice that since q_k is non-decreasing and bounded, see Theorem 6, switching will always occur for some finite k^*. Since for the steepest-descent algorithm the measure ν_k converges to a two-point measure supported at m and M, after switching the optimum 2-gradient algorithm is in a position where its convergence is fast provided that k^* is large enough for t_{k^*} to be small (it would converge in one iteration if the measure were supported on exactly two points, i.e. if t_{k^*} were zero). Moreover, the 3-point measure ν_{k^*} is invariant in two iterations of optimum 2-gradient, that is, $\nu_{k^*+2j} = \nu_{k^*}$ for any j, so that this fast convergence is maintained from iterations k^* to n. This can be formulated more precisely as follows.

Theorem 10. *When $\epsilon = \epsilon^*(n) = C\log(n)/n$ in Algorithm A, with C an arbitrary positive constant, the global rate of convergence*

$$R_n(x_0) = \left[\frac{f(x_n)}{f(x_0)}\right]^{1/n} = \left(\prod_{k=0}^{n-1} r_k\right)^{1/n}, \tag{3.39}$$

with r_k given by (3.14), satisfies

$$\limsup_{n\to\infty} \frac{\log[R_n(x_0)]}{\log(n)} < -\frac{1}{2}\,.$$

Proof. We know from Theorem 6 that the sequence (q_k) is non-decreasing and bounded. For steepest descent $s = 1$ and the bound (3.24) is $q_1^* = (M-m)^2/4$. Since $q_0 > 0$, it implies $k^* < (M-m)^2/(4\epsilon)$. Also, direct calculation gives $q_{k+1} - q_k = |\mathbf{M}_{2,0}^k|/q_k^2 = |\mathbf{M}_{2,0}^k|/|\mathbf{M}_{1,0}^k|^2$ and, using (3.36),

$$q_{k+1} - q_k = \frac{|\mathbf{M}_{2,0}^k|}{|\mathbf{M}_{2,1}^k|}\,\frac{1}{|\mathbf{M}_{1,-1}^k| - \mu_1 \frac{|\mathbf{M}_{2,-1}^k|}{|\mathbf{M}_{2,1}^k|}} > \frac{|\mathbf{M}_{2,0}^k|}{|\mathbf{M}_{2,1}^k|}\,\frac{1}{|\mathbf{M}_{1,-1}^k|}\,,$$

so that $q_{k^*+1} - q_{k^*} < \epsilon$ implies

$$\frac{|\mathbf{M}_{2,0}^{k^*}|}{|\mathbf{M}_{2,1}^{k^*}|} < \epsilon |\mathbf{M}_{1,-1}^{k^*}|\,. \tag{3.40}$$

For steepest descent, $r_k = |\mathbf{M}_{1,-1}^k|/(\mu_{-1}^k |\mathbf{M}_{1,1}^k|) = 1 - 1/(\mu_1^k \mu_{-1}^k) < R_1^*$, so that $|\mathbf{M}_{1,-1}^{k^*}| = \mu_1^{k^*}\mu_{-1}^{k^*} - 1 < R_1^*/(1-R_1^*) = (M-m)^2/(4mM)$ and (3.40) gives

$$\frac{|\mathbf{M}_{2,0}^{k^*}|}{|\mathbf{M}_{2,1}^{k^*}|} < \epsilon\,\frac{(M-m)^2}{4mM}\,. \tag{3.41}$$

The first iteration of the optimum 2-gradient algorithm has the rate

$$r_{k^*} = \frac{f(x_{k^*+1})}{f(x_{k^*})} = \frac{|\mathbf{M}_{2,-1}^{k^*}|}{\mu_{-1}^{k^*}|\mathbf{M}_{2,1}^{k^*}|} = \frac{|\mathbf{M}_{2,0}^{k^*}|}{mM\lambda\mu_{-1}^{k^*}|\mathbf{M}_{2,1}^{k^*}|}\,,$$

and using $\mu_{-1}^{k^*} > 1/M$ and (3.41) we get $r_{k^*} < B\epsilon$ with $B = (M-m)^2/[4M\lambda m^2]$. Since $d = 3$, $\nu_{k^*+2j} = \nu_{k^*}$ and $r_{k^*+2j} = r_{k^*}$ for $j = 1, 2, 3\ldots$

Now, for each iteration of steepest descent we bound r_k by R_1^*; for the optimum 2-gradient we use $r_k < B\epsilon$ for $k = k^* + 2j$ and $r_k < R_2^*$ for $k = k^* + 2j + 1$, $j = 1, 2, 3\ldots$ We have

$$\log[R_n(x_0)] = \frac{1}{n}\left\{\log\left[\prod_{k=0}^{k^*-1} r_k\right] + \log\left[\prod_{k=k^*}^{n-1} r_k\right]\right\}.$$

Since $R_2^* < R_1^*$, $B\epsilon < R_2^*$ for ϵ small enough, and $k^* < \bar{k} = (M-m)^2/(4\epsilon)$ we can write

$$\log[R_n(x_0)] < L_n(\epsilon) = \frac{1}{n}\left\{\bar{k}\log R_1^* + (n-\bar{k})\frac{1}{2}\left[\log(B\epsilon) + \log(R_2^*)\right]\right\}.$$

Taking $\epsilon = C\log(n)/n$ and letting n tend to infinity, we obtain

$$\lim_{n\to\infty} L_n/\log(n) = -1/2\,.$$

□

Algorithm A requires to fix the number n of iterations a priori and to choose ϵ as a function of n. The next algorithm does not require any such prior choice and uses alternatively a fixed number of iterations of steepest descent and optimum 2-gradient.

Algorithm B

Step 1 ($s = 1$): Use steepest descent for $m_1 \geq 1$ iterations; go to Step 2.
Step 2 ($s = 2$): Use the optimum 2-gradient algorithm for $2m_2$ iterations, $m_2 \geq 1$; return to Step 1.

Its performance satisfies the following.

Theorem 11. *For any choice of m_1 and m_2 in Algorithm B, the global rate (3.39) satisfies $R_n(x_0) \to 0$ as the number n of iterations tends to infinity.*

Proof. Denote $k_j = (j-1)(m_1+2m_2)+m_1$, $j = 1, 2 \ldots$, the iteration number for the jth switching from steepest descent to optimum 2-gradient. Notice that $\nu_{k_j-1} = \nu_{k_j+2m_2-1}$ since any three-point measure is invariant in two steps of the optimum 2-gradient algorithm. The repeated use of Step 2, with $2m_2$ iterations each time, has thus no influence on the behaviour of the steepest-descent iterations used in Step 1. Therefore, $\epsilon_j = q_{k_j} - q_{k_j-1}$ tends to zero as j increases. Using the same arguments and the same notation as in the proof of Theorem 10 we thus get $r_{k_j} < B\epsilon_j$ for the first of the $2m_2$ iterations of the optimum 2-gradient algorithm, with $B = (M-m)^2/[4M\lambda m^2]$. For large n, we write

$$j = \left\lfloor \frac{n}{m_1 + 2m_2} \right\rfloor$$

$n' = n - j(m_1 + 2m_2) < (m_1 + 2m_2)$. For the last n' iterations we bound r_k by R_1^*; for steepest-descent iterations we use $r_k < R_1^*$; at the jth call of Step 2 we use $r_{k_j} < B\epsilon_j$ for the first iteration of optimum 2-gradient and $r_k < R_2^*$ for the subsequent ones. This yields the bound

$$\log[R_n(x_0)] < \frac{1}{j(m_1 + 2m_2) + n'} \times \left\{ n' \log(R_1^*) + jm_1 \log(R_1^*) + j(2m_2 - 1) \log(R_2^*) + \sum_{i=1}^{j} \log(B\epsilon_i) \right\} .$$

Finally, we use the concavity of the logarithm and write

$$\frac{\sum_{i=1}^{j} \log(B\epsilon_i)}{j} < \log\left(B \frac{\sum_{i=1}^{j} \epsilon_i}{j} \right) = \log\left(B \frac{q_{k_j} - q_{m_1-1}}{j} \right) < \log\left(B \frac{q_1^*}{j} \right)$$

with $q_1^* = (M-m)^2/4$, see (3.24). Therefore, $\log[R_n(x_0)] \to -\infty$ and $R_n(x_0) \to 0$ as $n \to \infty$. □

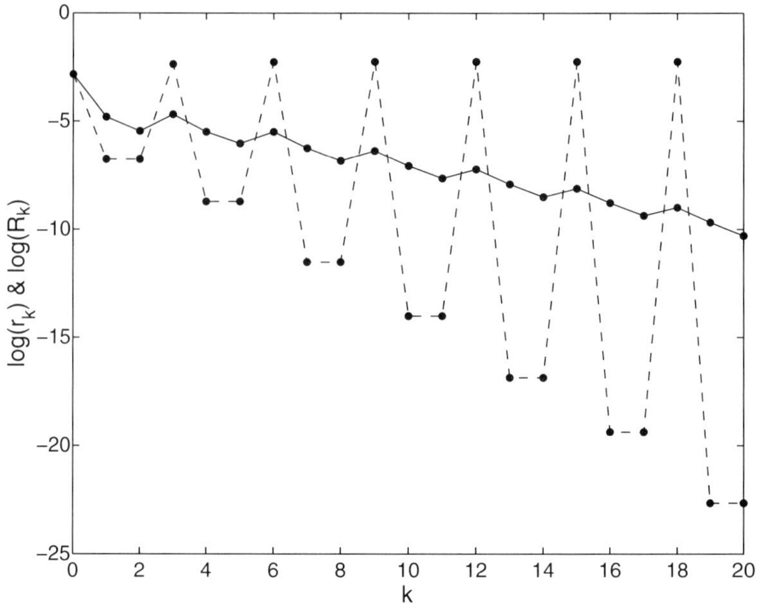

Fig. 3.7. Typical evolution of the logarithms of the rate of convergence r_k, in *dashed line*, and of the global rate R_k defined by (3.39), in *solid line*, as functions of k in Algorithm B

Figure 3.7 presents a typical evolution of $\log(r_k)$ and $\log(R_k)$, with R_k the global rate of convergence (3.39), as functions of the iteration number k in Algorithm B when $m_1 = m_2 = 1$ (z_0 is a random point on the unit sphere $\mathcal{S}_3$ and A has the eigenvalues 1, 3/2 and 2). The rate of convergence r_k of the steepest-descent iterations is slightly increasing, but this is compensated by the regular decrease of the rate for the pairs of optimum 2-gradient iterations and the global rate R_k decreases towards zero.

Remark 7. In Theorems 10 and 11, n counts the number of iterations. Let n now denote the number of gradient evaluations (remember that one iteration of the optimum 2-gradient algorithm corresponds to two steps of the conjugate gradient method, see Remark 1, and thus requires two gradient evaluations), and define

$$N_n(x_0) = \left[\frac{f[x(n)]}{f(x_0)}\right]^{1/n} \tag{3.42}$$

with $x(n)$ the value of x generated by the algorithm after n gradient evaluations. Following the same lines as in the proof of Theorem 10 we get for Algorithm A

$$\log[N_n(x_0)] = \frac{1}{n}\left\{\log\left[\prod_{k=0}^{k^*-1} r_k\right] + \log\left[\prod_{k=k^*}^{n-1}\sqrt{r_k}\right]\right\}$$
$$< L'_n(\epsilon) = \frac{1}{n}\left\{\bar{k}\log R_1^* + (n-\bar{k})\frac{1}{4}\left[\log(B\epsilon) + \log(R_2^*)\right]\right\}.$$

Taking $\epsilon = C\log(n)/n$ and letting n tend to infinity, we then obtain $\lim_{n\to\infty} L'_n/\log(n) = -1/4$.

Similarly to Theorem 11 we also have $N_n(x_0) \to 0$ as $n \to \infty$ in Algorithm B.

The importance of the results in Theorems 10 and 11 should not be overemphasized. After all, the optimum 3-gradient converges in one iteration in $\mathbb{R}^3$! It shows, however, that an algorithm with fast convergence can be obtained from the combination of two algorithms with rather poor performance, which opens a promising route for further developments. As a first step in this direction, next section shows that the combination used in Algorithm B still has a good performance in dimensions $d > 3$, with a behaviour totally different from the regular one observed in $\mathbb{R}^3$.

3.5.2 Switching Algorithms in $\mathbb{R}^d$, $d > 3$

We suppose that A is diagonalized with eigenvalues $m = \lambda_1 < \lambda_2 < \cdots < \lambda_d = M = \varrho m$ and that x_0 is such that ν_0 has $n_0 > 3$ support points. The behaviour of Algorithm B is then totally different from the case $d = 3$, where convergence is superlinear. Numerical simulations (see Sect. 3.5.2) indicate that the convergence is then only linear, although faster than for the optimum 2-gradient algorithm for suitable choices of m_1 and m_2. A simple interpretation is as follows. Steepest descent tends to force ν_k to be supported on m and M only. If m_1 is large, when switching to optimum 2-gradient, say at iteration $k_j = j(m_1 + 2m_2) + m_1$, the first iteration has then a very small rate r_{k_j}. Contrary to the case $d = 3$, $\nu_{k_j+2} \neq \nu_{k_j}$, so that the rate r_k quickly deteriorates as k increases and ν_k converges to a measure with three or four support points. However, when switching back to steepest descent at iteration $(j+1)(m_1 + 2m_2) = k_j + 2m_2$, the rate is much better than the bound R_1^* since $\nu_{k_j+2m_2}$ is far from a two-point measure. This alternation of phases where ν_k converges towards a two-point measure and then to a three or four-point measure renders the behaviour of the algorithm hardly predictable (Sect. 3.5.2 shows that a direct worst-case analysis is doomed to failure). On the other hand, each switching forces ν_k to jump to regions where convergence is fast. The main interest of switching is thus to prevent the renormalized gradient z_k from approaching its limit set where convergence is slow (since r_k is non-decreasing), and we shall see in Sect. 3.5.2 that choosing $m_1 = 1$ and $1 \leq m_2 \leq 5$ in Algorithm B is suitable.

3.5.2.1 The Limits of a Worst-Case Analysis

One of the simplest construction for a switching algorithm is as follows.

Algorithm C

> Use the optimum 2-gradient algorithm if its rate of convergence is smaller than some value $R < R_2^*$ and steepest descent otherwise.

When the state of the algorithm is given by the measure ν_k, denote $r_k = r(\nu_k)$ (respectively, $r'_k = r'(\nu_k)$) the rate of convergence (3.14) if a steepest-descent (respectively, an optimum 2-gradient) iteration is used. Despite the simplicity of its construction, the performance of Algorithm C resists to a worst-case analysis when one tries to bound the rate of convergence at each iteration of the algorithm. Indeed, one can easily check that the measures ν_s^* associated with the worst rates R_s^* for $s = 1, 2$, see (3.17), satisfy $r'(\nu_1^*) = 0$ and $r(\nu_2^*) = \sqrt{r'(\nu_2^*)} = \sqrt{R_2^*} = N_2^* > R_2^*$, see (3.21). This implies that when the state of the algorithm is given by the measure ν_2^* the rate of convergence equals R_2^* for an optimum 2-gradient iteration and is larger than R_2^* for a steepest-descent iteration; it thus ruins any hope to improve the performance of optimum 2-gradient at each iteration. The situation is not better when measuring the performance per gradient evaluation: the rate of convergence then equals N_2^* for both a steepest-descent and an optimum 2-gradient iteration for the measure ν_2^*. Therefore, the only possibility for obtaining an improvement over optimum 2-gradient is in the long run behaviour of the algorithm: when iterations with a slow rate of convergence occur they are compensated by a fast rate at some other iterations. This phenomenon is difficult to analyse since it requires to study several consecutive iterations of steepest descent and/or optimum 2-gradient, which is still an open issue. Some encouraging simulations results are presented in Sect. 3.5.2.

Although improving the value R_2^* for the rate of convergence is doomed to failure, it is instructive to investigate the possible choices for R in Algorithm C.

Consider a steepest-descent iteration. Define $L_k = \mu_{-1}^k \mu_1^k$, so that $L_k = 1/(1 - r_k)$. Since r_k is non-decreasing, L_k is non-decreasing too, and bounded by $1/(1 - R_1^*) = (\varrho + 1)^2/(4\varrho)$. Direct calculation gives

$$L_{k+1} - L_k = \frac{\mu_1^k |\mathbf{M}_{2,-1}^k|}{|\mathbf{M}_{1,0}^k|^2} .$$

When the optimum 2-gradient is used, the rate of convergence r'_k satisfies

$$r'_k = \frac{|\mathbf{M}_{2,-1}^k|}{\mu_{-1}^k |\mathbf{M}_{2,1}^k|} ,$$

so that using (3.36)

$$L_{k+1} - L_k = \mu_1^k \mu_{-1}^k r_k' \frac{|\mathbf{M}_{2,1}^k|}{|\mathbf{M}_{1,0}^k|^2} = \mu_1^k \mu_{-1}^k r_k' \frac{1}{|\mathbf{M}_{1,-1}^k| - \mu_1^k \frac{|\mathbf{M}_{2,-1}^k|}{|\mathbf{M}_{2,1}^k|}}$$
$$> \frac{\mu_1^k \mu_{-1}^k r_k'}{|\mathbf{M}_{1,-1}^k|} = \frac{r_k'}{r_k}. \qquad (3.43)$$

Now,

$$L_{k+1} - L_k = \frac{r_{k+1} - r_k}{(1 - r_{k+1})(1 - r_k)} < \frac{r_{k+1} - r_k}{(1 - R_1^*)^2}$$

and $r_k < R_1^*$ so that $r_{k+1} - r_k < F r_k'$ where

$$F = \frac{(1 - R_1^*)^2}{R_1^*} = \frac{16\varrho^2}{(\varrho - 1)^2(\varrho + 1)^2}$$

(and $F < 1$ if $\varrho > 2 + \sqrt{5}$). In Algorithm C, by construction the rate of convergence is less than R when optimum 2-gradient is used. When steepest descent is used, it means that $r_k' \geq R$, and therefore

$$r_k < r_{k+1} - FR < \bar{r}_1(R) = R_1^* - FR. \qquad (3.44)$$

Choosing R such that the bounds on the rates coincide for steepest-descent and optimum 2-gradient iterations, that is, such that $\bar{r}_1(R) = R$, gives

$$R = \bar{R} = \frac{(\varrho - 1)^4}{\varrho^4 + 14\varrho^2 + 1}, \qquad (3.45)$$

which is larger than R_2^* for any $\varrho > 1$ and therefore cannot be used in Algorithm C. Figure 3.8 presents R_1^* (dotted line), R_2^* (dashed line), $\bar{R}$ (dash-dotted line) and $\bar{r}_1(R_2^*)$, $\bar{r}_1(R_2^*/2)$ (solid lines, top for $\bar{r}_1(R_2^*/2)$), as functions of ϱ. The trade-off value $\bar{R}$ is clearly larger than R_2^*, taking $R = R_2^*$ gives a bound $\bar{r}_1(R_2^*)$ already close to R_1^* and getting even closer as R decreases.

The situation can be slightly improved by constructing a better bound than (3.44). We write

$$L_{k+1} - L_k = \frac{r_{k+1} - r_k}{(1 - r_{k+1})(1 - r_k)} < \frac{R_1^* - r_k}{(1 - R_1^*)(1 - r_k)}$$

so that for $r_k' \geq R$ (3.43) gives

$$(R_1^* - r_k)r_k > R(1 - R_1^*)(1 - r_k). \qquad (3.46)$$

The quadratic equation $(R_1^* - x)x = R(1 - R_1^*)(1 - x)$ has two roots $\underline{r}_2(R) < \bar{r}_2(R)$ if and only if

$$R > \bar{R}_\varrho = \frac{2 - R_1^* + 2\sqrt{1 - R_1^*}}{1 - R_1^*}$$

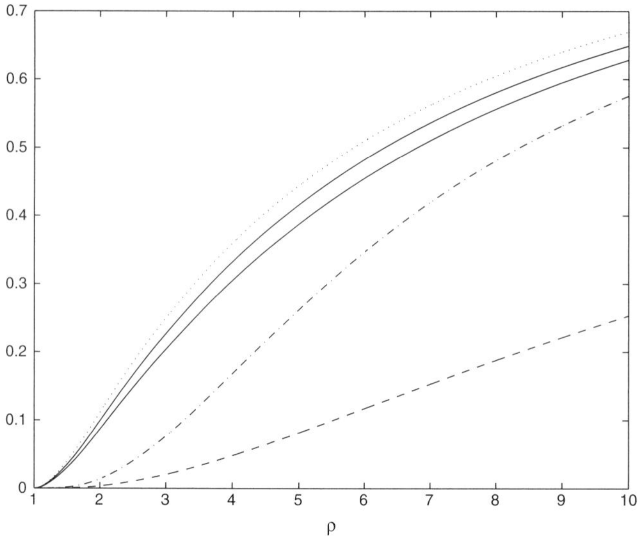

Fig. 3.8. *Dotted line*: R_1^*, *dashed line*: R_2^*, *dash-dotted line*: $\bar{R}$ defined by (3.45), *solid lines*: $\bar{r}_1(R_2^*) < \bar{r}_1(R_2^*/2)$, with $\bar{r}_1(R)$ defined by (3.44), as functions of ϱ

or

$$R < \underline{R}_\varrho = \frac{2 - R_1^* - 2\sqrt{1 - R_1^*}}{1 - R_1^*} = \frac{(\varrho + 1 - 2\sqrt{\varrho})^2}{4\varrho},$$

(and they are then positive since their product equals $R(1-R_1^*)$ and their sum is $R_1^* + R(1 - R_1^*)$). One may easily check that $R_2^* < \underline{R}_\varrho$ and $\bar{R}_\varrho > 4$ for any $\varrho > 1$. Choosing $R \leq R_2^*$ thus ensures that the two roots $\underline{r}_2(R), \bar{r}_2(R)$ exist and, from (3.46), r_k satisfies $\underline{r}_2(R) < r_k < \bar{r}_2(R)$. We have $\bar{r}_2(R) < \bar{r}_1(R) = R_1^* - FR$, which thus improves (3.44) (notice that the equation $\bar{r}_2(R) = R$ has now no solution in R). Figure 3.9 presents R_1^* (dotted line), $\sqrt{R_2^*}$ (dashed line), $\bar{r}_1(R_2^*) < \bar{r}_1(R_2^*/2)$ (solid lines) and $\bar{r}_2(R_2^*)$, $\bar{r}_1(R_2^*/2)$ (dash-dotted lines, top for $\bar{r}_2(R_2^*/2)$), as functions of ϱ. The improvement of $\bar{r}_2(R)$ over $\bar{r}_1(R)$ is clear for R near R_2^* (although $\bar{r}_2(R_2^*)$ remains quite close to R_1^*) but is negligible for $R = R_2^*/2$.

Remark 8. The bounds $\bar{r}_1(R)$ and $\bar{r}_2(R)$ are rather pessimistic, as evidenced by Fig. 3.9: the constraint $r'_k \geq R_2^*$ implies that ν_k is the measure ν_2^* given by (3.17) in Theorem 4, and therefore $r_k = \sqrt{R_2^*}$. The gap between $\sqrt{R_2^*}$ and the lower curves in solid line and dash-dotted line illustrates the pessimism of $\bar{r}_1(R_2^*)$ and $\bar{r}_2(R_2^*)$, respectively.

An exact bound could be obtained, at least numerically, for any value of $R \in [0, R_2^*]$ since the maximization of the rate r_k of the steepest-descent algorithm under the constraint that the rate r'_k of the optimum 2-gradient algorithm satisfies $r'_k \geq R$ for some $R \in [0, R_2^*]$ corresponds to a D_s-optimum design problem under constraint. Indeed, similarly to the proof of Theorem 4,

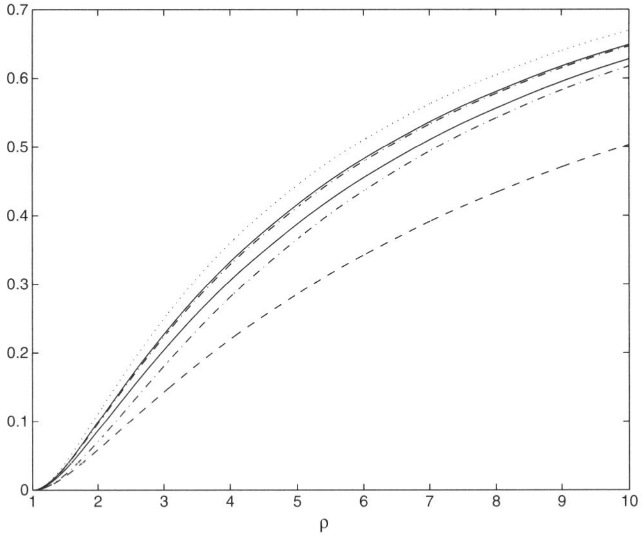

Fig. 3.9. *Dotted line:* R_1^*, *dashed line:* $\sqrt{R_2^*}$, *solid lines:* $\bar{r}_1(R_2^*) < \bar{r}_1(R_2^*/2)$, with $\bar{r}_1(R)$ defined by (3.44), *dash-dotted lines:* $\bar{r}_2(R_2^*) < \bar{r}_2(R_2^*/2)$, as functions of ϱ

the maximum value for r_k is obtained for the measure $\bar{\nu}$ supported on $[m, M]$ that minimizes the variance of the estimator of θ_0 in the regression model $\theta_0 + \theta_1 x$ with i.i.d. errors (a convex function of $\bar{\nu}$) under the restriction that the variance of the estimator of θ_0 in the model $\theta_0 + \theta_1 x + \theta_2 x^2$ is smaller than $1/R$ (which defines a convex constraint on $\bar{\nu}$). Numerical algorithms for solving such convex design problems can be constructed following, e.g. the ideas presented in Molchanov and Zuyev (2001).

As an attempt to consider several consecutive iterations to circumvent the limits of the worst-case analysis above, one may consider the following algorithm.

Step 1 ($s = 1$): Use steepest descent while $L_{k+1} - L_k \geq \epsilon$ for some $\epsilon > 0$, with $L_k = 1/(1 - r_k) = \mu_{-1}^k \mu_1^k$. When $L_{k+1} - L_k < \epsilon$, go to Step 2.
Step 2 ($s = 2$): Use the optimum 2-gradient algorithm while the rate of convergence r' is smaller than αR_2^* for some α, $\epsilon R_1^*/R_2^* < \alpha < 1$. When $r' > \alpha R_2^*$, return to Step 1.

The idea of the algorithm is that the rate of convergence of the first iteration of Step 2 is very good when ϵ is small enough. Indeed, when switching from $s = 1$ to $s = 2$, $L_{k+1} - L_k < \epsilon$ and (3.43) imply $r'_k < \epsilon r_k < \epsilon R_1^* < \alpha R_2^*$ (note that this switching necessarily occurs since L_k is not decreasing and bounded by $L^* = 1/(1 - R_1^*)$). We did not manage to improve the results above, however. The reason is that a worst-case analysis leads to consider cycles where the optimum 2-gradient algorithm is used for one iteration only, with a comeback to a sequence of n steepest-descent iterations with associated rates bounded by $1 - 1/[L^* - (n-1)\epsilon], 1 - 1/[L^* - (n-2)\epsilon], \ldots, 1 - 1/[L^* - \epsilon], 1 - 1/L^*$. The logarithm of the global rate of convergence over such a cycle of $n+1$ iterations can then be bounded by

$$\begin{aligned}
\log R_{\max} &= \left(\frac{1}{n+1}\right)\left[\sum_{i=0}^{n} \log\left(1 - \frac{1}{L^* - i\epsilon}\right) + \log(R_1^*\epsilon)\right] \\
&< \left(\frac{n}{n+1}\right) \log\left[\frac{1}{n}\sum_{i=0}^{n}\left(1 - \frac{1}{L^* - i\epsilon}\right)\right] + \frac{\log(R_1^*\epsilon)}{n+1} \\
&< \left(\frac{n}{n+1}\right) \log\left[1 - \frac{1}{n(L^*)^2}\sum_{i=0}^{n}(L^* + i\epsilon)\right] + \frac{\log(R_1^*\epsilon)}{n+1} \\
&= \left(\frac{n}{n+1}\right) \log\left[1 - \frac{1}{L^*} - \frac{(n-1)\epsilon}{2(L^*)^2}\right] + \frac{\log(R_1^*\epsilon)}{n+1}
\end{aligned}$$

which should be maximized with respect to n (to bound the worst-case cycle) and then minimized with respect to ϵ to optimize the bound. A careful analysis shows that the optimum is always attained for ϵ as large as possible and n small. The situation is then similar to that considered for Algorithm C: when $n = 1$ one should choose ϵ such that both the steepest-descent and the optimum 2-gradient iterations have a rate of convergence smaller than R_2^*, which is impossible.

3.5.2.2 Some Simulation Results

We apply Algorithm B with $m_1 = 1$, $m_2 = 4$ to a series of 1,000 problems in $\mathbb{R}^d$ with $d = 1{,}000$. For each problem, the eigenvalues of A are randomly generated with the uniform distribution in $[1, \varrho]$ and the initial renormalized gradient z_0 is also randomly generated, with the uniform distribution on the unit sphere $\mathcal{S}_{1000}$.

The algorithm is run for 100 iterations (which means 12 steepest-descent iterations and 88 iterations of the optimum 2-gradient algorithm, and thus 188 gradient evaluations). The results in terms of global rates R_n, see (3.39) and N_n, see (3.42), are summarized in Table 3.1 for the case $\varrho = 100$. For

Table 3.1. Global rates R_{100}, N_{188} and their logarithms for Algorithm B ($m_1 = 1$, $m_2 = 4$), averaged over 1,000 random problems in $\mathbb{R}^{1000}$ with $\varrho = 100$, together with their standard deviations, minimum and maximum values over the 1,000 problems

	Mean	Std. deviation	Minimum	Maximum
R_{100}	0.5538	0.0157	0.5047	0.6138
$\log(R_{100})$	−0.5914	0.0284	−0.6838	−0.4881
N_{188}	0.7394	0.0105	0.7061	0.7781
$\log(N_{188})$	−0.3020	0.0142	−0.3481	−0.2508

Table 3.2. Global rates R_{100}, N_{200} and their logarithms for the optimum 2-gradient algorithm, averaged over 1,000 random problems in $\mathbb{R}^{1000}$ with $\varrho = 100$, together with their standard deviations, minimum and maximum values over the 1,000 problems, and theoretical maxima

	Mean	Std. deviation	Minimum	Maximum	Theoretical max.
R_{100}	0.8199	0.0101	0.7766	0.8399	$R_2^* \simeq 0.8548$
$\log(R_{100})$	−0.1986	0.0123	−0.2528	−0.1745	$\log(R_2^*) \simeq -0.1569$
N_{200}	0.9055	0.0056	0.8812	0.9164	$N_2^* \simeq 0.9245$
$\log(N_{200})$	−0.0993	0.0062	−0.1264	−0.0873	$\log(N_2^*) \simeq -0.0785$

Table 3.3. Global rates R_{100}, N_{1888} and their logarithms for Algorithm B ($m_1 = 1$, $m_2 = 4$), averaged over 1,000 random problems in $\mathbb{R}^{1000}$ with $\varrho = 100$, together with their standard deviations, minimum and maximum values over the 1,000 problems

	Mean	Std. deviation	Minimum	Maximum
R_{1000}	0.5203	0.0079	0.4953	0.5467
$\log(R_{1000})$	−0.6535	0.0151	−0.7026	−0.6039
N_{1888}	0.7184	0.0055	0.7008	0.7365
$\log(N_{1898})$	−0.3307	0.0076	−0.3555	−0.3059

comparison, $R_1^* \simeq 0.9608$ for steepest descent, $R_2^* \simeq 0.8548$ and $N_2^* \simeq 0.9245$ for the optimum 2-gradient. The rate of convergence of the steepest-descent algorithm is known to be always close to its maximum value, see, e.g. Pronzato et al. (2001, 2006). Table 3.2 indicates that this is true also for the optimum 2-gradient algorithm: in that table, Algorithm B is run with $m_1 = 0$, that is, all iterations correspond to the optimum 2-gradient algorithm. One may notice that on average Algorithm B requires $0.3020/0.0993 \simeq 3$ times less gradient evaluations than the optimum 2-gradient algorithm to reach a given precision on the squared norm of the gradient. Even if one considers the very pessimistic situation that corresponds to the worst performance for Algorithm B and the best one for the optimum 2-gradient algorithm, the ratio is $0.2508/0.1264 \simeq 2$. To obtain similar performance for $\varrho = 100$ with an optimum s-gradient algorithm, that is, $N_s^* < 0.78$, one must take $s \geq 9$.

Tables 3.3 and 3.4 give the same information as Tables 3.1 and 3.2, respectively, but for the case when the algorithm is run for 1,000 iterations (which means 2,000 gradient evaluations for the optimum 2-gradient algorithm and 1,888 for Algorithm B with $m_1 = 1$ and $m_2 = 4$). The performance of the optimum 2-gradient algorithm are worse in Table 3.4 than in Table 3.2 (which comes as no surprise since the rate of convergence of the algorithm is non-decreasing), but those of Algorithm B are better when the number of iterations increases. This is confirmed by Fig. 3.10 that shows the global rates R_k from iteration 1 to iteration k, see (3.39), averaged over 1,000 random problems, for the optimum 2-gradient and Algorithm B as functions of k. Figure 3.11

Table 3.4. Global rates R_{1000}, N_{2000} and their logarithms for the optimum 2-gradient algorithm, averaged over 1,000 random problems in $\mathbb{R}^{1000}$ with $\varrho = 100$, together with their standard deviations, minimum and maximum values over the 1,000 problems, and theoretical maxima

	Mean	Std. deviation	Minimum	Maximum	Theoretical max.
R_{1000}	0.8484	0.0018	0.8403	0.8521	$R_2^* \simeq 0.8548$
$\log(R_{1000})$	−0.1644	0.0022	−0.1740	−0.1601	$\log(R_2^*) \simeq -0.1569$
N_{2000}	0.9211	0.0010	0.9167	0.9231	$N_2^* \simeq 0.9245$
$\log(N_{2000})$	−0.0822	0.0011	−0.0870	−0.0800	$\log(N_2^*) \simeq -0.0785$

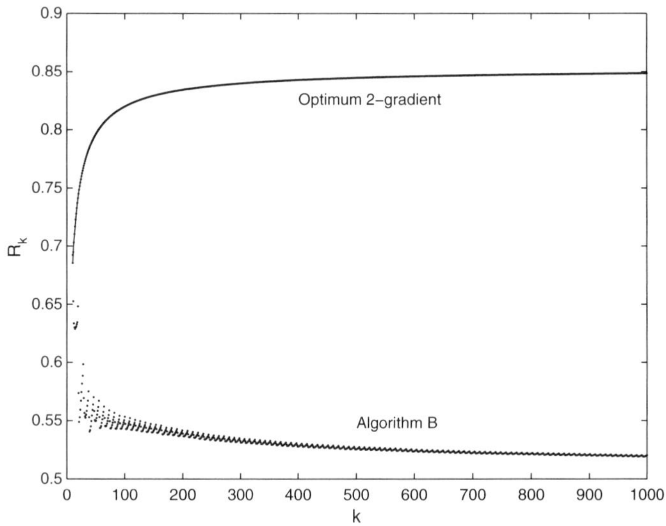

Fig. 3.10. Global rates R_k from iteration 1 to iteration k, (averaged over 1,000 random problems) for the optimum 2-gradient and Algorithm B with $m_1 = 1$, $m_2 = 4$ as functions of k

presents the rate of convergence r_k of both algorithms, averaged over the 1,000 random problems, as a function of k. The regular increase of r_k is clear for the optimum 2-gradient algorithm, whereas Algorithm B exhibits a rather specific pattern: the dots above the full line correspond to the steepest-descent iterations and those below to optimum 2-gradient iterations, the (averaged) rates of which tend to follow $m_2 = 4$ rather well-identified pairs of trajectories.

Table 3.5 gives the same information as Table 3.1 but for the case $\varrho = 1,000$ (which gives $R_2^* \simeq 0.9842$ and $N_2^* \simeq 0.9920$). To obtain $R_s^* < 0.93$ for $\varrho = 1,000$ with an optimum s-gradient algorithms one must take $s \geq 5$ and to get $N_s^* < 0.956$ one must take $s \geq 13$.

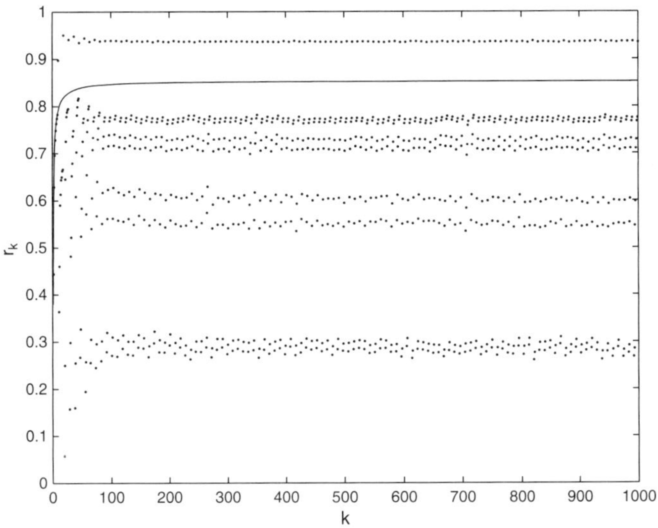

Fig. 3.11. Rate r_k at iteration k, averaged over 1,000 random problems, for the optimum 2-gradient (*solid line*) and Algorithm B with $m_1 = 1$, $m_2 = 4$ (*dots*) as a function of k

Table 3.5. Global rates R_{100}, N_{188} and their logarithms for Algorithm B ($m_1 = 1$, $m_2 = 4$), averaged over 1,000 random problems in $\mathbb{R}^{1000}$ with $\varrho = 1,000$, together with their standard deviations, minimum and maximum values over the 1,000 problems

	Mean	Std. deviation	Minimum	Maximum
R_{100}	0.8724	0.0182	0.8042	0.9154
$\log(R_{100})$	−0.1368	0.0209	−0.2179	−0.0884
N_{188}	0.9320	0.0099	0.8940	0.9554
$\log(N_{188})$	−0.0705	0.0106	−0.1121	−0.0457

References

Akaike, H. (1959). On a successive transformation of probability distribution and its application to the analysis of the optimum gradient method. *Annals of the Institute Statistical Mathematics Tokyo*, **11**, 1–16.

Elaydi, S. (2005). *An Introduction to Difference Equations*. Springer, Berlin. Third Edition.

Fedorov, V. (1972). *Theory of Optimal Experiments*. Academic Press, New York.

Forsythe, G. (1968). On the asymptotic directions of the s-dimensional optimum gradient method. *Numerische Mathematik*, **11**, 57–76.

Hoel, P. and Levine, A. (1964). Optimal spacing and weighting in polynomial prediction. *Annals of Mathematical Statistics*, **35**(4), 1553–1560.

Kantorovich, L. and Akilov, G. (1982). *Functional Analysis.* Pergamon Press, London. Second edition.

Kiefer, J. and Wolfowitz, J. (1959). Optimum designs in regression problems. *Annals of Mathematical Statistics*, **30**, 271–294.

Luenberger, D. (1973). *Introduction to Linear and Nonlinear Programming.* Addison-Wesley, Reading, Massachusetts.

Meinardus, G. (1963). Über eine Verallgemeinerung einer Ungleichung von L.V. Kantorowitsch. *Numerische Mathematik*, **5**, 14–23.

Molchanov, I. and Zuyev, S. (2001). Variational calculus in the space of measures and optimal design. In A. Atkinson, B. Bogacka, and A. Zhigljavsky, editors, *Optimum Design 2000*, chapter 8, pages 79–90. Kluwer, Dordrecht.

Nocedal, J., Sartenaer, A., and Zhu, C. (1998). On the accuracy of nonlinear optimization algorithms. Technical Report Nov. 1998, ECE Department, Northwestern University, Evanston, Il 60208.

Nocedal, J., Sartenaer, A., and Zhu, C. (2002). On the behavior of the gradient norm in the steepest descent method. *Computational Optimization and Applications*, **22**, 5–35.

Pronzato, L., Wynn, H., and Zhigljavsky, A. (2000). *Dynamical Search.* Chapman & Hall/CRC, Boca Raton.

Pronzato, L., Wynn, H., and Zhigljavsky, A. (2001). Renormalised steepest descent in Hilbert space converges to a two-point attractor. *Acta Applicandae Mathematicae*, **67**, 1–18.

Pronzato, L., Wynn, H., and Zhigljavsky, A. (2005). Kantorovich-type inequalities for operators via D-optimal design theory. *Linear Algebra and its Applications (Special Issue on Linear Algebra and Statistics)*, **410**, 160–169.

Pronzato, L., Wynn, H., and Zhigljavsky, A. (2006). Asymptotic behaviour of a family of gradient algorithms in $\mathbb{R}^d$ and Hilbert spaces. *Mathematical Programming*, **A107**, 409–438.

Sahm, M. (1998). Optimal designs for estimating individual coefficients in polynomial regression. Ph.D. Thesis, Fakultät für Mathematik, Ruhr Universität Bochum.

Silvey, S. (1980). *Optimal Design.* Chapman & Hall, London.

4

Bivariate Dependence Orderings for Unordered Categorical Variables

A. Giovagnoli, J. Marzialetti and H.P. Wynn

Summary. Interest in assessing the degree of association between two or more random variables has a long history in the statistical literature. Rather than *measuring* association, we want ways of *comparing* it. Restricting the attention in this chapter to unordered categorical random variables, we point at some possible definitions of dependence orderings which employ matrix theory and, to a lesser extent, group theory. This approach allows a unified investigation of the most common indicators in the statistical literature.

One very special type of association is the amount of agreement among different observers that classify the same group of statistical units: in the medical field this has led to widespread use of *Cohen's Kappa*. Starting with an axiomatic definition of agreement, we show its formal properties. Some criticism of Cohen's Kappa and other measures of agreement in use will ensue.

4.1 Introduction

Several statistical concepts (such as location, dispersion, concentration and dependence) can be studied via order and equivalence relations. The concept to be defined can be described by means of a partial ordering or a pre-ordering among the variables of interest and the relative measures are thus order-preserving functions. Bickel and Lehmann (1975) were among the first authors to introduce this approach to statistics. Many well-known properties of established statistical measures can be derived from the ordering representation.

Stochastic orderings (i.e. order relations among unidimensional or multidimensional random variables) and the relative order-preserving functions have a long history: an introduction is in Chap. 9 of the book by Ross (1995) and fundamental works are by Shaked and Shanthikumar (1994) and Müller and Stoyan (2002). Applications are to general statistical theory, in particular testing, reliability theory, and more recently risk and insurance. In this chapter it is not our intention to cover the basic material but rather to describe and investigate concepts related to association of random variables, as

L. Pronzato, A. Zhigljavsky (eds.), *Optimal Design and Related Areas in Optimization and Statistics*, Springer Optimization and its Applications 28,
DOI 10.1007/978-0-387-79936-0_4,

in Goodman and Kruskal (1979), namely the degree of dependence among the components of a multivariate variable. We stress that different concepts of association are possible: for example *interdependence* (all the variables have an exchangeable role) and *dependence* of one variable on the others. Furthermore, as pointed out in Chap. 11 of Bishop et al. (1975), a special case of association among variables is *inter-observer agreement*, which has important applications in several fields. In this case, there is one characteristic of interest observed on the same statistical units by different observers who may partly agree and partly disagree in their classifications or their scores, and the multidimensional random variables to be compared from the agreement viewpoint express the judgements of the "raters" in vector form.

So far studies of dependence and interdependence through orderings have focussed mainly on real random variables (see Joe (1997)): in the tradition of Italian statistics, Forcina and Giovagnoli (1987) and Giovagnoli (2000a, b) have defined some dependence orderings for bivariate *nominal* random variables, i.e. with values in unordered categories. This is the type of variables that we restrict ourselves to in this chapter. We review the existing results and introduce further developments in Sect. 4.2. In particular, we are not aware of a theory of *agreement orderings* so far, apart from a brief hint in the already mentioned paper by Giovagnoli (2002b). A possible definition and some new results are the topic of Sect. 4.3. Section 4.4 points at directions for research.

For reasons of simplicity and space in this chapter we only look at nominal variables in two dimensions. A further development which deals with an agreement ordering for a different type of multivariate variables, namely discrete or continuous, is the object of another paper by the same authors (Giovagnoli et al., 2006).

We end this introduction with some terminology. An *equivalence* in a set $\mathcal{S}$ is a reflexive, symmetric, transitive relation. A *pre-order* $\trianglelefteq$ in $\mathcal{S}$ is a reflexive and transitive relation (anti-symmetry is not required). To every pre-order $\trianglelefteq$ there corresponds an equivalence relation $\triangleq$: if $x, y \in \mathcal{S}$ and $x \trianglelefteq y$, $y \trianglelefteq x$ then we say that $x \triangleq y$. All the one-to-one maps φ of $\mathcal{S}$ onto $\mathcal{S}$ such that $\varphi(x) \triangleq x$ for all $x \in \mathcal{S}$ form a set $\mathbf{G}_{\mathrm{I}}$, called the *invariance* set of $\trianglelefteq$. All the maps ψ of $\mathcal{S}$ into $\mathcal{S}$ such that $x \trianglelefteq y$ implies $\psi(x) \trianglelefteq \psi(y)$ $\forall\, x, y \in \mathcal{S}$ form the *equivariance* set $\mathbf{G}_{\mathrm{E}}$ of $\trianglelefteq$. All the maps ϕ of $\mathcal{S}$ into $\mathcal{S}$ such that $\phi(x) \trianglelefteq x$ for all $x \in S$ form the *contraction* set $\mathbf{G}_{\mathrm{K}}$ of $\trianglelefteq$. As well as the contractions, we can define the *expansions*: all the maps $\tilde{\phi}$ of $\mathcal{S}$ into $\mathcal{S}$ such that $x \trianglelefteq \tilde{\phi}(x)$ for all $x \in \mathcal{S}$. Clearly $\mathbf{G}_{\mathrm{I}} \subset \mathbf{G}_{\mathrm{E}}$ and $\mathbf{G}_{\mathrm{I}} \subset \mathbf{G}_{\mathrm{K}}$; $\mathbf{G}_{\mathrm{I}}$ is a group, whereas $\mathbf{G}_{\mathrm{E}}$ and $\mathbf{G}_{\mathrm{K}}$ are semigroups.

A function $f : \mathcal{S} \to \mathbb{R}$ is *order-preserving* if $x \trianglelefteq y$ implies $f(x) \leq f(y)$. A trivial remark: if f is order-preserving and $g : \mathbb{R} \to \mathbb{R}$ is non-decreasing, then $g \circ f$ too is order-preserving. Clearly order-preserving functions must be invariant w.r.t. $\mathbf{G}_{\mathrm{I}}$.

4.2 Dependence Orderings for Two Nominal Variables

4.2.1 S-Dependence and D-Dependence of One Variable on the Other

Let X and Y be categorical variables with a finite number of nominal categories, which we shall denote by $x_1,\ldots,x_r$ and $y_1,\ldots,y_c$, respectively, just to label them, without the labels implying any order among the categories, i.e. x_1 does not "come before" or "is less than" x_2, etc. We are interested in the joint (frequency or probability) distribution of (X,Y), identified from now on by a table

$$P_{r\times c} = \begin{pmatrix} p_{11} & p_{12} & \cdots & p_{1c} \\ p_{21} & p_{22} & \cdots & p_{2c} \\ \vdots & \vdots & & \vdots \\ p_{r1} & p_{r2} & \cdots & p_{rc} \end{pmatrix} = (p_{ij}),$$

$i = 1,\ldots,r$; $j = 1,\ldots,c$, where $p_{ij} \geq 0$, $\sum_{ij} p_{ij} = 1$. An alternative description is by means of the conditional and marginal distributions, i.e. either $(P^*, \mathbf{p}_r)$, where $P^* = (p_{ij}/p_{i+})$ and $\mathbf{p}_r = (p_{1+},\ldots,p_{r+})^t$, or $(P^{**}, \mathbf{p}_c)$, where $P^{**} = (p_{ij}/p_{+j})$ and $\mathbf{p}_c = (p_{+1},\ldots,p_{+c})^t$. It is sometimes useful to include tables with null rows and/or columns, in which case P^* or P^{**} are defined by setting all zeroes in the corresponding row and/or column.

To describe the dependence of Y on X, the following order relation $\leq_S$ (we call it $S-$*dependence*) was defined in Forcina and Giovagnoli (1987).

Definition 1. *Let P and Q be two bivariate tables. Then $Q \leq_S P$ if and only if there exists a stochastic matrix S such that $Q = S^t P$.*

The relation $\leq_S$ implies that the column margins of P and Q are equal: $\mathbf{p}_c = \mathbf{q}_c$.

Proposition 1 (Forcina and Giovagnoli (1987)). *$Q \leq_S P$ is equivalent to the following two conditions holding simultaneously:*

$$(i) \quad Q^* = \dot{S}P^* \qquad\qquad (ii) \quad \mathbf{p}_r = \tilde{S}^t \mathbf{q}_r$$

with $\tilde{S}$ another stochastic matrix.

Forcina and Giovagnoli (1987) showed that $\leq_S$ satisfies some intuitive requirements for a dependence ordering.

We call *S-equivalence* the equivalence relation $\simeq_S$ defined by $\leq_S$. Since permutation matrices are doubly stochastic, a permutation of the rows of P gives an S-equivalent table; the converse is not true, namely not all S-equivalent tables can be obtained by permutation as the following result shows.

Proposition 2.

1. *Row-aggregation, i.e. replacing one of two rows by their sum and the other by a row of zeroes, leads to a distribution with less S-dependence.*
2. *When two rows are proportional, both row-aggregation and row-splitting (namely the inverse operation to row-aggregation) imply S-equivalence.*

Proof. The matrices

$$S_1 = \left(\begin{array}{cc|c} 1 & 0 & \multirow{2}{*}{0} \\ 1 & 0 & \\ \hline \multicolumn{2}{c|}{0^t} & \mathrm{I}_{r-2} \end{array}\right) \quad \text{and} \quad S_2 = \left(\begin{array}{cc|c} \alpha & 1-\alpha & \multirow{2}{*}{0} \\ 1 & 0 & \\ \hline \multicolumn{2}{c|}{0^t} & \mathrm{I}_{r-2} \end{array}\right)$$

are stochastic. Pre-multiplication by S_1^t gives aggregation of the first two rows. On the other hand, if the second row is zero, pre-multiplication by S_2^t splits the first row into two proportional ones. □

There is another type of dependence of Y on X.

Definition 2. *We define D-dependence as*

$$Q \leq_D P \overset{\text{def}}{\Longleftrightarrow} Q = PD \tag{4.1}$$

with D a T-matrix, i.e. a product of T-transforms, namely matrices of the form $T_\alpha = (1-\alpha)I + \alpha\Pi^{(2)}$, $0 \leq \alpha \leq 1$ and $\Pi^{(2)}$ a permutation matrix that exchanges only 2 elements; such a D is doubly-stochastic.

This can be thought of as a model for errors in the Y variable: since perfect dependence is obtained when to each x-category there corresponds precisely only one y-category, then α stands for the probability (frequency) of mistakenly exchanging two y-categories. This ordering is known in the literature as chain-majorization (see Marshall and Olkin (1979)). The equivalence relation $\hat{=}_D$ defined by $\leq_D$ is permutation of the columns of P.

Clearly these two orderings can be combined.

Definition 3. *Define SD-dependence as follows*

$$Q \leq_{SD} P \overset{\text{def}}{\Longleftrightarrow} Q = S^{\mathrm{t}} PD \tag{4.2}$$

where S is a stochastic matrix and D a product of T-transforms.

The $\leq_{SD}$ ordering was defined in Forcina and Giovagnoli (1987), who however did not carry out a proper investigation of its properties. Note that (4.2) can be written as $\mathrm{vec}(Q) = (S \otimes D)^t \, \mathrm{vec}(P)$ and means that there exists a bivariate distribution table R such that $R \leq_S P$ and $Q \leq_D R$, and also that there exists $\tilde{R}$ such that $\tilde{R} \leq_D P$ and $Q \leq_S \tilde{R}$. Clearly both $\leq_S$ and $\leq_D$ are special cases of $\leq_{SD}$. On the other hand, $\leq_{SD}$ allows more comparisons, in particular matrices P and Q no longer need to have identical row or column margins.

Relative to maximal and minimal elements w.r.t. the ordering $\leq_{SD}$, the following results hold true:

Proposition 3.

1. *All the tables with independent rows and a given marginal distribution of Y are S-equivalent and are smaller w.r.t. $\leq_S$ than all the other tables with the same Y-margin.*
2. *All the tables giving exact dependence of Y on X with the same Y-margin are S-equivalent and are greater w.r.t. $\leq_S$ than all the other tables with the same Y-margin.*

The proof is easily obtainable by the same techniques employed in Theorem 3 of Forcina and Giovagnoli (1987).

Observe that Proposition 3 is not true for the ordering $\leq_D$. However from Proposition 3 there follows

Corollary 1.

1. *The independence table with uniform margins $(1/rc)J_{r\times c}$, where J stands for the matrix of all ones, is smaller w.r.t. $\leq_{SD}$ than any other $r \times c$ joint probability table.*
2. *All the tables giving exact dependence of Y on X are SD-equivalent and are greater w.r.t. $\leq_{SD}$ than all the other $r \times c$ tables.*

Let us now look at ways of transforming the order relation.

Proposition 4. *The invariance group and contraction set of $\leq_S$, $\leq_D$ and $\leq_{SD}$ are as follows, where the pair (A, B) denotes two matrices of dimension $r \times r$ and $c \times c$, respectively, acting on P by pre- and by post-multiplication, respectively.*

1. $\mathbf{G}_{\mathrm{I}}(\leq_S) = \{(\Pi_1, I_c); I_c \text{ the identity, } \Pi_1 \text{ a permutation matrix}\}$
   $\mathbf{G}_{\mathrm{I}}(\leq_D) = \{(I_r, \Pi_2);\ I_r \text{ the identity, } \Pi_2 \text{ a permutation matrix}\}$
   $\mathbf{G}_{\mathrm{I}}(\leq_{SD}) = \{(\Pi_1, \Pi_2);\ \Pi_1,\ \Pi_2 \text{ permutation matrices}\} =$
   $= \mathbf{G}_{\mathrm{I}}(\leq_S) \bigvee \mathbf{G}_{\mathrm{I}}(\leq_D)$;
2. $\mathbf{G}_{\mathrm{K}}(<_S) = \{(S^t, I_c);\ I_c \text{ the identity, } S \text{ stochastic}\}$
   $\mathbf{G}_{\mathrm{K}}(\leq_D) = \{(I_r, D);\ I_r \text{ the identity, } D \text{ a } T\text{-matrix}\}$
   $\mathbf{G}_{\mathrm{K}}(\leq_{SD}) = \{(S^t, D);\ S \text{ stochastic, } D \text{ a } T\text{-matrix}\} =$
   $= \mathbf{G}_{\mathrm{K}}(\leq_S) \bigvee \mathbf{G}_{\mathrm{K}}(\leq_D)$;
3. $\mathbf{G}_{\mathrm{E}}(\leq_S) \supseteqq \{(S_1^t, S_2);\ S_1 \text{ stochastic and of full rank, } S_2 \text{ stochastic}\}$
   $\mathbf{G}_{\mathrm{E}}(\leq_D) \supseteqq \{(S^t, D);\ S \text{ stochastic, } D \text{ a } T\text{-matrix}\}$
   $\mathbf{G}_{\mathrm{E}}(\leq_{SD}) = \mathbf{G}_{\mathrm{K}}(\leq_S) \bigcap \mathbf{G}_{\mathrm{K}}(\leq_D)$.

Clearly, if we were interested in comparing bivariate tables as regards the dependence of X on Y we would consider the "transpose" order of $\leq_{SD}$, namely

$$Q \leq_{SD}^{t} P \overset{\text{def}}{\Longleftrightarrow} Q = DPS \qquad \text{where } S \text{ stochastic, } D \text{ a } T\text{-matrix.} \tag{4.3}$$

4.2.2 Measures of the Dependence of Y on X

Measures of dependence are usually requested to take on their minimum value when X and Y are independent, and usually such a minimum is assumed to be equal to 0. Furthermore they should have their maximum value when the dependence of one variable on the other is perfect. Usually such a maximum value is used to standardize the indicator so that it varies between 0 and 1. According to the approach of this chapter all the indicators of the dependence of Y on X (possibly before standardization) must preserve the $\leq_{SD}$-ordering, which amounts to preserving both orderings $\leq_S$ and $\leq_D$. In particular, they must take the same value on tables that are SD-equivalent. Thus all measures of SD-dependence must be invariant with respect to permutations of rows and of columns, row aggregation of proportional rows and row splitting when one row is zero.

The dependence indicators used in the literature, in general, seem to fall into three main types:

Type I) Measures that compare optimal prediction of Y given X with optimal prediction of Y when X is unknown, i.e.

$$\Phi_{Y\cdot X} = \frac{\Delta(\mathbf{p}_c) - \sum_i p_{i+}\Delta(\mathbf{p}_i^*)}{\Delta(\mathbf{p}_c)}$$

where $\Delta : \mathbb{R}^r \to \mathbb{R}$ stands for a measure of dispersion/heterogeneity of the distribution of Y, or of minimal expected loss in predicting Y.

Type II) Weighted averages of measures of the mean information of Y given X relative to the unconditional information of Y, i.e.

$$\Psi_{Y\cdot X} = g\left(\sum_i p_{i+} d(\mathbf{p}_i^*,\ \mathbf{p}_c)\right)$$

where $g : \mathbb{R} \to \mathbb{R}$ is increasing and $d(\mathbf{u}, \mathbf{v})$ is a measure of the "distance" or "diversity" of a distribution $\mathbf{u}$ on the set of unordered categories from another distribution $\mathbf{v}$, or of information gain from prior $\mathbf{v}$ to posterior $\mathbf{u}$.

Type III) Weighted averages of the "distances" $d(\cdot,\cdot)$ between all pairs of distributions of Y conditional on X

$$\Lambda_{Y\cdot X}(P) = g\left(\sum_{i=1}^{r}\sum_{i'=1}^{r} p_{i+}p_{i'+} d(\mathbf{p}_i^*, \mathbf{p}_{i'}^*)\right)$$

where $g(\cdot)$ and $d(\cdot,\cdot)$ are as in Type II.

Proposition 5. *The order $\leq_{SD}$ is preserved by Type I indices when Δ is convex and permutation invariant.*

Examples are:

1. Guttman's λ
$$\lambda(P) = \frac{\sum_i \max_j p_{ij} - \max_j p_{+j}}{1 - \max_j p_{+j}}$$
where $\Delta(\mathbf{u}) = 1 - \max_i u_i$.

2. Goodman and Kruskal (1954)'s τ
$$\tau(P) = \frac{\sum_i \sum_j p_{ij}^2/p_{i+} - \sum_j p_{+j}^2}{1 - \sum_j p_{+j}^2}$$
where $\Delta(\mathbf{u}) = 1 - \sum_i u_i^2$.

3. Theil's index η
$$\eta(P) = \frac{-\sum_i \sum_j p_{ij} \log(p_{ij}/p_{i+}p_{+j})}{\sum_j p_{+j} \log p_{+j}},$$
where Δ is Shannon's entropy: $\Delta(\mathbf{u}) = -\sum_i u_i \log(u_i)$.

Proposition 6. *The order $\leq_{SD}$ is preserved by Type II indices when g is convex, $d(\mathbf{u}, \mathbf{v})$ is a convex function on $\mathbb{R}^r \times \mathbb{R}^r$, invariant under permutations of $y_1, \ldots, y_c$.*

Examples are:

1. Gini's connection index G
$$G(P) = \frac{1}{2} \sum_i \sum_j |p_{ij} - p_{i+}p_{+i}| . \tag{4.4}$$
Here $d(\mathbf{u}, \mathbf{v}) = \sum_i |u_i - v_i|$ is the city-block distance.

2. Good's class of measures J
$$J_\lambda(P) = \sum_i \sum_j \frac{p_{ij}^\lambda}{p_{i+}^{\lambda-1} p_{+i}^{\lambda-1}}$$
(which is Pearson's χ^2 when $\lambda = 2$). Here $d(\mathbf{u}, \mathbf{v}) = \sum_i \frac{u_i^\lambda}{v_i^{\lambda-1}} - 1$.

3. Halphen's *modulus of dependence* H
$$H(P) = \sum_i \sum_j p_{ij} \log p_{ij} - \sum_i p_{i+} \log p_{i+} - \sum_j p_{+j} \log p_{+j} . \tag{4.5}$$
Here $d(\mathbf{u}, \mathbf{v}) = \sum_i u_i \log(u_i/v_i)$.

Proposition 7. *The order $\leq_{SD}$ is preserved by Type III indices if $g(\cdot)$ is an increasing real function, $d(\cdot,\cdot)$ is convex in both components and invariant under permutations of $y_1, \ldots, y_c$.*

For example, Goodman and Kruskal's indicator is obtained by letting $d(\cdot,\cdot)$ be the Euclidean distance and $g(\cdot)$ the square root.

Propositions 5 and 6 were proved by Forcina and Giovagnoli (1987), Proposition 7 is proved in Giovagnoli (2002a).

4.2.3 Interdependence Orderings

Interdependence between two variables may be defined as some type of "distance" of their joint distribution from the reference situation of independence, which corresponds to no association. A natural requirement is again invariance with respect to permutations of the rows and of the columns. Several ways of introducing orders of bilateral association between bivariate distributions, which make use of the heuristic arguments presented in Forcina and Giovagnoli (1987) for the dependence case, have been suggested in Giovagnoli (2002a) but we are not aware of any thorough investigation carried out on these order relations. For distributions with the same margins, Cifarelli and Regazzini (1986) and Scarsini (1991) also define an association order. Alternatively, another way is to combine the dependence ordering $\leq_{SD}$ of Y on X with its transpose (dependence of X on Y). Thus we can define an ordering $\leq_{SD-bil}$ of bilateral association if X is SD-dependent on Y **and** Y is SD-dependent on X.

Definition 4.

$$Q \leq_{SD-\text{bil}} P \overset{\text{def}}{\iff} Q = S_1^t P D_1 \quad \text{and} \quad Q = D_2^t P S_2 \tag{4.6}$$

for some stochastic matrices S_1 and S_2 and some T-matrices D_1 and D_2.

This is clearly a pre-ordering, invariant under any permutation of the rows and any permutation of the columns. As an example take:

$$P = \begin{pmatrix} 0.4 & 0.2 \\ 0.1 & 0.3 \end{pmatrix},\ Q = \begin{pmatrix} 0.25 & 0.35 \\ 0.25 & 0.15 \end{pmatrix},\ S_1 = \begin{pmatrix} 0.4 & 0.6 \\ 0.9 & 0.1 \end{pmatrix},\ S_2 = \begin{pmatrix} 0.25 & 0.75 \\ 0.75 & 0.25 \end{pmatrix},$$

$D_1 = D_2 = I_2$.

Proposition 8.

1. *For all P, let $C = P\mathbf{1}\mathbf{1}^t P$ be the table with the same margins as P whose margins are independent. Then $C \leq_{SD-bil} P$.*
2. *Table $(1/rc)J_{r\times c}$, standing for the independence distribution with uniform margins, is smaller w.r.t. $\leq_{SD-bil}$ than any other $r \times c$ joint probability table.*

3. *Exact dependence of* Y *on* X *and* X *on* Y *is possible only when* $r = c$. *When this happens, tables with exact dependence of rows on columns and column on rows are* $SD - bil$*-maximal.*

Note that if both P and Q have uniform margins, then S_1 and S_2 must be doubly stochastic for (4.6) to hold.

A special case of (4.6) is when there exist T-matrices D_1 and D_2 such that

$$Q = D_2 P D_1, \text{ i.e } \operatorname{vec}(Q) = (D_2 \otimes D_1)^t \operatorname{vec}(P). \tag{4.7}$$

Since T-matrices are doubly stochastic, so is $D_2 \otimes D_1$, thus (4.7) is also a special case of majorization of the vecs, namely the association order $\overset{m}{<}$ defined by Joe (1985)

$$Q \overset{m}{<} P \iff \operatorname{vec}(Q) = D \operatorname{vec}(P)$$

with D an $rc \times rc$ doubly stochastic matrix.

4.2.4 Measures of Interdependence Between X and Y

As to the order-preserving functions w.r.t. $\leq_{SD-\text{bil}}$, all the measures of SD-dependence seen in Sect. 4.2.2 which are symmetric, i.e. invariant under transposition of the matrix P ($\Phi_{Y \cdot X}(P) = \Phi_{Y \cdot X}(P^t)$) are clearly measures of bilateral SD-dependence. Among them there are:

1. Good's measures J_λ and more in general measures of the form

$$\sum_i \sum_j p_{i+} p_{+j} \, g\left(\frac{p_{ij}}{p_{i+} p_{+j}}\right),$$

with $g(\cdot)$ convex on $[0, \infty)$, as mentioned in Scarsini (1991).
2. Gini's indicator (4.4).
3. Halphen's indicator (4.5).
4. Type III indicators with $d(\mathbf{u}, \mathbf{v}) = d(\mathbf{v}, \mathbf{u})$.

In the literature one is advised to consider as a measure of reciprocal dependence of X and Y some type of average of two measurements, dependence of X on Y and dependence of Y on X, calculated by the same indicator:

$$M_{XY}(P) = M\left(\Phi_{Y \cdot X}(P), \Phi_{X \cdot Y}(P)\right)$$

which clearly are $SD - $ bil preserving if $\Phi_{Y \cdot X}$ is SD-preserving.

4.3 Inter-Raters Agreement for Categorical Classifications

4.3.1 How to Compare Agreement Among Raters

Classification and rating are basic in all scientific fields, and often there is the need to test the reliability of a classification process by assessing the level of agreement between two or more different observers (the raters) who classify the same group of statistical units (the subjects). In this section, we want to apply stochastic orderings to define what we mean by "agreement" among a group of observers who rate the same units on a categorical scale. Note that identical arguments apply if, instead, the same group of observers classify distinct sets of individuals or items, or the same observers rate the same items at different times.

In the literature, the extent of inter-rater agreement has been studied mainly by means of indicators. Cohen's Kappa is the most popular index of agreement for the case of two raters on a categorical scale of measurement (Cohen, 1960). A different approach is by means of statistical models which describe the structure of the relationship, mainly log-linear and latent class models, see Banerjee et al. (1999).

Our approach is different: given the sample space (the subjects) and the set $\mathcal{S}$ of all the possible ratings (unordered classes), we talk about the agreement among a group of observers as a type of dependence among the jointly distributed random variables with values in $\mathcal{S}$ representing the ratings of the various observers (Bishop et al., 1975). The question "when does one group or raters show more agreement than another?" is answered defining an order relation among multidimensional random variables (see also Giovagnoli (2002b)). Indicators of agreement will be all the real-valued functions preserving the ordering under consideration.

We start off with a set of "reasonable" requirements for any agreement ordering. Assume for now that d raters classify just one statistical unit (a subject) into one of m unordered classes. The way in which agreement of raters relative to that unit is expressed should satisfy the following set of axioms:

A-0 The agreement is a maximum when all the raters classify the subject into the same class.

A-1 The extent of agreement is independent of how we order (label) the classes.

A-2 The extent of agreement is independent of how we order (label) the raters.

We believe these axioms are sufficient to characterize agreement in the case of just two raters, whereas further properties, may be needed when their number is greater than two. In particular, we want to express the fact that agreement increases if one of the raters in a minority group changes his/her classification of a particular subject (statistical unit) to a class with larger consensus among the other raters. If we define the class distribution of that subject to be the frequency distribution of the way in which the statistical unit is diversely classified by the raters, we state that

A-3 The extent of agreement cannot decrease when the class distribution of the subject among the raters becomes more concentrated.

We point out that Axiom A-3 is in accordance with the widespread approach in the literature (Armitage et al. (1966); Fleiss (1971); Davies and Fleiss (1982)) that measures agreement by counting pairs of raters who agree on the classification of a single unit.

4.3.2 An Agreement Ordering for the Case of Two Raters

For simplicity in this chapter we consider the case of just two raters. Our aim is to define an agreement ordering for bivariate distributions in such a way that Axioms A-0 to A-2 hold true, so that these properties are automatically satisfied by any agreement measure that preserves the ordering. This approach will help us clarify the behaviour of some commonly used indicators. Let $P = (p_{ij})$ with $i, j = 1, \ldots, m$, denote the joint classification probabilities of the two raters. By Axioms A-1 and A-2, for every $m \times m$ table P and permutation matrix Π, we require P to be equivalent to $\Pi^t P \Pi$ and also to be equivalent to P^t, so that

$$P \doteqdot_{\text{agr}} P, \quad P \doteqdot_{\text{agr}} \Pi^t P \Pi, \quad P \doteqdot_{\text{agr}} P^t, \quad P \doteqdot_{\text{agr}} \Pi^t P^t \Pi \tag{4.8}$$

where we write $\doteqdot_{\text{agr}}$ to mean equivalence under agreement. Furthermore, by Axiom A-3 if we let

$$\mathbf{p}_i^t = (p_{i1}, p_{i2}, \ldots, p_{ij}, \ldots, p_{im})$$

the agreement increases by replacing this row with

$$\tilde{\mathbf{p}}_i^t = (p_{i1}, p_{i2}, \ldots, p_{ii} + \delta, .., p_{ij} - \delta, \ldots, p_{im})$$

where $0 < \delta \leq p_{ij}$, and similarly for any column. This can be formalized as follows. Define $E_{ij} = (e_{hk}^{ij})$ with $i \neq j$ the $m \times m$ "shift" matrix where

$$e_{hk}^{ij} = 1 \text{ if } h = k = i, \; e_{hk}^{ij} = -1 \text{ if } h = i, k = j, \; e_{hk}^{ij} = 0 \text{ otherwise.}$$

Definition 5. *Given two tables $P = (p_{ij})$ and $Q = (q_{ij})$, $i, j = 1, \ldots, m$, the agreement of P is lower than that of Q if P is obtained from Q by means of a finite number of "shifts" on the rows or columns, i.e.*

$$Q = P + \sum_{i \neq j} \delta_{ij} E_{ij} + \sum_{i \neq j} \tilde{\delta}_{ij} E_{ij}^t \qquad \textit{where } \; \delta_{ij} \geq 0, \;\; \tilde{\delta}_{ij} \geq 0, \;\; \delta_{ij} + \tilde{\delta}_{ij} \leq p_{ij} \, .$$

We call this relation "Δ-ordering" and write $\leq_\Delta$. It is easy to check that:

Proposition 9. *The Δ-ordering is a partial order.*

It is also easy to show that the Δ-ordering is consistent with the axioms. A-0 is clearly true. Furthermore:

Proposition 10. *Given two $m \times m$ tables P and Q*

$$P \leq_\Delta Q \implies \Pi^t P \Pi \leq_\Delta \Pi^t Q \Pi \quad \text{and} \quad P^t \leq_\Delta Q^t . \tag{4.9}$$

Remark 1. It is easy to see that given $m \times m$ tables P, Q and T

$P \leq_\Delta T$ and $T \doteq_{\text{agr}} Q \implies$ there exists R such that $P \doteq_{\text{agr}} R$ and $R \leq_\Delta Q$.

This enables us to widen the definition of agreement ordering to the following relation $\leq_{\text{agr}}$.

Definition 6. *$P \leq_{\text{agr}} Q$ if and only if there exists T such that $P \leq_\Delta T$ and $T \doteq_{\text{agr}} Q$.*

Proposition 11. *The relation $P \leq_{\text{agr}} Q$ is a pre-order. Furthermore, $P \leq_{\text{agr}} Q$ and $Q \leq_{\text{agr}} P$ if and only if $P \doteq_{\text{agr}} Q$, with $\doteq_{\text{agr}}$ defined as in (4.8).*

Proof. By virtue of Remark 1, transitivity of $\leq_{\text{agr}}$ holds. □

The invariance, equivalence and contraction sets of $\leq_{\text{agr}}$ are implicitly defined by Definitions (7), (8) and 5, respectively.

4.3.3 Order Preserving Indicators with Respect to the Agreement Ordering

We now want to check how the definition of $\leq_{\text{agr}}$ fits in with existing measures of agreement. A detailed description of such measures is given in Shoukri (2004).

To preserve $\leq_{\text{agr}}$, an indicator must be a function of table P which

1. Is invariant under any permutation of the rows and the same permutation of the columns of P
2. Is invariant by transposition of P
3. Preserves $\leq_\Delta$

Measures that are invariant under any permutation of the rows and any permutation of the columns do not appear to be suitable as potential measures of agreement. This applies for instance to Pearson's well-known Chi-squared indicator.

We recall that $Q \leq_{\text{agr}} P$ implies that there exists a permutation π such that

1. $q_{ii} \leq p_{\pi(i)\pi(i)}$, for all $i = 1, \ldots, m$
2. Either $q_{ij} \geq p_{\pi(i)\pi(j)}$, or $q_{ij} \geq p_{\pi(j)\pi(i)}$, for all $i = 1, \ldots, m$

Note that when π is the identity, property 1 above defines the $\leq_{\text{NAIF}}$ order relation of Giovagnoli (2002b).

4.3.3.1 Total Proportion of Agreement

The total proportion of agreement is the indicator TPA $= \sum_i p_{ii}$. Clearly TPA is order-preserving with respect to $\leq_{\text{agr}}$.

4.3.3.2 Cohen's Kappa

When the TPA measure is chance-corrected, namely the amount of agreement obtained for the effect of chance alone is subtracted, and the result is normalized with respect to the maximum value it can assume, it gives rise to the Kappa measure introduced by Cohen (1960):

$$\kappa(P) = \frac{\sum_i p_{ii} - \sum_i p_{i+}p_{+i}}{1 - \sum_i p_{i+}p_{+i}}.$$

Corollary 2. *For tables with the same margins, Cohen's Kappa preserves the agreement ordering* $\leq_{\text{agr}}$.

However the following counterexample, in which $P \leq_{\text{agr}} Q \leq_{\text{agr}} R$ and $\kappa(R) \leq \kappa(Q) \leq \kappa(P)$, shows that this result does not hold in general for any two bivariate distribution tables:

$$P = \begin{pmatrix} 0.10 & 0.20 & 0.50 \\ 0.01 & 0.00 & 0.01 \\ 0.16 & 0.01 & 0.01 \end{pmatrix} \qquad \kappa = -0.2970,$$

$$Q = \begin{pmatrix} 0.20 & 0.10 & 0.50 \\ 0.01 & 0.00 & 0.01 \\ 0.16 & 0.01 & 0.01 \end{pmatrix} \qquad \kappa = -0.2989,$$

$$R = \begin{pmatrix} 0.30 & 0.00 & 0.50 \\ 0.01 & 0.00 & 0.01 \\ 0.16 & 0.01 & 0.01 \end{pmatrix} \qquad \kappa = -0.3014.$$

Remark 2.

(i) This is an undesirable behaviour of the Kappa indicator. However, it can be shown to take place only under special circumstances, and only when its values are negative, which does not occur very frequently in actual practice (a detailed discussion can be found in Marzialetti (2006)).

(ii) It can also be shown that for $m = 2$ Cohen's Kappa always preserves the Δ-ordering and thus the $\leq_{\text{agr}}$-ordering (Marzialetti, 2006).

4.3.3.3 Farewell and Sprott (1999)'s Index

Another measure that clearly is invariant under a permutation of the sub-fixes and their exchange is

$$\sum_{i<j} \log \frac{p_{ii}p_{jj}}{p_{ij}p_{ji}}.$$

Because of properties 1. and 2. this measure preserves $\leq_{\text{agr}}$.

4.3.3.4 Rogot and Goldberg (1966)'s Indices

The following measures are defined only for $m = 2$, and are both order-preserving w.r.t. $\leq_{\text{agr}}$:

(i) $A_1 = \dfrac{p_{11}}{p_{1+} + p_{+1}} + \dfrac{p_{22}}{p_{2+} + p_{+2}}$.

A_1 is clearly invariant under permutations and transposition. Considering just one row shift for simplicity, we must show that

$$\frac{p_{11}}{p_{1+} + p_{+1}} + \frac{p_{22}}{p_{2+} + p_{+2}} \leq \frac{p_{11} + \delta}{p_{1+} + p_{+1} + \delta} + \frac{p_{22}}{p_{2+} + p_{+2} - \delta}.$$

This is always true since

$$\frac{p_{11}}{p_{1+} + p_{+1}} \leq \frac{p_{11} + \delta}{p_{1+} + p_{+1} + \delta} \quad \text{and} \quad \frac{p_{22}}{p_{2+} + p_{+2}} \leq \frac{p_{22}}{p_{2+} + p_{+2} - \delta}.$$

Hence A_1 preserves $\leq_{\text{agr}}$.

(ii) $A_2 = \dfrac{p_{11}}{p_{1+}} + \dfrac{p_{11}}{p_{+1}} + \dfrac{p_{22}}{p_{2+}} + \dfrac{p_{22}}{p_{+2}}$.

A_2 is clearly invariant under permutations and transposition. The rest of the proof retraces the steps of the previous one.

4.3.3.5 Jacquard's Coefficient

This is given by $J = p_{11}/(p_{11} + p_{21} + p_{22})$. This measure of agreement is not permutation invariant and does not appear to fit in our set-up, since one of the two categories, namely the first one, is taken to be of special interest, and not interchangeable with the second one.

In conclusion, we have shown that in most cases the order relation $\leq_{\text{agr}}$ defined by Definition 6 seems to capture the essence of the concept of agreement between two observers measured by the most commonly used indicators.

4.4 Conclusions and Further Research

There are many ways of defining association for multivariate nominal variables, and in particular agreement among observers seen as a special case of the degree of association of their joint classification probabilities. We hope to have shown that stochastic orderings sometimes combined with group theory provide a powerful mathematical environment for investigation of measures of association and agreement. The group theoretical ideas need further development. Moreover, the idea of matrix G-majorization (defined in Giovagnoli and Wynn (1985)) could be extended to cover some of the order relations defined in this chapter, in particular the order relation of Definition 3.

We believe that a research programme in this area should address the following questions, which are only partly answered in the present chapter.

1. (a) When does a multidimensional random variable show the same "association structure" as another one? In other words, how can we describe the equivalence relation "two multidimensional variables are equally associated"?
   (b) Which operations on a joint probability distribution do not affect the dependence structure?
2. (a) When is one multidimensional random variable more associated than another?
   (b) Which operations on a set of jointly distributed variables increase or decrease dependence among them?
3. What are suitable criteria for representing the suggested ordering on a numerical scale, i.e. what is the class of all order preserving functions (consistent indicators) for that type of association?

Acknowledgements

Section 4.3 is part of second author's PhD thesis (Marzialetti, 2006), jointly supervised by the other authors of the present work. This research was completed when the third author was a Senior Fellow of the I.S.A. (Institute of Advanced Study) of the University of Bologna. Thanks are also due to Professor Antonio Forcina of Perugia University for stimulating discussions.

References

Armitage, P., Blendis, L.M., and Smyllie, H.C. (1966). The measurement of observer disagreement in the recording of signs. *Journal of the Royal Statistical Society A*, **129**, 98–109.

Banerjee, M., Capozzoli, M., McSweeney, L., and Sinha, D. (1999). Beyond kappa: A review of interrater agreement measures. *The Canadian Journal of Statistics*, **27**(1), 3–23.

Bickel, P.J. and Lehmann, E.L. (1975). Descriptive statistics for nonparametric models. i: Introduction. *The Annals of Statistics*, **3**, 1038–1044.

Bishop, Y.M.M., Fienberg, S.E., and Holland, P.W. (1975). *Discrete multivariate analyses: Theory and practice*. MIT Press, Cambridge, MA.

Cifarelli, D.M. and Regazzini, E. (1986). Concentration function and its role in descriptive statistics. *Proceedings of the 33rd Scientific Meeting of the Italian Statistical Society*, **2**, 347–352.

Cohen, J. (1960). A coefficient of agreement for nominal scales. *Educational and Psychological Measurement*, **20**, 37–46.

Davies, M. and Fleiss, J. (1982). Measuring agreement for multinomial data. *Biometrics*, **38**, 1047–1051.

Farewell, V.T. and Sprott, D.A. (1999). Conditional inference for predictive agreement. *Statistics in Medicine*, **18**, 1435–1449.

Fleiss, J.L. (1971). Measuring nominal scale agreement among many raters. *Psychological Bulletin*, **76**, 378–381.

Forcina, A. and Giovagnoli, A. (1987). Mathematical properties of cross-classification matrices. *Bollettino Unione Matematica Italiana*, **VII(1-B)**(2), 523–544.

Giovagnoli, A. (2002a). Differenti aspetti della connessione: un approccio attraverso la teoria degli ordinamenti. In *Studi in onore di Angelo Zanella.* Ed. Vita e Pensiero, 329–334.

Giovagnoli, A. (2002b). Stochastic orderings and their use in statistics: the case of association between two variables. In *Proceedings of the XLI Scientific Meeting of the Italian Statistical Society*, 5-7 June 2002, Milan, 95–104.

Giovagnoli, A. and Wynn, H.P. (1985). G-majorization with applications to matrix orderings. *Linear Algebra and its Applications*, **67**, 111–135.

Giovagnoli, A., Marzialetti, J., and Wynn, H.P. (2006). A new approach to inter-rater agreement through stochastic orderings: the discrete case. Manuscript.

Goodman, L.A. and Kruskal, W.H. (1954). Measures of association for cross classifications. *Journal of the American Statistical Association*, **49**, 732–764.

Goodman, L.A. and Kruskal, W.H. (1979). *Measures of association for cross classification.* Springer-Verlag, New York.

Joe, H. (1985). An ordering of dependence for contingency tables. *Linear Algebra and its Applications*, **70**, 89–103.

Joe, H. (1997). *Multivariate models and dependence concepts.* Chapman & Hall, London.

Marshall, A.W. and Olkin, I. (1979). *Inequalities: Theory of Majorization and Its Applications.* Academic Press, London.

Marzialetti, J. (2006). *Lo Studio dell'Agreement mediante gli Ordinamenti.* Ph.D. thesis, Department of Statistical Sciences, University of Bologna.

Müller, A. and Stoyan, D. (2002). *Comparison Methods for Stochastic Models and Risks.* Wiley, London.

Rogot, E. and Goldberg, I.D. (1966). A proposed index for measuring agreement in test-retest studies. *Journal of Chronical Disease*, **19**, 991–1006.

Ross, S. (1995). *Stochastic Processes.* Wiley, London.

Scarsini, M. (1991). An ordering of dependence. In *Topics in Statistical Dependence*, pages 403–414. Block, H.W., Sampson A.R., Savits, T.H. eds, Institute of Mathematical Statistics (Hayward).

Shaked, M. and Shanthikumar, J.G. (1994). *Stochastic Orders and their Applications.* Academic Press, New York, London.

Shoukri, M.M. (2004). *Measures of Interobserver Agreement.* Chapman & Hall, Boca Raton.

5

Methods in Algebraic Statistics for the Design of Experiments

G. Pistone, E. Riccomagno and M.P. Rogantin

Summary. We present a brief review of classical experimental design in the spirit of algebraic statistics. Notion of identifiability, aliasing and estimability of linear parametric functions, confounding are expressed in relation to a set of polynomials identified by the design, called the *design ideal.* An effort has been made to indicate the classical linear algebra counterpart of the objects of interest in the polynomial space, and to indicate how the algebraic statistics approach generalizes the classical theory. In the second part of this chapter we address new questions: a seemingly limitation of the algebraic approach is discussed and resolved in the ideas of minimal and maximal fan designs, again generalizing classical notions; an algorithm is provided to switch between two major representations of a design, one of which uses Gröbner bases and the other one uses indicator functions. Finally, all the theory in the chapter is applied and extended to the class of mixture designs which present a challenging structure and questions.

5.1 Introduction

The approach to the design of experiments (DOE) via computational commutative algebra (CCA) was originally proposed in Pistone and Wynn (1996) and developed in Pistone et al. (2001). Relevant references to polynomial ring algebra can be found in the textbooks, Cox et al. (1997, 2005) and Kreuzer and Robbiano (2000, 2005). For a summary of the algebraic theory used in DOE, see in particular Pistone et al. (2001, Chap. 2).

Unless otherwise specified in this chapter, a *design* $\mathcal{D}$ is a finite subset of k^m, where k is a number field, for example $\mathbb{Q}$ or $\mathbb{R}$ or $\mathbb{C}$. The (numeric) codes of the factor levels are values in k, and m is the number of factors. We denote by X_j the projection of the design on the jth factor. By saying that a design is a set, we mean that each point has a single replicate. Designs with replicate points are dealt with in particular in Sect. 5.5.

The reader will find a short summary of the algebraic background for this chapter in Sect. 5.2. The design definition given above is clean and simple, and is also justified by the success of the algebraic methods that follow from

L. Pronzato, A. Zhigljavsky (eds.), *Optimal Design and Related Areas in Optimization and Statistics*, Springer Optimization and its Applications 28,
DOI 10.1007/978-0-387-79936-0_5,

this approach. However, this approach needs to be considered in the larger perspective of the various approaches to the statistics of experiments. This is done in Sect. 5.3 as a methodological discussion. The reader who is interested above all in new algebraic methods could skip this part.

Hierarchical monomial models are discussed in Sect. 5.4. Section 5.5 is devoted to a short review of the use of indicator polynomials for qualitative fractional factorial designs. In particular, the use of a special coding with the roots of the unity is suggested as the best mathematical setting for the so-called multiplicative notation of aliasing relations. In Sect. 5.6. we present new results on how to transform the description of a factorial design using generating equations of the design ideal into the description of the same design using its indicator polynomial, and backward. The extension of this to mixture designs is discussed in Sect. 5.7. The main feature of the mixture design case is their interpretation as finite subsets of a projective space.

5.2 Background

The symbol $k[x_1,\ldots,x_m]$ indicates the set of all polynomials in the unknowns (also called indeterminates) $x_1,\ldots,x_m$, and with coefficients in the number field k. The set $k[x_1,\ldots,x_m]$ is a ring, i.e. it is closed under addition and multiplication; it is also a k-vector space. The set of all polynomials vanishing on the design $\mathcal{D}$ has the algebraic structure of an *ideal*, i.e. it is closed under addition and multiplication and also under multiplication by any $f \in k[x_1,\ldots,x_m]$.

We call such an ideal the *design ideal* $I(\mathcal{D})$. According to the fundamental Hilbert basis theorem, cf. Kreuzer and Robbiano (2000, Chap. XX), there exists a finite set of polynomials $f_1,\ldots,f_r \in k[x_1,\ldots,x_m]$ such that

$$I(\mathcal{D}) = \left\{ \sum_{i=1}^{r} s_i f_i : s_i \in k[x_1,\ldots,x_m] \right\} .$$

Such a set is called a *generating set* of $I(\mathcal{D})$ and we write

$$I(\mathcal{D}) = \langle f_1,\ldots,f_r \rangle .$$

Each generating set $f_1,\ldots,f_r$ defines a system of polynomial equations:

$$\begin{cases} f_1(x_1,\ldots,x_m) = 0 \\ \qquad\qquad\vdots \\ f_1(x_1,\ldots,x_m) = 0 \end{cases} \tag{5.1}$$

whose set of solutions is $\mathcal{D}$. The generating set is not uniquely defined nor is its length r. We will see in what follows that special generating sets exist with a weak form of uniqueness.

In the statistical picture, a point in $\mathcal{D}$ is a treatment that produces a numerical response value. A (numerical) response on $\mathcal{D}$ is a function on $\mathcal{D}$ taking values from some super-field K, $K \supseteq k$. A common choice is to use rational codings for levels, $k = \mathbb{Q}$, and to consider real responses, $K = \mathbb{R}$, while other options are possible. The set of all responses is a vector space of dimension equal to the number of points of $\mathcal{D}$, denoted by $\#\mathcal{D}$. We are interested here in finding those linear bases of responses whose elements are monomials.

Apart from being associated to the system polynomial equation (5.1), the design ideal $I(\mathcal{D})$ actually characterizes all *aliasing relations* of polynomial form, in the sense that if $f, g \in k[x_1, \ldots, x_m]$ then $f(a) = g(a)\ \forall a \in \mathcal{D}$ if, and only if, $f - g \in I(\mathcal{D})$. An aliasing relation is an equivalence relation, defined on $k[x_1, \ldots, x_m]$ by setting two polynomials f, g equivalent if $f - g \in I(\mathcal{D})$. Such an equivalence relation defines a new ring, called a quotient ring, denoted by $k[x_1, \ldots, x_m]/I(\mathcal{D})$. This quotient ring is a representation of the k-valued responses on $\mathcal{D}$. Two equivalent polynomials in fact take the same value on each point of $\mathcal{D}$ and each response can be interpolated on $\mathcal{D}$ by a polynomial.

The quotient ring is a k-vector space of dimension $\#\mathcal{D}$ and it admits many linear bases of practical interest. It should be noticed that, given a linear basis of the k-valued responses, the same basis works for any numerical extension of the number field k, in particular for real valued responses. In fact, the canonical basis e_a, $a \in \mathcal{D}$, is common for all response spaces.

Here, we are mainly interested in bases whose elements are monomials, i.e. of the form x^α, where α is a multi-exponent, $\alpha = (\alpha_1, \ldots, \alpha_m)$ and $x^\alpha = x_1^{\alpha_1} \cdots x_m^{\alpha_m}$. In regression models, a monomial is associated to a simple factor when it is univariate, e.g. $X^{(2,0)} (= X_1^2)$, or an interaction if it involves more than one indeterminate, e.g. $X^{(2,1)} (= X_1^2 X_2)$. Here and whenever possible we use this notational convention: capital letters denote responses, while indeterminates, number fields and values, are denoted by small letters.

We say that a set of monomials $S = \{x^\alpha : \alpha \in N\}$ is *hierarchical* if it is closed under factorization. For example, if $x_1 x_2^2 \in S$, then $1, x_1, x_2, x_1 x_2, x_2^2 \in S$. A hierarchical monomial basis (HMB) is a hierarchical set of monomials which is a linear basis of the response space.

A linear basis of monomials

$$B = \{x^\alpha \mid \alpha \in N\}$$

is hierarchical, if and only if, the complement set N^c of N in $\mathbb{Z}_+^m$ is closed under component-wise sum with an arbitrary element of $\mathbb{Z}_+^m$. In fact, if x^β divides some $x^\alpha \in B$, then $\beta \notin N^c$ because otherwise $\beta + (\alpha - \beta) = \alpha$ would be in N^c.

The set of all HMBs is called the *statistical fan* of the design. According to some authors, linear models based on HMBs are especially fit for modelling because they assume that the model should contain all the terms that are simpler than all the other terms. HMBs and their fans are discussed in Sect. 5.4 below.

Example 1. The first example is given by *factorial designs*, where

$$\mathcal{D} = \mathcal{D}_1 \times \cdots \times \mathcal{D}_j \times \cdots \times \mathcal{D}_m$$

where $\mathcal{D}_j$ is a set of n_j points in k. The generating set of the factorial design ideal $I(\mathcal{D})$ is the set of m univariate polynomials of degree n_j, $j = 1 \ldots m$

$$I(\mathcal{D}) = < d_1(x_1), \ldots, d_j(x_j), \ldots, d_m(x_m) > ,$$

where $d_j(x_j) = x_j^{n_j} - r_j(x_j)$, $r_j(x_j)$ is a polynomial of degree less than n_j and $d_j(x_j)$ vanishes on each element of $\mathcal{D}_j$.

The standard expression of the Vandermonde determinant shows that there is only one HMB and it is the same whatever the codes of the factor levels are. The set of the exponents is

$$L = \{0, 1, \ldots, n_1 - 1\} \times \cdots \times \{0, 1, \ldots, n_m - 1\} .$$

It should be noticed that such a basis consists of all the monomials that are not divided by any of the leading terms $x_j^{n_j}$ of the polynomials $d_j(x_j)$. We will see below that this statement, however involved, generalizes to any design.

In many cases, the design of interest will be a subset of a larger and simpler design. When the larger design is a factorial design, we call the subset a *fractional factorial design* or simply a *fraction*, $\mathcal{F}$.

A special generating set of a fractional factorial design is obtained by adding a set of *generating polynomials*

$$g_1, \ldots, g_p \in k[x_1, \ldots, x_m]$$

to the generators of the factorial design $d_1(x_1), \ldots, d_m(x_m)$. The common zeros of the generating equations that are points of the factorial designs are the points of $\mathcal{F}$.

There are many ways of determining the g_i's. For example, if the points are known, the algorithm in (Abbott et al., 2000) can be used to compute the generating set of $I(\mathcal{D})$.

There exists a particular set of generating polynomials formed by only one equation that is based on the indicator function of the fraction. The *indicator function* F of a fraction $\mathcal{F}$ (without replicates) is a particular response function on $\mathcal{D}$ such that

$$F(\zeta) = \begin{cases} 1 & \text{if } \zeta \in \mathcal{F}, \\ 0 & \text{if } \zeta \in \mathcal{D} \setminus \mathcal{F}. \end{cases}$$

A polynomial function F is an indicator function of some fraction $\mathcal{F}$ if, and only if, $F^2 - F = 0$ on $\mathcal{D}$. If F is the indicator function of the fraction $\mathcal{F}$, then $F - 1 = 0$ is a generating equation of the same fraction.

Another interesting case is when the design ideal is a subset of a larger ideal such that the set of zeros is a continuous variety. This is the case of *mixture designs* where all the points satisfy the equation

$$\sum_{j=1}^{m} x_j = 1 \, .$$

A mixture design is be specified by adding a set of generating polynomials to the polynomial $x_1 + \cdots + x_m - 1$.

Example 2. The simplex lattice design was introduced in Scheffé (1958). A $\{m, n\}$ simplex lattice design is the intersection of the simplex in $\mathbb{R}^m$ and the full factorial design in m factors and with the $n+1$ uniformly spaced levels $\{0, 1/n, \ldots, 1\}$. A generating set of the polynomial ideal of $\{m, n\}$ is

$$\prod_{j=0}^{n} (x_1 - j/n), \ldots, \prod_{j=0}^{n} (x_m - j/n), \sum_{i=1}^{m} x_i - 1 \, .$$

The proof of the existence of at least one HMB for each $\mathcal{D}$ is based on the theory of a special class of generating sets of an ideal I. This theory in fact depends on a way of listing the terms of a polynomial in decreasing order, and thus of identifying the leading term of each multivariate polynomial, in the same way we do for univariate polynomials by listing the terms according to the degree. The key is the following definition.

Definition 1. *A* term-ordering τ *on the ring* $k[x_1, \ldots, x_m]$ *is a total order* $\prec$ *on the set of monomials that satisfies the following properties:*

1. $1 \prec x^\alpha$ for $\alpha \neq 0$;
2. if $x^\alpha \prec x^\beta$ then $x^{\alpha+\gamma} \prec x^{\beta+\gamma}$.

The reader can refer to any of the cited commutative algebra monographs for a detailed discussion. The practical fact is that such monomial orders exist and they are implemented in symbolic software such as CoCoA or Maple. The most common term-orderings are called reverse lexicographic, total degree reverse lexicographic and elimination ordering, see, e.g. Pistone et al. (2001, Sect. 2.3).

In turn, given a ring $k[x_1, \ldots, x_m]$ and a term-ordering τ, it is possible to define special generating sets of the ideal I as follows.

Definition 2. *A subset* $\{g_1, \ldots, g_t\}$ *of an ideal* I *is a* Gröbner basis *of* I *with respect to a term-ordering* τ *if and only if*

$$\langle \mathrm{LT}_\tau(g_1), \ldots, \mathrm{LT}_\tau(g_r) \rangle = \langle \mathrm{LT}(I) \rangle \, ,$$

where $\mathrm{LT}_\tau(f)$ *denotes the leading term of the polynomial* f *according to* τ, *and* $\mathrm{LT}_\tau(I) = \{\mathrm{LT}_\tau(f) \ : f \in I\}$.

The key result is the following.

Proposition 1. *Let* $I \subset k[x_1, \dots, x_m]$ *be an ideal,* $\langle g_1, \dots, g_t \rangle_\tau$ *a Gröbner basis for* I *w.r.t a term-ordering* τ*, and* $f \in k[x_1, \dots, x_m]$*. Therefore, there exists a* unique *remainder* $r \in k[x_1, \dots, x_m]$ *and a polynomial* $g \in I$ *such that*

1. $f = g + r$*, and*
2. *No term of* r *is divisible by any of* $\mathrm{LT}_\tau(g_1), \dots, \mathrm{LT}_\tau(g_t)$.

The unique remainder r is called *normal form* of f in I (w.r.t. τ) and is denoted by $\mathrm{NF}_\tau(f)$. A polynomial f, such that $f = \mathrm{NF}_\tau(f)$, is said to be reduced (w.r.t. the ideal I and the term-ordering τ).

The set of monomials that does not divide any leading term of the Gröbner basis is a basis for the k-vector space of responses. This implies that for each $\mathcal{D}$ there exists at least an HMB formed by monomials. Such an HMB is called the set of estimable monomials and is denoted by $\mathrm{Est}_\tau(\mathcal{D})$. However, not all HMBs are obtained in this way. For a more detailed account of this matter, see Pistone et al. (2001, Chap. 3).

The following example gives an illustration of these ideas.

Example 3. We consider a 2^{3-1} regular fraction $\mathcal{F}$ of a binary factorial design $\mathcal{D}$, $\mathcal{D} = \{-1, 1\}^3$. A defining equation of $\mathcal{F}$ is:

$$xyz = 1$$

and, for each term-ordering, a generating set for the design ideal $I(\mathcal{F})$ is the Gröbner basis:

$$I(\mathcal{F}) =< x^2 - 1, y^2 - 1, z^2 - 1, x - yz, y - xz, z - xy >$$

and $\mathrm{Est}(\mathcal{F}) = \{1, x, y, z\}$.

5.3 Generalized Confounding and Polynomial Algebra

In this section, we briefly discuss the mathematics of the basic DOE theory, as it is presented in many textbooks. Our aim is to properly include the algebraic polynomial theory for DOE in the standard literature on the design of experiments. Some recent textbooks on DOE are Wu and Hamada (2000), which is more application oriented, and Hinkelmann and Kempthorne (2005), which is devoted to a formal mathematical development.

5.3.1 Classical Theory of Designs

Given a set of *units* $\mathcal{U}$ and a set of treatments $\mathcal{T}$, a protocol for a designed experiment without repetitions is a function $\varphi : \mathcal{U} \longrightarrow \mathcal{T}$. In general, *design* $\mathcal{D}$ is the image set of φ together with the multiplicity (number of replications) of each treatment, so that the dependence on the unit is hidden.

Typically, but not necessarily, the set of treatments has a factorial structure:

$$\mathcal{T} = \mathcal{F}_1 \times \mathcal{F}_2 \times \cdots \times \mathcal{F}_m .$$

This equation highlights the practical usefulness of identifying some marginal (sets of) factors, which are usually defined by the real problem being studied. Often each $\mathcal{F}_j$ is a subset of a number field, for example if $\mathcal{F}_j$ is associated to some categorical type of data, and a numeric coding is chosen for the categories. The coding is in part dictated by the real problem being studied and in part it is arbitrary. Moreover, we can assume all the $\mathcal{F}_j$'s to be one dimensional. Each of them is the rank of the one dimensional projections on the factors X_j of Sect. 5.1.

We can assume that the factor level sets, $\mathcal{F}_j$ for $j = 1, \ldots, m$, belong to a number field k. We require k to be a field of characteristic zero, that is, a field containing $\mathbb{Q}$. This depends on the requirement of having responses in $\mathbb{R}$ or $\mathbb{C}$ represented as polynomials, which implies that the number field k must be a sub-field of $\mathbb{R}$. We can assume that each design is the zero set of a system of polynomial equations with coefficients in the chosen field and in as many indeterminates as there are uni-dimensional factors, as in Sect. 5.2.

The *response* of an experiment is usually a multivariate real valued function, which depends on both the units and the treatments. We almost always consider univariate responses. In practice, the responses are modelled as random quantities.

The *parameters of interest*, in contrast to *nuisance parameters*, are assumed to be functions only of the treatments and not of the experimental units. Thus, we assume that for $i = 1, 2, \ldots, n$ the ith design point produces an experimental value which is a random real number distributed according to a given parametric family:

$$Y_i \sim \mathcal{P}(f(t_i), \sigma_i) ,$$

where, clearly, $f(t_i) = f(t_j)$ if $\phi(u_i) = \phi(u_j)$; the σ_i parameter set is the nuisance parameter and it depends on both the unit and the treatment; the function $f : \mathcal{D} \to \mathbb{R}$ is supposed to belong to a family of models V which is a subset of the set of real functions on $\mathcal{D}$. The function value $f(t_i)$ is called the *true effect of the treatment* t_i. If we denote the set of real linear functions defined on the treatment by $\mathcal{L}(\mathcal{D})$, then V is a linear subspace of $\mathcal{L}(\mathcal{D})$. To underline the assumption that the statistical model has a vector space structure over $\mathbb{R}$ we use the symbol M instead of V. We also assume that $f(t_i)$ is the mean value of Y_i, and write $\mathbb{E}(Y(t_i)) = f(t_i)$ where the expected value is with respect to the model distribution. Gaussian regression models and generalized linear models with an identity link function are classical examples of the previously described setting.

Linear monomial models are classes of statistical models of particular interest. The linear space V of these models is spanned by a set of monomial functions $\{X^\alpha,\ \alpha \in L\}$ where L is a finite subset of $\mathbb{Z}_+^m$. The response to the treatment t is written as

$$f(t) = \sum_{\alpha \in L} \theta_\alpha X^\alpha(t)\,,$$

where $t = (t_1, \ldots, t_m) \in \mathcal{F}_1 \times \cdots \times \mathcal{F}_m$ and $X^\alpha(t) = t_1^{\alpha_1} \cdots t_m^{\alpha_m}$. In this case, the true effect function f is a polynomial and the coefficient θ_α is called the *true effect of the simple factor or interaction* X^α under the model f.

More generally, let $v_1, \ldots, v_r$ be a set of vectors spanning M,

$$M = \operatorname{span}(v_1, \ldots, v_r)\,.$$

For $\theta = (\theta_1, \ldots, \theta_r)^t \in \mathbb{R}^r$ let $f(t) = \sum_{i=1}^r \theta_i v_i(t) \in M$ be a generic true effect. Let $M_\mathcal{D}$ be the model matrix. We use the notation

$$M_\mathcal{D} = [v_j(t_i)]_{i,j}\,,$$

where i is the row index and j is the column index. In some cases, the admissible coefficients θ could be restricted by linear constraints $C\theta = 0$. A typical example is the standard analysis of variance. Another class of statistical models we shall consider are linear models whose vector space basis is formed by polynomials v_j which are not monomials. In this case, the model often cannot be reduced to a linear monomial model. For example, for $v_1 = 1$ and $v_2 = X + X^2$, $f(t) = \theta_1 + \theta_2(X + X^2)(t)$ is not a linear monomial model.

Once a vector space basis v_j, $j = 1, \ldots, r$ of M has been chosen, the parameters of interest are given by the vector θ of the coefficients and in standard matrix notation we write

$$\mathbb{E}_\theta(Y) = M_\mathcal{D}\,\theta$$

where Y is the vector of responses $[Y_i]_{i=1,\ldots,n}$. The sub-θ in the expectation highlights that the unknown component of interest of the distribution is θ only.

The linear combination $c^t\theta$ of the parameters θ_j's is said to be *estimable* if it admits an unbiased linear estimator, that is, if there exists $s \in \mathbb{R}^n$ such that:

$$\mathbb{E}_\theta(s^tY) = c^t\,\theta \quad \text{for all } \theta\,.$$

It is well known, see, e.g. Raktoe et al. (1981), that this is equivalent to the condition

$$c \in \operatorname{span}\left(M_\mathcal{D}^t\right)\,,$$

that is, c^t belongs to the row space of the design matrix. Thus, for all vector $s \in \mathbb{R}^n$ such that $s^tM_\mathcal{D} = c^t$, the statistics s^tY is an unbiased linear estimator of the parametric function $c^t\theta$. Clearly, s^tY is an unbiased estimator of the same linear combination of the treatment's true effects $\mathbb{E}_\theta(s^tY)$:

$$\mathbb{E}_\theta(s^t Y) = \sum_{i=1}^{n} s_i v(t_i) .$$

The vector s is unique up to elements of $\ker M_{\mathcal{D}}^t$; then it is unique if, and only if, the rows of the model matrix are linearly independent. The true effect $f(t) = \sum_{i=1}^{r} \theta_i v_i(t)$ in the statistical family M is said to be estimable if the corresponding parameter set θ is estimable.

The set $\{c^t\theta : c \in \mathbb{R}^r\}$ is called the set of parametric functions for θ. All parametric functions $c^t\theta$ are estimable if, and only if, r rows of the model matrix $M_{\mathcal{D}}$ are linearly independent, or if, and only if, the r columns of the model matrix $M_{\mathcal{D}}$ are linearly independent. This is equivalent to the existence of a sub-design, i.e. a sub-set of $\mathcal{D}$, where the model is saturated.

5.3.2 Confounding, Aliasing and Polynomial Algebra

So far we have discussed a framework for linear regression models and the notion of estimable models. Two related notions of *confounding* and *aliasing* are introduced in the classical theory of DOE, see, e.g. Raktoe et al. (1981) and Kobilinsky (1997). Different definitions of aliasing and confounding can be found in literature. In this section, we discuss two formal definitions that are natural in our framework and clarify their meaning and use in practice.

The possibly most important observation derived from the polynomial algebra approach to DOE is that the set of all univariate responses over a design is a set of equivalence classes which is isomorphic to an object which is well known in computational commutative algebra and is called a quotient space. We recall its definition from Sect. 5.2. Two polynomials f and g in $k[x_1, \ldots, x_m]$ are equivalent with respect to a design $\mathcal{D}$ if $(f - g)$ equals zero on each point of $\mathcal{D}$, or if $(f - g)$ belongs to the design ideal $I(\mathcal{D})$. In this case f and g are said to be *aliased* on the design $\mathcal{D}$, see Pistone and Wynn (1996); Pistone et al. (2001) where a slightly different nomenclature was used. The term "confounding" refers here the relation between parameters of models which are equal on all points of the design. A consequence of performing the computation in the algebraic environment is that there is an algorithm to determine a monomial vector space basis of M. First, we comment further on confounding with the aid of a simple example.

Example 4. Consider the fraction $\mathcal{F}$ of the full factorial design $\mathcal{D} = \{-1, 1\}^2$ defined by $xy = 1$, namely $\mathcal{F} = \{(-1, -1), (1, 1)\}$. For the two models

$$m_1 = \theta_{00} + \theta_{10}x + \theta_{01}y + \theta_{11}xy \quad \text{and} \quad m_2 = (\theta_{00} + \theta_{11}) + (\theta_{10} + \theta_{01})x$$

it is easy to verify that $m_1(t) = m_2(t)$ for $t \in \mathcal{F}$, that is, m_1 and m_2 are confounded. There is a relatively simple algorithmic procedure in computational commutative algebra to verify whether two models are confounded. In algebraic terms, m_1 and m_2 are polynomials with coefficients in the field of rational functions of the parameters $\theta_{00}, \theta_{10}, \theta_{01}, \theta_{11}$, see Cox et al. (1997, Chap. 5, § 5) for proper algebraic definitions and Pistone et al. (2001, § 3.8) for some examples. Let $k(\theta_{00}, \theta_{10}, \theta_{01}, \theta_{11})$ be the field of rational functions in $\theta_{00}, \theta_{10}, \theta_{01}, \theta_{11}$ with coefficients in the same number field in which the treatment codes are. For example, let us choose $k = \mathbb{Q}$, the choice made in practice. Let $K = \mathbb{Q}(\theta_{00}, \theta_{10}, \theta_{01}, \theta_{11})[x, y]$ be the ring of polynomials with the indeterminates x, y and whose coefficients are rational functions in $\mathbb{Q}(\theta_{00}, \theta_{10}, \theta_{01}, \theta_{11})$. Clearly $\mathbb{Q}[x, y] \subset K$ and therefore

$$I(\mathcal{D}) = \langle x^2 - 1, y^2 - 1, xy - 1 \rangle,$$

which is an ideal in $\mathbb{Q}[x, y]$, and also an ideal of the ring K. The "confounding" of the two models m_1 and m_2 is translated into the fact that

$$\begin{aligned} &- m_1 + m_2 = \\ &\qquad - (\theta_{00} + \theta_{10}x + \theta_{01}y + \theta_{11}xy) + ((\theta_{00} + \theta_{11}) + (\theta_{10} + \theta_{01})x) = \\ &\qquad\qquad = \theta_{11} + \theta_{01}x - \theta_{01}y - \theta_{11}xy = \\ &\qquad\qquad\qquad = \theta_{11}(1 - xy) + \theta_{01}(x - y) \ \in \ I(\mathcal{D}) \end{aligned}$$

because $x - y = -x(xy - 1) + y(x^2 - 1) \in I(\mathcal{D})$.

There are algorithms to test whether a polynomial belongs to a polynomial ideal. The ideal membership problem in computational commutative algebra for polynomial ideals is one of the first applications of the Gröbner basis techniques.

A further notion of confounding of two factors or interactions was introduced in Galetto et al. (2003), based on the orthogonal spaces of the ANOVA decomposition of the full factorial design. This approach is not discussed here.

The use of the word confounding for two models whose difference is in $I(\mathcal{D})$ has produced some misunderstandings in the statistical milieu, where confounding usually is a property of model parameters and their estimators. In what follows, we try address the source of this confusion by giving a formalized definition of both confounding and aliasing. At the root of this confusion is the fact that in the set-up of statistical models described in this section confounding and aliasing are related concepts: confounding relates to parameters and aliasing to functions on the design.

5.3.3 Estimability and Confounding of Linear Parametric Functions

Let us denote the row space of the model matrix $M_{\mathcal{D}}$ by W,

$$W = \text{span}\left(M_{\mathcal{D}}^t\right)$$

and its dimension by p. The parametric function $c^t\theta$ is not estimable if $c \notin W$. In such a case, the space $\mathbb{R}^r$ splits into two subspaces W and W^c, $W \cap W^c =$

$\{\emptyset\}$, and uniquely each vector c, $c \in \mathbb{R}^r$, is uniquely decomposed in the two components on W and W^c:

$$c = c_W + c_{W^c} \ .$$

A special case is when $W^c = W^\perp$. More generally, given any splitting space W^c, a scalar product on $\mathbb{R}^r$ exists that makes W and W^c orthogonal.

Let the parametric function $c^t_W\theta$ be estimable by s^tY. Therefore, as

$$c^t_W\theta = c^t\ \theta - c^t_{W^c}\ \theta \ ,$$

$c^t\theta$ is estimable under the constraint $c^t_{W^c}\theta = 0$. In fact, s exists so that $s^t M_{\mathcal{D}}\ \theta = c^t_W\theta$ for all θ. Therefore,

$$s^t M_{\mathcal{D}}\ \theta = c^t\theta - c^t_{W^c}\theta \quad \text{for all } \theta$$

and $c^t\theta$ is estimable for those θ that satisfy the constraint $c^t_{W^c}\theta = 0$.

We denote a matrix whose rows generate W^c by L. The dimension of L is $(r-p) \times r$. If all the constraints $L\theta = 0$ are satisfied, then each parametric function $c^t\theta$ is estimable.

It should be noticed that W depends on the design and the model, while c_W, c_{W^c}, and s depend on the choice of the complementary space W^c, which is only restricted by the complementary assumption. Vice-versa, given c and s, one vector is derived in the relevant $W^{c\prime}$.

Confounding is defined by (Raktoe et al., 1981, Sect. 7.2) as follows.

Definition 3. *Assume that the parametric function $c_1^t\ \theta$ is not estimable. In such a case, $c_1^t\ \theta$ is said to be* confounded *with the parametric function $c_2^t\ \theta$, where c_1 and c_2 are algebraically independent, if there exists a non-zero real number δ such that*

$$(c_1 + \delta c_2)^t\ \theta$$

is estimable.

If the definition is satisfied, as

$$c_1 + \delta c_2 \in W \qquad \text{then} \qquad c_{1_{W^c}} + \delta c_{2_{W^c}} = 0 \ .$$

It follows that the general form of c_2 such that $c_2^t\theta$ is confounded with $c_1^t\theta$ is:

$$c_2 = c + \gamma c_{1_{W^c}} \quad \text{for all } c \in W \ .$$

Estimability and confounding of parametric functions have been described in the row space W of the model matrix $M_{\mathcal{D}}$. There is a dual notion in the column space, called *aliasing*. We now define aliasing and discuss the relationship between the two.

The estimable parametric function space is identified with W, $W = \text{span}\,(M^t_{\mathcal{D}})$ and

$$\operatorname{span}\left(M_{\mathcal{D}}^t\right) = \left(\ker(M_{\mathcal{D}})\right)^{\perp} . \tag{5.2}$$

For all vectors θ, such that $M_{\mathcal{D}}\,\theta = 0$ and all vectors c orthogonal to θ, $c \in W$, so that $c^t\theta$ is estimable.

A rephrasing of the same statement is: for all vectors θ such that

$$\sum \theta_j v_j(t) = 0 \qquad \forall\, t \in \mathcal{D}\,,$$

and all vectors c orthogonal to θ, then $c^t\theta$ is estimable. In this way, we emphasize a sub-vector space of the ideal of the design $I(\mathcal{D})$.

$$\left\{f = \sum \theta_j v_j \mid f(t) = 0\, \forall\, t \in \mathcal{D}\right\} = \left\{f = \sum \theta_j v_j(t)\right\} \cap I(\mathcal{D}). \tag{5.3}$$

Vice-versa, if $c^t\theta = 0$ for all the estimable function $c^t\theta$, then $\theta \in \ker(M_{\mathcal{D}})$, i.e. $\sum \theta_j v_j(t) = 0$.

Definition 4. *The function f of the model V, $f = \sum \alpha_j v_j(t)$, is an* alias *of the function g, $g = \sum \beta_j v_j(t)$, if they are equal for all the treatments of the design $\mathcal{D}$, i.e.*

$$\sum (\alpha_j - \beta_j) v_j(t) = 0 \quad \forall\, t \in \mathcal{D}, \quad \text{i.e. } (f - g) \in I(\mathcal{D})\ .$$

Equalities between members of a model of a given design are sometimes called *aliasing relations* of the model on the design.

It should be noticed that the design ideal $I(\mathcal{D})$ is generated by a finite number of polynomials $g_1, \ldots, g_k$ and each equation $g_j = 0$ is an aliasing between g_j and zero, in some model. As the model is sometimes not actually specified, e.g. regular fractions of factorial designs, we suggest, in such cases, to use the term alias as synonymous of member of the design ideal.

The design ideal $I(\mathcal{D})$ contains all the polynomials that are zero on $\mathcal{D}$. This suggests an algebraic way of identifying all aliasing relations via the membership ideal algorithms. In fact, any available ideal membership algorithm, such as those described in Cox et al. (1997) or Kreuzer and Robbiano (2000), finds the members of the model which are aliased with zero.

If we apply the argument of formula (5.2) to the constraint matrix L, then the kernel of L, $\ker L = \{\theta \mid L\theta = 0\}$, is the same as the orthogonal space of the complementary space W^c.

If $W^c = W^{\perp}$, then L defines both constraints on the parameters

$$L\theta = 0$$

and aliasing relations with 0 given by

$$LM = 0\,,$$

where M is the vector of the model functions. The reverse is true: if $L\theta = 0$ and $LM = 0$, then $W^c = W^{\perp}$. Any of these two equivalent form can be used to derive the other, if one is willing to use orthogonal constraints. The example used to introduce this section was of this type; other examples follow.

Example 5. We consider a 2^{3-1} regular fraction $\mathcal{F}$ of a binary factorial design $\mathcal{D}$, $\mathcal{D} = \{-1, 1\}^3$. A defining equation of $\mathcal{F}$ is

$$xyz = 1$$

and a generating set for the design ideal $I(\mathcal{F})$ is the Gröbner basis:

$$I(\mathcal{F}) =< x^2 - 1, y^2 - 1, z^2 - 1, x - yz, y - xz, z - xy > \; .$$

We consider the model without the three-way interaction:

$$\mathbb{E}_\theta(Y) = \theta_{000} + \theta_{100}x + \theta_{010}y + \theta_{001}z + \theta_{110}xy + \theta_{101}xz + \theta_{011}yz \in V \; .$$

Because of the generating equations, the model on the fraction is equal to:

$$\theta_{000} + (\theta_{100} + \theta_{011})x + (\theta_{010} + \theta_{101})y + (\theta_{001} + \theta_{110})z \, .$$

As in (5.3), the model is identically zero on $\mathcal{F}$ if, and only if, $K\theta = 0$ where:

$$K = \begin{bmatrix} 1\,0\,0\,0\,0\,0\,0 \\ 0\,1\,0\,0\,0\,0\,1 \\ 0\,0\,1\,0\,0\,1\,0 \\ 0\,0\,0\,1\,1\,0\,0 \end{bmatrix} \; .$$

Therefore $W = \text{span}\,(K^t)$ and the kernel of the matrix K is the space $W^\perp$.

Among all the matrices with the same kernel, the matrix K has a special relevance, because it has independent rows and was derived through an algebraic and statistically meaningful procedure.

When necessary, a basis of the kernel itself could be computed in a number of numerical ways, for example the software CoCoA gives:

$$L = \begin{bmatrix} 0\,1\,0\,0 & 0 & 0 & -1 \\ 0\,0\,1\,0 & 0 & -1 & 0 \\ 0\,0\,0\,1 & -1 & 0 & 0 \end{bmatrix} \; .$$

A special set of constraints that makes all the parametric functions $c^t\theta$ estimable is $L\theta = 0$.

Moreover,

$$\begin{bmatrix} 0\,1\,0\,0 & 0 & 0 & -1 \\ 0\,0\,1\,0 & 0 & -1 & 0 \\ 0\,0\,0\,1 & -1 & 0 & 0 \end{bmatrix} \begin{bmatrix} 1 \\ x \\ y \\ z \\ xy \\ xz \\ yz \end{bmatrix} = 0$$

are the aliasing relations that define the fraction $\mathcal{F} \subset \mathcal{D}$ as we have chosen the orthogonal splitting $W^c = W^\perp$. It should be noticed that in this case such aliasing relations coincide with the three generating equations of the fraction in the Gröbner basis.

We denote the map from the parameter space Θ, contained in $\mathbb{R}^r$, to the linear parametric model V, contained in $\mathcal{L}(\mathcal{D})$, by m:

$$m : \begin{cases} \mathbb{R}^r \supset \Theta \longrightarrow V \subset \mathcal{L}(\mathcal{D}) \\ (\theta_1, \dots, \theta_r) \longmapsto \theta_1 v_1 + \dots + \theta_r v_r \,. \end{cases}$$

Therefore, the kernel of the map m contains the coefficients of the polynomial models belonging to the ideal of the design:

$$\ker(m) = \left\{ \theta \in \Theta \ : \ \sum \theta_j v_j \in I(\mathcal{D}) \right\} \,.$$

Let $(Y_t)_{t \in \mathcal{F}}$, $\mathcal{F} \subseteq \mathcal{D}$, be a random sample. Let

$$s : \quad (y_t)_{t \in \mathcal{F}} \longmapsto \mathcal{L}(\mathcal{D})$$

be a generic statistic.

We say that $s\,(Y_t)_{t \in \mathcal{F}}$ is a *linear estimator* if

$$\begin{aligned} \mathbb{E}_\theta \left(s\,(Y_t)_{t \in \mathcal{F}} \right) &= s\,(\mathbb{E}_\theta\,(Y_t) \ : \ t \in \mathcal{F}) \\ &= s \left(\sum_{j=1}^{r} \theta_j \ v_j(t) \ : \ t \in \mathcal{F} \right) \\ &= \sum_{j=1}^{r} \theta_j \ s\,(v_j(t) \ : \ t \in \mathcal{F}) \,. \end{aligned} \tag{5.4}$$

If $\mathcal{F} = \{t_1, \dots, t_n\}$ is finite, then $s\,(y_1, \dots, y_n) = \sum_{i=1}^n s_i y_{t_i}$ and

$$\mathbb{E}_\theta \left(\sum_{i=1}^{n} s_i Y_{t_i} \right) = \sum_{i=1}^{n} s_i \mathbb{E}_\theta\,(Y_{t_i}) = \sum_{i=1}^{n} s_i \sum_{j=1}^{r} \theta_j v_j(t_i) = \sum_{j=1}^{r} \theta_j \sum_{i=1}^{n} s_i v_j(t_i) \,.$$

The previous Definition 5.4 also applies to continuous cases. If $\mathcal{D}$ is continuous, an example is

$$s\left((y_t)_{t \in \mathcal{F}} \right) = \int_{\mathcal{F}} K(y_t) \ \mu(\mathrm{d}t) \,,$$

where $K(y) \in V$ and μ is a measure on $\mathcal{D}$.

Example 6. We consider a three-way ANOVA model without interaction when the factors are coded by 0 and 1. It follows that $\mathcal{D} = \{0,1\}^3$ and

$$I(\mathcal{D}) = < x^2 - x, y^2 - y, z^2 - z > .$$

The usual (over-parameterized) model is

$$\mathbb{E}_\theta(Y_{ij}) = \mu + \alpha_i + \beta_j + \gamma_k , \qquad i, j, k = 1, 2,$$

with $\theta = (\mu, \alpha_0, \alpha_1, \beta_0, \beta_1, \gamma_0, \gamma_1) \in \mathbb{R}^7$.

The model matrix $M_\mathcal{D}$ consists of the constant and of the indicators of $x = 0$, $x = 1$, $y = 0$,$y = 1$, $z = 0$ and $z = 1$, as columns:

$$M_\mathcal{D} = \begin{bmatrix} 1\,1\,0\,1\,0\,1\,0 \\ 1\,1\,0\,1\,0\,0\,1 \\ 1\,1\,0\,0\,1\,1\,0 \\ 1\,1\,0\,0\,1\,0\,1 \\ 1\,0\,1\,1\,0\,1\,0 \\ 1\,0\,1\,1\,0\,0\,1 \\ 1\,0\,1\,0\,1\,1\,0 \\ 1\,0\,1\,0\,1\,0\,1 \end{bmatrix} .$$

The map from the parameter space to the model is

$$\begin{aligned} m:\ \theta = (\mu, \alpha_0, \alpha_1, \beta_0, \beta_1, \gamma_0, \gamma_1) \longmapsto \\ \mu + \alpha_0 1_{[x=0]} + \alpha_1 1_{[x=1]} + \beta_0 1_{[y=0]} + \beta_1 1_{[y=1]} + \gamma_0 1_{[z=0]} + \gamma_1 1_{[z=1]} . \end{aligned}$$

The indicators are represented on $\mathcal{D}$, as reduced polynomials by $1 - x$, x, $1 - y$, y,$1 - z$, z, respectively. The coefficients of the ANOVA polynomial model therefore belong to $\ker(m)$ if, and only if,

$$(\mu + \alpha_0 + \beta_0 + \gamma_0) + (\alpha_1 - \alpha_0)x + (\beta_1 - \beta_0)y + (\gamma_1 - \gamma_0)z \in I(\mathcal{D}) . \quad (5.5)$$

It is known, both from the algebraic statistics theory and from the usual check of the linear independence of the matrix model vectors, that 1, x, and y are linearly independent on $\mathcal{D}$. Therefore, the membership relation (5.5) is equivalent to the system of linear equations:

$$\mu + \alpha_0 + \beta_0 + \gamma_0 = 0, \quad \alpha_1 - \alpha_0 = 0, \quad \beta_1 - \beta_0 = 0, \quad \gamma_1 - \gamma_0 = 0 . \quad (5.6)$$

This reminds the well-known ANOVA re-parametrization, as implemented in statistical software, such as SAS, Minitab, that assume the constant and the indicators of $x = 1$, $y = 1$ and $z = 1$ on the design as full matrix model. The re-parametrization is:

$$\delta = K\theta \qquad \text{where} \qquad K = \begin{bmatrix} 1 & 1 & 0 & 1 & 0 & 1 & 0 \\ 0 & -1 & 1 & 0 & 0 & 0 & 0 \\ 0 & 0 & 0 & -1 & 1 & 0 & 0 \\ 0 & 0 & 0 & 0 & 0 & -1 & 1 \end{bmatrix} .$$

The system (5.6) is equivalent to $K\theta = 0$. We obtain:

$$W = \text{span}\left(K^t\right) \qquad \text{and} \qquad \ker(m) = \ker(M_{\mathcal{D}}) = \ker(K)\,.$$

For example, a basis of the kernel can be computed using the CoCoA software which gives:

$$L = \begin{bmatrix} 1 & 0 & 0 & -1 & -1 & 0 & 0 \\ 0 & 1 & 1 & -1 & -1 & 0 & 0 \\ 0 & 0 & 0 & -1 & -1 & 1 & 1 \end{bmatrix}\,.$$

However, the corresponding set of constraints:

$$\begin{cases} \mu = \beta_0 + \beta_1 \\ \alpha_0 + \alpha_1 = \beta_0 + \beta_1 \\ \gamma_0 + \gamma_1 = \beta_0 + \beta_1 \end{cases}$$

gives unusual estimators of the parameters. A more common choice is $L'\,\theta = 0$, with:

$$L' = \begin{bmatrix} 0 & 1 & 1 & 0 & 0 & 0 & 0 \\ 0 & 0 & 0 & 1 & 1 & 0 & 0 \\ 0 & 0 & 0 & 0 & 0 & 1 & 1 \end{bmatrix}\,.$$

The rows of L' generate a space W^c which corresponds to a non orthogonal splitting of $\mathbb{R}^7$.

As an example of non-estimable linear parametric function we consider the parameter μ:

$$\mu = (1, 0, 0, 0, 0, 0, 0)\,\theta\,.$$

The decomposition of c in c_W and c_{W^c} is

$$(1,0,0,0,0,0,0) = \left(1, \frac{1}{2}, \frac{1}{2}, \frac{1}{2}, \frac{1}{2}, \frac{1}{2}, \frac{1}{2}\right) - \left(0, \frac{1}{2}, \frac{1}{2}, \frac{1}{2}, \frac{1}{2}, \frac{1}{2}, \frac{1}{2}\right)$$

and c_W can be written as

$$c_W = k_1 + \frac{1}{2}k_2 + \frac{1}{2}k_3 + \frac{1}{2}k_4\,,$$

where k_i is the ith row of the matrix K.

A linear estimator of a generic parametric function is s^tY:

$$s^tY = s_{000}Y_{000} + s_{001}Y_{001} + s_{010}Y_{010} + s_{011}Y_{011} \\ + s_{100}Y_{100} + s_{101}Y_{101} + s_{110}Y_{110} + s_{111}Y_{111}\,.$$

Therefore,

$$\mathbb{E}_\theta(s^tY) = \sum_{xyz} s^t M_{\mathcal{D}}\,\theta = (\mu + \alpha_0 + \beta_0 + \gamma_0)\sum_{xyz} s_{xyz} + \\ (\alpha_1 - \alpha_0)\sum_{xyz} s_{xyz}\;x + (\beta_1 - \beta_0)\sum_{xyz} s_{xyz}\;y + (\gamma_1 - \gamma_0)\sum_{xyz} s_{xyz}\;z \\ = (\mu + \alpha_0 + \beta_0 + \gamma_0)s_{+++} + (\alpha_1 - \alpha_0)s_{1++} + (\beta_1 - \beta_0)s_{+1+} + (\gamma_1 - \gamma_0)s_{++1}\,.$$

We derive estimators of the parametric functions $k_1\theta$, $k_2\theta$, $k_3\theta$ and $k_4\theta$.

The function $k_1\theta$, $k_1\theta = \mu + \alpha_0 + \beta_0 + \gamma_0$, is estimable if $s_{+++} = 1$, $s_{1++} = 0$, $s_{+1+} = 0$ and $s_{++1} = 0$; a solution is

$$s_1^t = \frac{1}{4}\left(2, 1, 1, 0, 1, 0, 0, -1\right) .$$

Estimators of the other three parametric functions corresponding to k_2, k_3 and k_4 are:

$$s_2^t = \frac{1}{4}\left(-1, -1, -1, -1, 1, 1, 1, 1\right) ,$$
$$s_3^t = \frac{1}{4}\left(-1, -1, 1, 1, 1, -1, -1, 1\right) ,$$
$$s_4^t = \frac{1}{4}\left(-1, 1, -1, 1, -1, 1, -1, 1\right) .$$

Therefore, a linear estimator of μ is

$$s_\mu = s_1 + \frac{1}{2}s_2 + \frac{1}{2}s_3 + \frac{1}{2}s_4 = \left(\frac{1}{8}, \frac{1}{8}, \frac{1}{8}, \frac{1}{8}, \frac{1}{8}, \frac{1}{8}, \frac{1}{8}, \frac{1}{8}\right) .$$

5.4 Models and Monomials

5.4.1 Hierarchical Monomial Basis HMB, Why and Where

We recall that a set of monomials $S = \{x^\beta : \beta \in N\}$ is *hierarchical* if it is closed under factorization. We have collected a number of equivalent formulations of this key property in the Proposition below. For the proof, see e.g. the discussion order ideal and Dickson's lemma in Cox et al. (1997), or the short presentation in Pistone et al. (2001, Sect. 4.5).

Proposition 2. *The monomial set S is hierarchical if, and only if, the following equivalent statements hold true:*

1. *The set of multi-exponents N is an echelon, i.e. is the union of boxed sets of the form $\{\alpha : 0 \leq \alpha \leq \beta\}$, where $\leq$ denotes the component-wise partial order. The minimal set of β's is called the* step set.
2. *N is a partially ordered set for the component-wise "$\leq$" relation, and it is such that if a point is included, all points below are also included, i.e. it is a lower interval. As a partially ordered set, it can be represented as a Hasse diagram.*
3. *For each variable x_i and each integer h, the section $S_{i,h} = \{x^\beta : \beta \in N, \beta_i = h\}$ is hierarchical and, moreover, decreases in number for increasing h.*

4. *S is the complement of a monomial ideal, i.e. a set of monomials such that it contains $x^{\gamma+\delta}$ whenever it contains x^γ. A monomial ideal has a unique minimal finite set of generators, which are monomials called cut-outs of S, because S consists of all the monomials that are not divided by any cut-out.*

For example, consider $S = \{1, x, y, z\}$.

Therefore the boxes are $(1, 0, 0), (0, 1, 0), (0, 0, 1)$; the Hasse diagram is

$$\begin{array}{c} 100\ 010\ 001 \\ \diagdown \ | \ \diagup \\ 000 \end{array} ;$$

the sections are $S_{10} = \{1, y, z\}$, $S_{11} = \{x\}$, $S_{20} = \{1, x, z\}$, $S_{21} = \{y\}$, $S_{30} = \{1, x, y\}$, $S_{31} = \{z\}$; the cut-outs are x^2, y^2, z^2, xyz.

We have already observed that, given a design $\mathcal{D}$, and a term-ordering τ, a k-vector basis of the quotient space $k[x_1 \ldots x_m]/I(\mathcal{D})$ consists of all the monomials x^α that are not divided by the leading term of the polynomials in a τ-Gröbner basis of the design ideal $I(\mathcal{D})$. By the very definition of Gröbner basis, the monomial ideal generated both by the leading terms of a G-basis and by the leading terms of all the polynomials in the ideal, it follows that the hierarchical monomial basis, HMB, obtained from a G-basis is non-reducible by any aliasing relation in the design ideal. However, not all monomial ideals are derived from a G-basis. As a consequence, there are HMBs that do not arise this way. The following example shows a simple case that is further elaborated in what follows.

Example 7. The design ideal $I(\mathcal{F})$ with

$$\mathcal{F} = \{(0, 0), (1, -1), (-1, 1), (0, 1), (1, 0)\}$$

with number field $\mathbb{Q}^2$ is generated by the three polynomials $x_1^3 - x_1$, $x_2^3 - x_2$, $(x_1 + x_2)(x_1 + x_2 - 1)$. The HMBs $\{1, x_1, x_2, x_2^2, x_1x_2\}$ and $\{1, x_1, x_2, x_1^2, x_1x_2\}$ are derived from a G-basis, while $\{1, x_1, x_2, x_1^2, x_2^2\}$ is not (B. Sturmfels, personal communication.).

A statistical model based on a HMB or, more generally, on a hierarchical monomial set satisfies the *marginal functionality property* MFP advocated by McCullagh and Nelder (1989). The equivalent property they actually use is related to Proposition 2(3):

Proposition 3. *MFP is defined recursively as follows.*

1. *A univariate polynomial model has the MFP if it contains all the terms up to some maximal degree.*
2. *An m-variate polynomial model has the MFP if, as a polynomial in an individual variable x_i it has MFP coefficients of decreasing complexity; that is, if the maximal degree of x_i is r_i, the polynomial can be written as:*

$$\sum_{k=0}^{r_i} \left(\sum_{\beta \in N_k} c_{k\beta}\, x^{\beta} \right) x_i^k \qquad \text{with } N_0 \supseteq N_1 \cdots \supseteq N_k$$

and each N_h *is an echelon.*

Another equivalent formulation of the same principle is the notion, due to Peixoto (1990), of "well-formulated" or "well-formed". Using the MFP, Peixoto shows that

Proposition 4. *A polynomial model is well-formed if, and only if, the model is invariant under change of scale and location of each of the covariates. The class of well-formed polynomial models coincides with the class of hierarchical monomial models.*

Well-formed polynomial models have the property that, if used as a response surface, the fit of the model is invariant under linear coding transformations of explanatory variables. In fact, an HMB or an HM set with N as the exponent set is transformed into a linear combination of the same HMB or HM set (with the same exponent set) under linear transformation of numerical codings. In a polynomial model, a term x^{β} is absent if, and only if, the β-derivative is zero at 0.

5.4.2 How many HMBs (with Hugo Maruri-Aguilar)

The procedure illustrated above to associate a set of saturated HMBs to a given design ideal depends to a great extent on the choice of a term ordering.

In Holliday et al. (1999), this dependence was exploited in favour of a robust analysis of a data set related to a real case study from automotive industry. Specifically, having chosen a total degree ordering, the HMB for all permutations of the factors were computed and the intersection of these were selected as a starting model for the subsequent model selection procedure. The intersection was an HMB, was not empty and actually contained a large number of terms. In Pistone et al. (2000), a blocked term ordering was used to take into account expert prior information on the relative minor importance, but potential relevance of some factors.

Definition 5. *Let* $\mathcal{D}$ *be a design with* n *distinct points. The set of all HMBs with* n *elements retrieved with the G-basis procedure is the algebraic fan of* $\mathcal{D}$.

Theorem 1. *Let* $\mathcal{D}$ *be a design. Its algebraic fan is finite.*

Proof. The set of all possible HMBs in m dimensions and with n elements is finite. The algebraic fan is non empty as, given a term-ordering, the G-basis procedure returns one of its elements. □

Table 5.1. Number of HMBs and corner cuts in 2 and 3 dimensions

HMB size	m=2 n. of HMB	m=2 n. of CC	m=3 n. of HMB	m=3 n. of CC
1	1	1	1	1
2	2	2	3	3
3	3	3	6	6
4	5	4	13	10
5	7	6	24	18
6	11	7	48	27
7	15	8	86	36
8	22	10	160	69
9	30	12	282	79
10	42	13	500	109
11	56	16	859	133
12	77	16	1,479	201
13	101	18	2,485	222
14	135	20	4,167	234

In Example 7, we have seen that there are HMBs that are not included in the algebraic fan and identified by the design. This has led us to the definition of the *statistical fan* of a design, namely the set of all HMBs of the same size as the design and for which the design model matrix is full rank.

In Onn and Sturmfels (1999), the set of HMBs retrievable with the G-basis method for some designs is called the set of *corner cuts*. It is shown that an HMB is a corner cut if it is separable from its complement by a hyper-plane. An indication of the relative sizes of the set of corner cuts and of all HMB models for $m = 2, 3$ and a number of terms up to 14 is given in Table 5.1. It is clear that the ratio between the size of the set of HMB with n monomials and retrievable with Gröbner basis methods and the total number of HMBs goes to zero as n increases. Detailed discussion can be found in Babson et al. (2003), Onn and Sturmfels (1999) and Maruri-Aguilar (2007, Chap. 5 and Appendix B2).

In Example 7, we show that the model $1, x_1, x_1^2, x_2, x_2^2$ is not a corner cut model. However, it is the most symmetric of the models in the statistical fan. In fact, to destroy symmetry is a feature of the Gröbner basis computation, as term orderings intrinsically do not preserve symmetries, which are often preferred in statistical modelling. Other advantageous uses of algebraic fans and corner cut ideas, coupled with the Hilbert function, in the design and analysis of experiments are described in Maruri-Aguilar (2007, Chaps. 5 and 6).

The above observations outline some features of the application of algebraic statistics to the design and analysis of experiments which seem to be negative. Next we show that they are actually not so negative.

As previously mentioned, the authors in Holliday et al. (1999) and Pistone et al. (2000) vary term-orderings to create a larger class of model supports that a statistician generally use to perform model selection. The systematic span of the set of HMB bases of the $k[x_1, \ldots, x_m]/I(\mathcal{D})$, and thus of the set of the identifiable models, is one of the key features of the application of computational algebraic methods in the analysis of data sets.

The computational ability that the algebraic theory offers of rewriting models in different k-vector space bases in $k[x_1, \ldots, x_m]/I(\mathcal{D})$ using the so-called rewriting rules is even more important. This issue depends, and is somehow computationally equivalent, to the notions of confounding and aliasing, as discussed in Sect. 5.3.2. These notions are embedded in the design ideal and do not depend on term-ordering which remains just a tool used to perform computations. It is natural to investigate the effects of term orderings not only on the HMBs retrieved, as before, but also on the identification of particular generating sets of the design ideal.

Thus, the whole design ideal is the set of confounding relationships imposed by the design on the set of all monomials T^m in m indeterminates. A generator set of the ideal "generates" those relationships. A Gröbner basis is a special generator with properties that are useful from a computational point of view, in the same way as the defining words can be used to generate the full alias table for a regular design. It should be noted that an alias table is a finite object as the highest degree of the interactions to be considered is set a priori. There is no mathematical impediment to extend an alias table to the whole of T^m. This is rather evident for the class of designs in Definition 6.

Definition 6. *A design is called a* minimal fan design *if its statistical fan contains only one saturated model.*

The algebraic fan of a minimal fan design is equal to the statistical fan. There is in fact, only one reduced Gröbner basis (the proof is obvious). This Gröbner basis is the closest possible analogy to the defining words for regular fractions. The normal form, as obtained by the rewriting rules, of the (unique) leading terms allows us to construct the alias table as illustrated in Example 8.

Example 8. The reduced Gröbner basis of the minimal fan design

$$\mathcal{D} = \{(0,0), (1,0), (2,0), (1,1), (0,2)\}$$

is $G = \{b^2 - b, a^2b - ab, a^3 - 3a^2 + 2a\}$ and its standard basis is $\{1, a, a^2, b, ab\}$.

The alias table of the design can therefore be constructed as in Table 5.2.

The elements of G label the first row on Table 5.2 and are interpreted as functions over $\mathcal{D}$, in particular $B^2 - B = 0$ when evaluated at each point of $\mathcal{D}$. The monomials in T^m are listed in the first column. In the rows relative to Step 1, each $a \in T^m$ multiplies the column labels. In Step 2, the result is written in the standard basis. For example, A times $A^3 = 3A^2 - 2A$ gives $A^4 = 3A^3 - 2A^2$ which on $\mathcal{D}$ equals $3(3A^2 - 2A) - 2A^2 = 7A^2 - 6A$. Thus one can read the polynomial in normal form confounded with each monomial in the body of Table 5.2. For example, the higher order monomial A^4 is written

Table 5.2. Alias table for Example 8

Step 1	$B^2 = B$	$A^2B = AB$	$A^3 = 3A^2 - 2A$
B	$B^3 = B^2$	$A^2B^2 = AB^2$	$A^3B = 3A^2B - 2AB$
B^n	$B^{n+2} = B^{n+1}$	$A^2B^{n+2} = AB^{n+1}$	$A^3B^n = 3A^2B^n - 2AB^n$
A	$AB^2 = AB$	$A^3B = A^2B$	$A^4 = 3A^2 - 2A^2$
A^2	$A^2B^2 = A^2B$	$A^4B = A^3B$	$A^5 = 3A^4 - 2A^3$
A^n	$A^nB^2 = A^nB$	$A^{n+2}B = A^{n+1}B$	$A^{n+3} = 3A^{n+2} - 2A^{n+1}$
AB	$AB^3 = AB^2$	$A^3B^2 = A^2B^2$	$A^4B = 3A^3B - 2A^2B$
A^nB^m		$\dots$	
Step 2	$B^2 = B$	$A^2B = AB$	$A^3 = 3A^2 - 2A$
B	$B^3 = B$	$A^2B^2 = AB$	$A^3B = AB$
B^n	$B^{n+2} = B$	$A^2B^{n+2} = AB$	$A^3B^n = AB$
A	$AB^2 = AB$	$A^3B = AB$	$A^4 = 7A^2 - 6A$
A^2	$A^2B^2 = AB$	$A^4B = AB$	$A^5 = 15A^2 - 14A$
A^n	$A^nB^2 = AB$	$A^{n+2}B = AB$	$A^{n+3} = \alpha_n A^2 - \beta_n A$
AB	$AB^3 = AB$	$A^3B^2 = AB$	$A^4B = AB$
A^nB^m	$A^nB^{m+2} = AB$	$A^{n+2}B^{m+1} = AB$	$A^{n+3}B^m = AB$

as a linear combination of elements in the standard basis as $7A^2 - 6A$. Here, $\alpha_{n+1} = \beta_{n+1} + 1$, $\beta_{n+1} = 2\alpha_n$ and $\alpha_n = 3w^{n-1} + \sum_{i=0}^{n-1} 2^i$, for n, m positive integers.

5.5 Indicator Function for Complex Coded Designs

In this section, we summarize the main results of Pistone and Rogantin (2008).

We consider mixed factorial designs with replicates and we code the n_j levels of a factor by the n_jth roots of the unity. The set of functions $\{X^\alpha, \ \alpha \in L\}$ is an *orthonormal* basis of the *complex responses* on the full factorial design $\mathcal{D}$.

5.5.1 Indicator Function and Counting Function

The indicator polynomial was originally introduced to treat the single replicate case. This case has special features, mainly because the equivalent description with generating equations is available. We introduced the new name *counting function* for the general case with or without replicates. The design with replicates associated to a counting function can be considered a multi-subset $\mathcal{F}$ of the design $\mathcal{D}$, or an array with repeated rows. In the following, we also use the name "fraction" in this extended sense.

The *counting function* R of a fraction $\mathcal{F}$ is a response defined on $\mathcal{D}$ so that for each $\zeta \in \mathcal{D}$, $R(\zeta)$ equals the number of appearances of ζ in the fraction.

A 0–1 valued counting function is the *indicator function* F of a single replicate fraction $\mathcal{F}$.

We denote the coefficients of the representation of R on $\mathcal{D}$ using the monomial basis by b_α:

$$R(\zeta) = \sum_{\alpha \in L} b_\alpha \; X^\alpha(\zeta) \,, \qquad \zeta \in \mathcal{D} \,.$$

Theorems 2 and 3 of Sect. 5.6 allow one to compute the indicator function, given the fraction polynomial generating equations. The following proposition shows how to compute the indicator function, given the fraction points.

Proposition 5.

1. *The coefficients b_α of the counting function of a fraction $\mathcal{F}$ are*

$$b_\alpha = \frac{1}{\#\mathcal{D}} \sum_{\zeta \in \mathcal{F}} \overline{X^\alpha(\zeta)} \;;$$

   in particular, b_0 is the ratio between the number of points of the fraction and those of the design.
2. *In a single replicate fraction, the coefficients b_α of the indicator function are related according to:*

$$b_\alpha = \sum_{\beta \in L} b_\beta \; b_{[\alpha - \beta]} \;.$$

5.5.2 Orthogonal Responses and Orthogonal Arrays

We discuss the general case of fractions $\mathcal{F}$ with or without replicates. We denote the mean value of a response f on the fraction $\mathcal{F}$, by $E_{\mathcal{F}}(f)$, $E_{\mathcal{F}}(f) = (1/\#\mathcal{F}) \sum_{\zeta \in \mathcal{F}} f(\zeta)$. We say that a response f is *centered* on a fraction $\mathcal{F}$ if $E_{\mathcal{F}}(f) = 0$ and we say that two responses f and g are *orthogonal* on $\mathcal{F}$ if $E_{\mathcal{F}}(f\,\overline{g}) = 0$, i.e. the response $f\,\overline{g}$ is centered.

It should be noticed that the term "orthogonal" refers to vector orthogonality with respect to a given Hermitian product. The standard practice in orthogonal array literature, however, is to define an array as orthogonal when all the level combinations appear equally often in relevant subsets of columns, e.g. (Hedayat et al., 1999, Def. 1.1). Vector orthogonality is affected by the coding of the levels, while the definition of an orthogonal array is purely combinatorial. A characterization of orthogonal arrays can be based on vector orthogonality of special responses. This section and the next one are devoted to discussing how the choice of complex coding makes such a characterization as straightforward as in the classical two-level case with coding −1,+1.

Proposition 6. *Let $R = \sum_{\alpha \in L} b_\alpha X^\alpha$ be the counting function of a fraction $\mathcal{F}$.*

1. *The term* X^α *is centered on* $\mathcal{F}$ *if, and only if,* $b_\alpha = b_{[-\alpha]} = 0$.
2. *The terms* X^α *and* X^β *are orthogonal on* $\mathcal{F}$ *if, and only if,* $b_{[\alpha-\beta]} = b_{[\beta-\alpha]} = 0$.
3. *If* X^α *is centered then, for each* β *and* γ *such that* $\alpha = [\beta - \gamma]$ *or* $\alpha = [\gamma - \beta]$*, then* X^β *is orthogonal to* X^γ.

Now we discuss the relations between the coefficients b_α, $\alpha \in L$, of the counting function and the property of being an orthogonal array. Let

$$\mathrm{OA}(n, s_1^{p_1}, \ldots, s_m^{p_m}, t)$$

be a mixed level orthogonal array with n rows and m columns, $m = p_1 + \cdots + p_m$, in which p_1 columns have s_1 symbols, ..., p_k columns have s_m symbols, and with strength t, as defined, e.g. in Wu and Hamada (2000, p. 260). Strength t means that, for any t column of the matrix design, all possible combinations of symbols appear equally often in the matrix.

Definition 7. *Let* I *be a non-empty subset of* $\{1, \ldots, m\}$*, and let* J *be its complementary set,* $J = I^c$*. Let* $\mathcal{D}_I$ *and* $\mathcal{D}_J$ *be the corresponding full factorial designs over the* I*-factors and the* J*-factors, so that* $\mathcal{D} = \mathcal{D}_I \times \mathcal{D}_J$*. Let* $\mathcal{F}$ *be a fraction of* $\mathcal{D}$ *and let* $\mathcal{F}_I$ *and* $\mathcal{F}_J$ *be its projections.*

1. *A fraction* $\mathcal{F}$ factorially projects *onto the* I*-factors if* $\mathcal{F}_I = s\ \mathcal{D}_I$*, that is, the projection is a full factorial design where each point appears* s *times.*
2. *A fraction* $\mathcal{F}$ *is a* mixed orthogonal array *of strength* t *if it factorially projects onto any* I*-factors with* $\#I = t$*.*

Proposition 7. *For each point* ζ *of* $\mathcal{F}$*, we consider the decomposition* $\zeta = (\zeta_I, \zeta_J)$ *and we denote the counting function restricted to* $\mathcal{F}_I$ *by* R_I*.*

1. *A fraction* factorially projects onto the I-factors *if and only if*

$$R_I(\zeta_I) = \#\mathcal{D}_J\ b_0 = \frac{\#\mathcal{F}}{\#\mathcal{D}_I} \qquad \text{for all } \zeta_I$$

   and if and only if all of the coefficients of the counting function involving only the I*-factors are 0.*
2. *A fraction is an* orthogonal array of strength t *if, and only if, all the coefficients of the counting function up to the order* t *are zero, except the constant term* b_0*.*

5.5.3 Regular Fractions

We consider a fraction *without replicates*. Let $n = \mathrm{lcm}\{n_1, \ldots, n_m\}$. We denote the set of the nth roots of the unity, by Ω_n, $\Omega_n = \{\omega_0, \ldots, \omega_{n-1}\}$. Let $\mathcal{L}$ be a subset of exponents, $\mathcal{L} \subset L = (\mathbb{Z}_n)^m$, containing $(0, \ldots, 0)$ and let l be its cardinality ($l > 0$). Let e be a map from $\mathcal{L}$ to Ω_n, $e : \mathcal{L} \to \Omega_n$.

Definition 8. *A fraction $\mathcal{F}$ is* regular *if*

1. *$\mathcal{L}$ is a sub-group of L,*
2. *e is a homomorphism, $e([\alpha+\beta]) = e(\alpha)\, e(\beta)$ for each $\alpha, \beta \in \mathcal{L}$,*
3. *The equations*
$$X^\alpha = e(\alpha)\,, \qquad \alpha \in \mathcal{L}\,,$$
define the fraction $\mathcal{F}$, i.e. they are a set of generating equations. The Equations $X^\alpha = e(\alpha)$ with $\alpha \in \mathcal{L}$ are called the defining *equations of $\mathcal{F}$.*

If $\mathcal{H}$ is a minimal generator of the group $\mathcal{L}$, then the equations $X^\alpha = e(\alpha)$ with $\alpha \in \mathcal{H} \subset \mathcal{L}$ are called minimal generating *equations.*

Proposition 8. *Let $\mathcal{F}$ be a fraction. The following statements are equivalent:*

1. *The fraction $\mathcal{F}$ is regular according to Definition 8.*
2. *The indicator function of the fraction has the form*
$$F(\zeta) = \frac{1}{l} \sum_{\alpha \in \mathcal{L}} \overline{e(\alpha)}\; X^\alpha(\zeta)\,, \qquad \zeta \in \mathcal{D}\,,$$
where $\mathcal{L}$ is a given subset of L and $e : \mathcal{L} \to \Omega_n$ is a given mapping.
3. *For each $\alpha, \beta \in L$ the parametric functions represented on $\mathcal{F}$ by the terms X^α and X^β are either orthogonal or totally confounded.*

5.6 Indicator Function vs. Gröbner Basis

Gröbner bases and indicator functions are powerful computational tools to deal with ideals of fractions and each outline different theoretical aspects of the ideals. For this reason, it is interesting to know how to move from one representation to the other.

When a list of treatment values is available, F can be computed using some form of interpolation formula, as performed in the aforementioned papers. Here we follow a new approach based on generating equations.

Theorem 2. *Let the ideals of the design and of the fraction be:*
$$I(\mathcal{D}) = < d_1, \ldots, d_p > \quad \text{and} \quad I(\mathcal{F}) = < d_1, \ldots, d_p, g_1 \ldots, g_q > \ .$$
Let G be a $\mathcal{D}$-reduced polynomial. A single equation $G = 0$ is a generating equation of $\mathcal{F}$, i.e.
$$< d_1, \ldots, d_p, g_1 \ldots, g_q > = < d_1, \ldots, d_p, G >$$
if and only if (1) and (2) below are both satisfied:

(1) There exist $h_j \in k[x_1, \ldots, x_m]$, $j = 1, \ldots, q$ such that
$$G - \sum_j h_j g_j \in I(\mathcal{D})\ ;$$

(2) For all g_i, there exists $s_i \in k[x_1, \ldots, x_m]$ such that

$$g_i - s_i \; G \in I(\mathcal{D}) \; .$$

Moreover, for $G = 1 - F$ if (1) and

(3) For all g_i, $F g_i \in I(\mathcal{D})$

hold, then F is the indicator function of $\mathcal{F}$ in $\mathcal{D}$.

Proof. Condition (1) is equivalent to $G \in I(\mathcal{F})$; (1) and (2) together are equivalent to $< d_1, \ldots, d_m, G >= I(\mathcal{F})$.

If $G = F - 1$, from condition (1) we obtain $1 - F(a) = 0$ for all $a \in \mathcal{F}$. For all $a \in \mathcal{D} \setminus \mathcal{F}$, condition (3) implies $F(a) g_i(a) = 0$ for all $i = 1, \ldots, q$. As not all $g_i(a)$ can be simultaneously equal to zero, then for all $a \in \mathcal{D} \setminus \mathcal{F}$ there exists $\bar{i}$ such that $g_{\bar{i}}(a) \neq 0$. □

Remark 1. The computation of the indicator polynomial from the generating equations could be reduced to the case of a single generating equation. In fact, if $\mathcal{F}_i$ is the fraction whose generating equation is g_i, with indicator polynomial F_i, then $\mathcal{F} = \cap_{i=1}^{k} \mathcal{F}_i$ and $F = \mathrm{NF}_{\mathcal{D}}\,(F_1 \cdots F_k)$.

Example 9. Let us consider the 3^2 factorial design with level set $\{-1, 0, +1\}$ In this case the design equations are $d_1(x) = x^3 - x$ and $d_2(y) = y^3 - y$. We consider the "cross" fraction with generating polynomial $g(x, y) = xy$. The system that corresponds to statements (1) and (2) of Theorem 2 is

$$\begin{cases} x^3 - x = 0 \, , \\ y^3 - y = 0 \, , \\ 1 - f - hxy = 0 \, , \\ fxy = 0 \, . \end{cases}$$

We want to eliminate h and determine f as function of x and y. From the third equation multiplied by x^2y^2 we obtain $x^2y^2 - fx^2y^2 - hx^3y^3 = 0$, and, using the other equations, we obtain $x^2y^2 - 1 + f = 0$. The indicator polynomial is $F(x, y) = 1 - x^2y^2$. The equivalent system of equations

$$\begin{cases} x^3 - x = 0 \, , \\ y^3 - y = 0 \, , \\ f + x^2y^2 - 1 = 0 \, , \\ hxy + f - 1 = 0 \, , \end{cases}$$

is in lower-triangular form with respect to the lexicographic order of monomials with $x \prec y \prec f \prec h$ and has the smallest leading terms. In other words, it is a Gröbner basis for the lexicographic order. This leads to Theorem 3.

Theorem 3. *We consider the ring $k[h_1, \ldots, h_q, f, x_1, \ldots, x_m]$. The lexicographic Gröbner basis of the elimination ideal*

$$< d_1, \ldots, d_m,\ (1-f) - \sum_j h_j g_j,\ f g_1, \ldots, f g_k > \cap\, k[f, x_1, \ldots, x_m]$$

contains a unique polynomial of the form $f - \sum_{\alpha \in L} b_\alpha x^\alpha$, so that the indicator function F is $\sum_{\alpha \in L} b_\alpha X^\alpha$.

Proof. The polynomial $f - \sum_{\alpha \in L} b_\alpha x^\alpha$ belongs to the elimination ideal because of theorem 2. It has a minimum leading term among the polynomials containing the indeterminates f and $x_1, \ldots, x_m$. □

If g is a response on the design $\mathcal{D}$, the mapping

$$\mathcal{L}(\mathcal{D}) \ni g \mapsto \mathrm{NF}_{\mathcal{D}}\,(gF) \in \mathcal{L}(\mathcal{D})$$

sets g to zero outside $\mathcal{F}$, and provides a distinguished representative of the equivalence class of responses which are aliased on the fraction. The computation of $\mathrm{NF}_{\mathcal{D}}\,(X^\alpha F)$, $\alpha \in L$ is a nice way of studying the aliasing structure of the fraction. Again, the computation is performed without computing the treatments.

Theorem 4. *We fix a term order on the set of exponents L. We compute the system of the Normal Forms on $\mathcal{D}$ of the polynomials $X^\alpha F$, $\alpha \in L$ with the previous order. We denote the matrix of the coefficients by R and their elements by $r_{\alpha\beta}$:*

$$\mathrm{NF}_{\mathcal{D}}\,(X^\alpha F) = \sum_\beta r_{\alpha\beta} X^\beta \ .$$

Any vector k in the left kernel of R, i.e. such that $k^t R = 0$, gives an aliasing relation among the monomials X^α:

$$\sum_{\alpha \in L} k_\alpha X^\alpha = 0 \qquad \text{on } \mathcal{F}.$$

A hierarchical monomial basis of the fraction can be found from the ker-matrix K. If $K_{L \setminus N}$ is a non singular sub-matrix of K, then X^α, with $\alpha \in N$, is a monomial basis of $\mathcal{F}$.

Proof. We write the ker-matrix in two vertical blocks as $K = [K_{L\setminus N} \mid K_N]$ and the matrix R in the two corresponding horizontal blocks $R = \left[\frac{R_{L\setminus N}}{R_N}\right]$. If $K_{L\setminus N}$ is non-singular, then $K_{L\setminus N}^{-1} K\ R = 0$ implies

$$R_N + K_{L\setminus N}^{-1}\ K_N\ R_N = 0$$

and X^α, with $\alpha \in N$, is a basis of the fraction. □

Example 10. Composite fraction with 2 factors. We consider a 2 factor design with 5 levels and the fraction of the following 9 points:

$$\{(-2,0),(-1,1),(-1,-1),(0,2),(0,0),(0,-2),(1,1),(1,-1),(2,0)\}$$

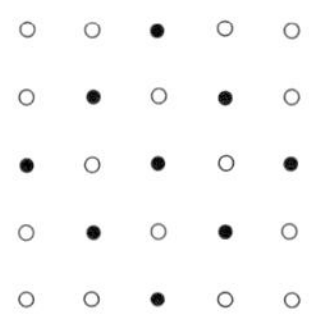

The equations of the full factorial design with level sets $\{-2,-1,0,1,2\}$ are

$$d_1 = x_1(x_1^2-1)(x_1^2-4) \quad \text{and} \quad d_2 = x_2(x_2^2-1)(x_2^2-4)$$

and a natural choice of generating polynomials of the composite design is

$$\begin{gathered} g_1 = x_1x_2(x_1^2-1)\,,\ g_2 = x_1x_2(x_2^2-1)\,, \\ x_1(x_1^2-4)(x_2^2-1)\,,\ x_2(x_2^2-4)(x_1^2-1)\,. \end{gathered}$$

An equivalent, reduced generating set, obtained by simple substitution, is given by:

$$g_1\ ,g_2\ ,g_3 = x_1^3+3x_1x_2^2-4x_1\,,\ g_4 = 3x_1^2x_2+x_2^3-4x_2\,.$$

According to the remark after Theorem 2, we choose to compute the indicator function of the fraction generated by each g_i separately for $i=1,\ldots,4$ and then compute the Normal Form of their product. Theorem 3 and the Buchberger algorithm, used to compute the Gröbner basis with the lexicographic order, provide the indicator function for each generating equation as

$$\begin{aligned}
F_1 &= \frac{1}{48}x_1^4x_2^4 - \frac{5}{48}x_1^4x_2^2 - \frac{1}{48}x_1^2x_2^4 + \frac{5}{48}x_1^2x_2^2 + 1 \\
F_2 &= \frac{1}{48}x_1^4x_2^4 - \frac{1}{48}x_1^4x_2^2 - \frac{5}{48}x_1^2x_2^4 + \frac{5}{48}x_1^2x_2^2 + 1 \\
F_3 &= \frac{19}{144}x_1^4x_2^4 - \frac{79}{144}x_1^4x_2^2 + \frac{1}{3}x_1^4 - \frac{67}{144}x_1^2x_2^4 + \frac{271}{144}x_1^2x_2^2 - \frac{4}{3}x_1^2 + 1 \\
F_4 &= \frac{19}{144}x_1^4x_2^4 - \frac{67}{144}x_1^4x_2^2 - \frac{79}{144}x_1^2x_2^4 + \frac{271}{144}x_1^2x_2^2 + \frac{1}{3}x_2^4 - \frac{4}{3}x_2^2 + 1
\end{aligned}$$

and

$$F = \frac{31}{144}x_1^4x_2^4 - \frac{127}{144}x_1^4x_2^2 - \frac{127}{144}x_1^2x_2^4 + \frac{1}{3}x_1^4 + \frac{511}{144}x_1^2x_2^2 + \frac{1}{3}x_2^4 - \frac{4}{3}x_1^2 - \frac{4}{3}x_2^2 + 1\,.$$

We can compute the Normal Forms of $X^\alpha F$, $\alpha \in L$ with the DegRevLex order. A left kernel K of the matrix R, computed by CoCoA, gives a good picture of the aliasing patterns. The display is divided into two parts:

$$
\begin{array}{ccccccccccccccc}
00 & 01 & 02 & 03 & 04 & 10 & 11 & 12 & 13 & 14 & 20 & 21 & 22 & 23 & 24
\end{array}
$$
$$
\begin{pmatrix}
0 & -4 & 0 & 1 & 0 & 0 & 0 & 0 & 0 & 0 & 0 & 0 & 0 & 0 & 0 \\
0 & 0 & -4 & 0 & 1 & 0 & 0 & 0 & 0 & 0 & 0 & 0 & 0 & 0 & 0 \\
0 & 0 & 0 & 0 & 0 & 0 & 1 & 0 & 0 & 0 & 0 & 0 & 0 & 0 & 0 \\
0 & 0 & 0 & 0 & 0 & 0 & 0 & 1 & 0 & -1 & 0 & 0 & 0 & 0 & 0 \\
0 & 0 & 0 & 0 & 0 & 0 & 0 & 0 & 1 & 0 & 0 & 0 & 0 & 0 & 0 \\
0 & 0 & 0 & 0 & 0 & 0 & 0 & 0 & 0 & 0 & 0 & 1 & 0 & 0 & 0 \\
0 & 0 & 0 & 0 & 0 & 0 & 0 & 0 & 0 & 0 & 0 & 0 & 1 & 0 & 0 \\
0 & 0 & 0 & 0 & 0 & 0 & 0 & 0 & 0 & 0 & 0 & 0 & 0 & 1 & 0 \\
0 & 0 & 0 & 0 & 0 & 0 & 0 & 0 & 0 & 0 & 0 & 0 & 0 & 0 & 1 \\
0 & 0 & 0 & 0 & 0 & -4 & 0 & 0 & 0 & 3 & 0 & 0 & 0 & 0 & 0 \\
0 & 0 & 0 & 0 & 0 & 0 & 0 & 0 & 0 & -1 & 0 & 0 & 0 & 0 & 0 \\
0 & 0 & 0 & 0 & 0 & 0 & 0 & 0 & 0 & 0 & 0 & 0 & 0 & 0 & 0 \\
0 & 0 & 0 & 0 & 0 & 0 & 0 & 0 & 0 & -1 & 0 & 0 & 0 & 0 & 0 \\
0 & 0 & 0 & 0 & 0 & 0 & 0 & 0 & 0 & 0 & -4 & 0 & 0 & 0 & 0 \\
0 & 0 & 0 & 0 & 0 & 0 & 0 & 0 & 0 & 0 & 0 & 0 & 0 & 0 & 0 \\
0 & 0 & 0 & 0 & 0 & 0 & 0 & 0 & 0 & 0 & 0 & 0 & 0 & 0 & 0
\end{pmatrix}
$$

$$
\begin{array}{cccccccccc}
30 & 31 & 32 & 33 & 34 & 40 & 41 & 42 & 43 & 44
\end{array}
$$
$$
\begin{pmatrix}
0 & 0 & 0 & 0 & 0 & 0 & 3 & 0 & 0 & 0 \\
0 & 0 & 0 & 0 & 0 & 0 & 0 & 3 & 0 & 0 \\
0 & -1 & 0 & 0 & 0 & 0 & 0 & 0 & 0 & 0 \\
0 & 0 & 0 & 0 & 0 & 0 & 0 & 0 & 0 & 0 \\
0 & -1 & 0 & 0 & 0 & 0 & 0 & 0 & 0 & 0 \\
0 & 0 & 0 & 0 & 0 & 0 & -1 & 0 & 0 & 0 \\
0 & 0 & 0 & 0 & 0 & 0 & 0 & -1 & 0 & 0 \\
0 & 0 & 0 & 0 & 0 & 0 & -1 & 0 & 0 & 0 \\
0 & 0 & 0 & 0 & 0 & 0 & 0 & -1 & 0 & 0 \\
1 & 0 & 0 & 0 & 0 & 0 & 0 & 0 & 0 & 0 \\
0 & 0 & 1 & 0 & 0 & 0 & 0 & 0 & 0 & 0 \\
0 & -1 & 0 & 1 & 0 & 0 & 0 & 0 & 0 & 0 \\
0 & 0 & 0 & 0 & 1 & 0 & 0 & 0 & 0 & 0 \\
0 & 0 & 0 & 0 & 0 & 1 & 0 & 3 & 0 & 0 \\
0 & 0 & 0 & 0 & 0 & 0 & -1 & 0 & 1 & 0 \\
0 & 0 & 0 & 0 & 0 & 0 & 0 & -1 & 0 & 1
\end{pmatrix}
$$

We can observe that there are 12 rows with 2 non-zero values (-1 and 1) and 23 zeros. We have the full aliasing of the corresponding monomials:

$$
\begin{aligned}
&X^{(31)} = X^{(11)},\ X^{(31)} = X^{(33)},\ X^{(31)} = X^{(13)}\,, \\
&X^{(41)} = X^{(21)},\ X^{(41)} = X^{(23)},\ X^{(41)} = X^{(43)}\,, \\
&X^{(14)} = X^{(12)},\ X^{(14)} = X^{(32)},\ X^{(14)} = X^{(34)}\,, \\
&X^{(42)} = X^{(22)},\ X^{(42)} = X^{(44)},\ X^{(42)} = X^{(24)}\,.
\end{aligned}
$$

Another 4 rows correspond to cases of partial aliasing:

$$
\begin{aligned}
&X^{(03)} = 4X^{(01)} - 3X^{(41)},\ X^{(04)} = 4X^{(02)} - 3X^{(42)}\,, \\
&X^{(30)} = 4X^{(10)} - 3X^{(14)},\ X^{(40)} = 4X^{(20)} - 3X^{(42)}\,.
\end{aligned}
$$

A non-singular sub-matrix of the previous kernel of the R matrix corresponds to $N = \{(\alpha_1, \alpha_2) \mid 0 \leq \alpha_i \leq 2\}$; the determinant of the matrix $K_{L \setminus N}$ is 1. In particular, the response surface with terms with total degree up to 2, is estimable.

It should be noticed that the previous discussion concerning the aliasing, based on the matrix K, depends on the very special form of the kernel matrix produced by CoCoA. This software tries to compute a "simple" matrix with integer entries and such that the number of non-zero entries of each row is low. A generic numerical software would have produced an floating point orthonormalized matrix that would be unsuitable for a statistical interpretation.

5.7 Mixture Designs

In classical DOE literature, mixture designs are considered as a subset of the simplex. Maruri-Aguilar et al. (2007) instead actually identify each point of the mixture design with a line through the origin. This allows them to consider a larger class of homogeneous models associated to the design. A mixture design $\mathcal{D}$ is analysed through its cone which is a projective variety associated to $\mathcal{D}$. In particular, homogeneous polynomials of a given degree are considered. We illustrate the results in Maruri-Aguilar et al. (2007) through analogy to the statements in Sect. 5.2.

Let $\mathcal{D}$ be a mixture design, that is, a set of n distinct points in $\mathbb{R}^m$, where $d = (d_1, \ldots, d_m)$ such that $0 \leq d_i \leq 1$ and $\sum_{i=1}^m d_i = 1$. The analogue of the design $\mathcal{D} \subset \mathbb{R}^m$ is the cone of $\mathcal{D}$

$$\mathcal{C}_\mathcal{D} = \{\alpha d : d \in \mathcal{D} \text{ and } \alpha \in \mathbb{R}\} \subset \mathbb{R}^k \ ;$$

namely, the union of lines through the origin of $\mathbb{R}^m$ and a point in $\mathcal{D}$. Each line in $\mathcal{C}_\mathcal{D}$ corresponds to a projective point in the $m-1$ projective space $\mathbb{P}^{m-1}$.

The analogue of the design ideal $I(\mathcal{D})$ is the ideal generated by $\mathcal{C}_\mathcal{D}$

$$\text{Ideal}(\mathcal{C}_\mathcal{D}) = \{f \in \mathbb{R}[x_1, \ldots, x_m] : f(d) = 0 \text{ for all } d \in \mathcal{C}_\mathcal{D}\} \ .$$

A polynomial is *homogeneous* if each one of its terms has the same total degree. An ideal is homogeneous if it is generated by a set of homogeneous polynomials, or, it admits a reduced Gröbner basis formed by homogeneous polynomials, for each term-ordering. Ideal$(\mathcal{C}_\mathcal{D})$ is the largest homogeneous ideal in $\mathbb{R}[x_1, \ldots, x_m]$ that vanishes on all the points of $\mathcal{D}$, see Maruri-Aguilar et al. (2007). Moreover, Ideal$(\mathcal{D})$ = Ideal$(\mathcal{C}_\mathcal{D}) + \langle \sum_{i=1}^m x_i - 1 \rangle$, that is, a polynomial vanishing on $\mathcal{D}$ can be written as a combination of homogeneous components vanishing on $\mathcal{D}$ and the sum to one condition. If G is a generator set of Ideal$(\mathcal{C}_\mathcal{D})$ then G and $\sum_{i=1}^m x_i - 1$ form a generator set of Ideal$(\mathcal{D})$. This means that the ideal of the mixture design can be obtained by cutting its cone at the simplex. A generating set of Ideal($\mathcal{C}_\mathcal{D}$) is obtained by homogenizing the Gröbner basis with respect to a graded term ordering of Ideal($\mathcal{D}$).

The mixture analogue of the full factorial design is the class of simplex lattice designs in Example 2. The analogue of HMBs is given by suitable subsets of the set of all homogeneous monomials of a given degree. A $\mathbb{R}$-vector space basis of $\mathbb{R}[x_1, \ldots, x_k]/\operatorname{Ideal}(\mathcal{D})$ can in fact be chosen to be formed by homogeneous polynomials of degree s. Moreover, given a term ordering and sufficiently large s, a basis of the quotient space is given by the monomials of degree s that are not divisible by the leading terms of the reduced Gröbner basis of the cone ideal. This procedure is illustrated in the following example.

Example 11. Let $\mathcal{D} = \{(0,1),(1,0),(1/2,1/2)\}$, with $\mathcal{C}_\mathcal{D} = \{(0,a),(b,0),(c,c)\}$ and with a, b, and c non-zero real numbers.

1. Determine a Gröbner basis of Ideal($\mathcal{C}_\mathcal{D}$), with respect to a term ordering; this is $\{x_1x_2^2 - x_1^2x_2\}$.
2. Compute the leading terms of each element of the Gröbner basis, for the example $x_1x_2^2$, for any term ordering for which $x_2 > x_1$.
3. Consider all the monomials of a sufficiently large total degree s; for example in $\mathbb{R}[x_1, x_2]$ there are four monomials of degree $s = 3$, namely $x_1^3, x_1^2x_2, x_1x_2^2, x_2^3$.
4. Determine all monomials of degree s not divisible by the leading terms of the Gröbner basis, in the example $x_1^3, x_1^2x_2, x_2^3$.

Let us recall some basic facts about projective varieties. Homogeneous polynomials and projective varieties are naturally associated. A homogeneous polynomial vanishing on a set of homogeneous coordinates of a projective point p in fact vanishes on all the homogeneous coordinates of p. Moreover, if V is a projective variety and $I(V)$ the largest homogeneous ideal containing V, then the variety generated by $I(V)$ is V itself. The Projective Strong Nullstellensatz theorem also states that if I is a homogeneous ideal and the variety generated by I, call it V, is not empty, then the ideal generated by V is the radical ideal of I.

Example 12. In $\mathbb{P}^2$ with homogeneous coordinates (x, y, z), the x-axis is defined by the ideal $I = \langle y \rangle$ but the ideal $\langle y^2, xy, yz \rangle$ also defines the x-axis. The radical ideals of I and J are equal but $J \subset I$. The difference is that I is a saturated ideal while J is not. I can be obtained from J with the saturation operation, see Cox et al. (1997).

Finally, we define an analogue of the indicator function of Sect. 5.6. It should be observed that a function f is well defined over a projective object if $f(a) = f(\lambda a)$ for all a homogeneous coordinates of a projective point and λ a non zero real number. Ratios of homogeneous polynomials have this property. Hence we give the following definition.

Definition 9. *Let $\mathcal{D}$ be a simplex lattice design or some other mixture design, p a point in $\mathcal{D}$ and $\mathcal{F}$ a fraction of $\mathcal{D}$.*

1. A separator *(polynomial) of the point* $p \in \mathcal{D}$ *is any (homogeneous) polynomial* S_p *such that* $S_p \notin \text{Ideal}(\mathcal{C}_{\{p\}})$ *and* $S_p \in \text{Ideal}(\mathcal{C}_{\mathcal{D}\setminus\{p\}})$.
2. *The* indicator function *of* $p \in \mathcal{D}$ *is*

$$S_{\{p\}} = \frac{S_p}{(\sum_{i=1}^{k} x_i)^{s_p}},$$

where s_p *is the degree of* S_p.
3. *A* separator *(polynomial) of* $\mathcal{F} \subset \mathcal{D}$ *is any (homogeneous) polynomial* $S_{\mathcal{F}}$ *such that* $S_{\mathcal{F}} \notin \text{Ideal}(\mathcal{C}_{\mathcal{F}})$ *and* $S_{\mathcal{F}} \in \text{Ideal}(\mathcal{C}_{\mathcal{D}\setminus\mathcal{F}})$.
4. *The indicator of the fraction* $\mathcal{F} \subset \mathcal{D}$ *is the rational polynomial function*

$$S_{\mathcal{F}} = \sum_{p\in\mathcal{F}} \frac{S_p}{(\sum_i x_i)^{s_p}}.$$

The previous Items 2. and 4. are representations of the unique indicator functions. The next theorem shows that the definition of the indicator function is well posed. Obviously the given definitions for a single point are redundant as a single point is a very simple fraction. However, we find it useful to distinguish the particular design fraction of size one.

Theorem 5. *An indicator function for* $\mathcal{F} \subset \mathcal{D}$*, where* $\mathcal{D}$ *is a mixture design, is the rational polynomial function obtained by averaging the indicator functions of the single points in the fraction, namely*

$$S_{\mathcal{F}} = \sum_{p\in\mathcal{F}} \frac{S_p}{(\sum_i x_i)^{s_p}},$$

where S_p *is a homogeneous polynomial of total degree* s_p *such that* $S_p(d) = 1$ *if* $d = p$ *and* 0 *if* $d \in \mathcal{D} \setminus \{p\}$.

Proof. In Abbott et al. (2000), it is shown that S_p exists for all p. Moreover, for $\alpha \in \mathbb{R}$ and $d \in \mathcal{D}$,

$$\begin{aligned} S_{\mathcal{F}}(\alpha d) &= \sum_{p\in\mathcal{F}} \frac{S_p(\alpha d)}{(\sum_i x_i)^{s_p}(\alpha d)} \\ &= \sum_{p\in\mathcal{F}} \frac{\alpha^{s_p} S_p(d)}{\alpha^{s_p}(\sum_i x_i)^{s_p}(d)} \\ &= \sum_{p\in\mathcal{F}} \frac{S_p(d)}{(\sum_i x_i)^{s_p}(d)} = S_{\mathcal{F}}(d) = \begin{cases} 1 \text{ if } d \in \mathcal{F}, \\ 0 \text{ otherwise}. \end{cases} \end{aligned}$$

□

Thus separators are homogeneous polynomials and indicators are in fact functions over the projective points. They take the same values when evaluated at any representative of the projective point in $\mathcal{D}$. In particular, for $p \in \mathcal{D}$, we obtain $S_p(a) = 0$ if $a \in \mathcal{D} \setminus \{p\}$ and $S_p(p) = (\sum_{i=1}^{m} p_i)^{s_p}$ where $p = (p_1 :$

$\ldots : p_k)$ and $S_p(p)$ is not zero when evaluated in any point of the simplex. Non significant differences occur if we divide S_p times $(\sum b_i x_i)^{s_p}$ for b_i real numbers, up to assuming that $\sum b_i x_i \notin \text{Ideal}(\mathcal{C}_{\{p\}})$ for all $p \in \mathcal{D}$.

When the coordinates of the points in the fraction and the ambient design are known, the separators and indicators can be directly computed from their definitions. An algorithm to do this efficiently is implemented in CoCoA under the macro called `IdealOfProjectivePoints`. In Examples 13 and 14, we compute the separators and indicators directly from the generating equations of the fraction and the design in analogy to Theorem 2.

Example 13. The following CoCoA script allows us to isolate point $(1, 1)$ in design $\mathcal{D}$ of Example 2. In Line (1), we inform the system that we use polynomials with rational coefficients and four indeterminates, f, h, x, y and the lexicographic term ordering with $f < h < z < y$. In Line (2), D describes the generator set of $\mathcal{C}_{\mathcal{D}}$ and $\mathcal{C}_{\mathcal{F}} = \langle xy^2 - x^2y, xy\rangle$. The polynomial P in Line (3) corresponds to $(\sum_{i=1}^{k} x_i)^{s_p}$ with $s_p = 2$ (this information is given by the Hilbert function and here we omit details). Theorem 2 is set-up in Line (4) and performed in Lines (6) and (7). Line (5) computes saturation whose need in illustrated in Example 12. Finally the indicator function of $\mathcal{F}$ in $\mathcal{D}$ is

$$S = \left(\frac{x - y}{x + y}\right)^2 .$$

```
(1) Use T::=Q[fhxy],Lex;
(2) D:=xy^2-x^2y;  G:=xy;
(3) P:=(x+y)^2;
(4) L:=[D,P-f -hG, fG];
(5) Id:=Saturation(Ideal(L),Ideal(x,y));
(6) GB:=ReducedGBasis(Id);
(7) S:=f -GB[1];S;
```

Example 14. Let $\mathcal{D}$ be the design whose cone ideal is generated by $y^2z - yz^2, x^2z - xz^2, x^2y - xy^2$ and $\mathcal{F}$ the fraction obtained by adjoining the generating equations $g_1 = xz - yz$ and $g_2 = -xy + yz$. First, we compute the indicator function for each generating equation as in Example 13, then we consider their product to determine the indicator function of the fraction

```
Use T::=Q[f h xyz], Lex;
Set Indentation;
D:=[y^2z - yz^2, x^2z - xz^2, x^2y - xy^2];
G:=[xz - yz, -xy + yz];
P :=(x+y+z)^3;
S:=[];
For I:=1 To Len(G) Do
 L:=ConcatLists([D,[P-f-hG[I],fG[I]]]);
 Id:=Saturation(Ideal(L),Ideal(x,y,z));
 GB:=ReducedGBasis(Id);
```

```
  S:=Concat( [f-GB[1]], S);
 EndFor;
 SF:=NF(Product(S), Ideal(D));SF;
```

The indicator function of the fraction obtained is

$$S = \frac{x^6 - 2xy^5 + 732xyz^4 - 2xz^5 + y^6 - 2yz^5 + z^6}{(x+y+z)^6}.$$

5.8 Conclusions

We have provided a review of the main application of algebraic statistics in experimental design by collecting the basic results available in literature and extending and completing them in various parts. The two main directions that the topic has taken, namely the study of the indicator function and the study of the Gröbner bases associated with a design, are illustrated with examples and algorithms. The interlink between them is developed in Sect. 6 allowing, at least from an algorithmic viewpoint, to take the full advantage of them both.

A comparison with common concepts in design of experiment is in Sect. 3, whilst the discussion on HMBs in Sect. 4 illustrates how this algebraic approach is a natural generalization of many of such concepts. Our discussion on HMBs is by no means final. The theory here is yet not fully developed. The importance of computational commutative algebra in the description of the mathematical structure of the response space associated to a design is clear and well understood. But the understanding of the implications of the interpretation of a design as an algebraic ideal on modelling, model interpretation and design is not fully mature yet. In particular, there is scope to investigate the full power of the notion of confounding which is embedded in a design ideal with the many possible representations of the design ideal itself. The study of corner cut models in Maruri-Aguilar (2007) goes in the direction.

Questions like which HMB in the algebraic/statistical fan of a design should be used to build a statistical model have so far been addressed resorting to usual statistical techniques (for example, Holliday et al. (1999)). But the combined use of corner cut models and Hilbert function seems promising in determining novel criteria to compare model associated with a design. More importantly, so far only few successful attempts have been made to tackle the inverse problem of determining a design suitable to identify a model and with some other desirable properties. Here the indicator function seems the most promising tool, see Fontana et al. (2000).

Acknowledgement

This chapter stems from many discussions the authors had between themselves and with many colleagues. The authors are grateful to them all. In particular they acknowledge the stimulus provided by Henry Wynn in first thinking some of the issues in this chapter and his impetus in the applications of algebraic statistics in design of experiment. Hugo Maruri-Aguilar has been mentioned as co-author in Sect. 5.4.2.

References

Abbott, J., Bigatti, A., Kreuzer, M., and Robbiano, L. (2000). Computing ideals of points. *Journal of Symbolic Computation*, **30**(4), 341–356.

Babson, E., Onn, S., and Thomas, R. (2003). The Hilbert zonotope and a polynomial time algorithm for universal Gröbner bases. *Advances in Applied Mathematics*, **30**(3), 529–544.

Cox, D., Little, J., and O'Shea, D. (1997). *Ideals, Varieties, and Algorithms.* Springer-Verlag, New York, second edition.

Cox, D., Little, J., and O'Shea, D. (2005). *Using Algebraic Geometry.* Springer, New York, second edition.

Fontana, R., Pistone, G., and Rogantin, M.P. (2000). Classification of two-level factorial fractions. *Journal of Statistical Planning and Inference*, **87**(1), 149–172.

Galetto, F., Pistone, G., and Rogantin, M.P. (2003). Confounding revisited with commutative computational algebra. *Journal of Statistical Planning and Inference*, **117**(2), 345–363.

Hedayat, A.S., Sloane, N.J.A., and Stufken, J. (1999). *Orthogonal Arrays.* Springer-Verlag, New York.

Hinkelmann, K. and Kempthorne, O. (2005). *Design and Analysis of Experiments. Vol. 2. Advanced Experimental Design.* John Wiley & Sons, Hoboken, NJ.

Holliday, T., Pistone, G., Riccomagno, E., and Wynn, H.P. (1999). The application of computational algebraic geometry to the analysis of designed experiments: a case study. *Computational Statistics* , **14**(2), 213–231.

Kobilinsky, A. (1997). *Les Plans Factoriels*, chapter 3, pages 879–883. ASU–SSdF, Éditions Technip.

Kreuzer, M. and Robbiano, L. (2000). *Computational Commutative Algebra 1.* Springer, Berlin-Heidelberg.

Kreuzer, M. and Robbiano, L. (2005). *Computational Commutative Algebra 2.* Springer-Verlag, Berlin.

Maruri-Aguilar, H. (2007). *Algebraic Statistics in Experimental Design.* Ph.D. thesis, University of Warwick, Statistics, March 2007 .

Maruri-Aguilar, H., Notari, R., and Riccomagno, E. (2007). On the description and identifiability analysis of mixture designs. *Statistica Sinica*, **17**(4), 1417–1440.

McCullagh, P. and Nelder, J.A. (1989). *Generalized Linear Models.* Chapman & Hall, London, second edition.

Onn, S. and Sturmfels, B. (1999). Cutting corners. *Advances in Applied Mathematics*, **23**(1), 29–48.

Peixoto, J.L. (1990). A property of well-formulated polynomial regression models. *The American Statistician*, **44**(1), 26–30.

Pistone, G. and Rogantin, M. (2008). Algebraic statistics of codings for fractional factorial designs. *Journal of Statistical Planning and Inference*, **138**(1), 234–244.

Pistone, G. and Wynn, H.P. (1996). Generalised confounding with Gröbner bases. *Biometrika*, **83**(3), 653–666.

Pistone, G., Riccomagno, E., and Wynn, H.P. (2000). Gröbner basis methods for structuring and analyzing complex industrial experiments. *International Journal of Reliability, Quality, and Safety Engineering*, **7**(4), 285–300.

Pistone, G., Riccomagno, E., and Wynn, H.P. (2001). *Algebraic Statistics: Computational Commutative Algebra in Statistics.* Chapman & Hall, London.

Raktoe, B.L., Hedayat, A., and Federer, W.T. (1981). *Factorial Designs.* John Wiley & Sons Inc., New York.

Scheffé, H. (1958). Experiments with mixtures. *Journal of the Royal Statistical Society B*, **20**, 344–360.

Wu, C.F.J. and Hamada, M. (2000). *Experiments. Planning, Analysis, and Parameter Design Optimization.* John Wiley & Sons Inc., New York.

6

The Geometry of Causal Probability Trees that are Algebraically Constrained

E. Riccomagno and J.Q. Smith

Summary. Algebraic geometry is used to study properties of a class of discrete distributions defined on trees and called algebraically constrained statistical models. This structure has advantages in studying marginal models as it is closed under learning marginal mass functions. Furthermore, it allows a more expressive and general definition of causal relationships and probabilistic hypotheses than some of those currently in use. Simple examples show the flexibility and expressiveness of this model class which generalizes discrete Bayes networks.

6.1 The Algebra of Probability Trees

We begin with some definitions. A tree is a directed graph $\mathcal{T} = (V(\mathcal{T}), E(\mathcal{T}))$ with vertex set $V(\mathcal{T})$ and edge set $E(\mathcal{T})$ such that there is only one vertex v_0, called the *root vertex*, with no edge into it and for any other vertex there exists exactly one edge into it. If there is no ambiguity, we could write $\mathcal{T} = (V, E)$.

Let $e = (v, v') \in E(\mathcal{T})$ be the directed edge *emanating* from the vertex v into the vertex v'. In the edge $e = (v, v')$, the vertex v is called the *parent* of v or the parent in e and v' the *child* of v or in e. Vertices with no children are called *leaves*. Non-leaf vertices are called *situations* and $S(\mathcal{T}) \subset V(\mathcal{T})$ is the set of situations. For $v \in S(\mathcal{T})$ let $E(v)$ be the set of edges emanating from v; $E(v)$ is in one-to-one correspondence with the children of v.

A *path* is an ordered sequence of edges where the child vertex in an edge is the parent in the subsequent edge. A *root-to-leaf path* or *atomic event* is a path where the first edge emanates from the root and the last edge ends into a leaf. The notation $(e_1(\lambda), \ldots, e_{n(\lambda)}(\lambda))$ gives the ordered lists of $n(\lambda)$ edges in the path λ and $(v_0(\lambda), \ldots, v_{n(\lambda)}(\lambda))$ is the ordered list of vertices in λ. Let $\Lambda(\mathcal{T})$ be the set of all root-to-leaf paths in $\mathcal{T}$ and N its cardinality. Atomic events are in one-to-one correspondence with leaves.

A tree $\mathcal{T}$ is made into a probability tree by associating a value $\pi(v, v') \in [0, 1]$ to each edge $(v, v') \in E(\mathcal{T})$ so that the simplex condition $\sum_{v'} \pi(v, v') = 1$ holds for all situations $v \in S(\mathcal{T})$. The value $\pi(v, v')$ is called a *primitive*

L. Pronzato, A. Zhigljavsky (eds.), *Optimal Design and Related Areas in Optimization and Statistics*, Springer Optimization and its Applications 28,
DOI 10.1007/978-0-387-79936-0_6,

probability and often indicated as $\pi(e)$ where $e = (v, v')$. Thus, the simplex condition formulas above become $\sum_{e \in E(v)} \pi(e) = 1$ for $v \in S(\mathcal{T})$. Let Π be the set of primitive probabilities, namely

$$\Pi = \{\pi(e) : e \in E(\mathcal{T})\} = \{\pi(e) : e \in E(v) \text{ and } v \in S(\mathcal{T})\}.$$

Without ambiguity we can write $\pi(v, v') = \pi_{v'}$. Note that naturally the value

$$p(\lambda) = \prod_{i=1}^{n(\lambda)} \pi(e_i(\lambda))$$

is associated to the root-to-leaf path $\lambda = (e_1(\lambda), \ldots, e_{n(\lambda)}(\lambda))$ and

$$\sum_{\lambda \in \Lambda(\mathcal{T})} p(\lambda) = 1.$$

This probability structure can be thought of as follows. For each situation $v \in \mathcal{T}$ a random variable, $X(v)$, sits on it. The support of $X(v)$ is the set of edges emanating from v, equivalently its children, and for $e = (v, v') \in E(v)$ we define $\text{Prob}(X(v) = e) = \text{Prob}(X(v) = v') = \pi(e)$. These random variables are assumed defined over the common probability space formed by the set of atomic events, its power set and the probability defined above, and are assumed mutually independent.

Example 1. A simple probability tree is given in Fig. 6.1.

The set of situation is $S(\mathcal{T}) = \{v_i : 0 \leq i \leq 4\}$ and the random variable $X(v_i)$ is associated to the situation $v_i \in S(\mathcal{T})$, for $0 \leq i \leq 4$. The random variables $X(v_i)$, $0 \leq i \leq 4$, have state spaces of dimensions $3, 2, 3, 2, 2$, respectively. The primitive probabilities are π_i for $i = 1, \ldots, 12$. The eight root-to-leaf paths λ of $\mathcal{T}$ can be indexed by the index of their leaf. So for example $\lambda_{12} = (v_0, v_1, v_4, v_{12})$. The probabilities $p_i = \text{Prob}(\lambda_i)$ of these atomic events λ_i, $5 \leq i \leq 12$ are given by the following monomials in the primitive probabilities

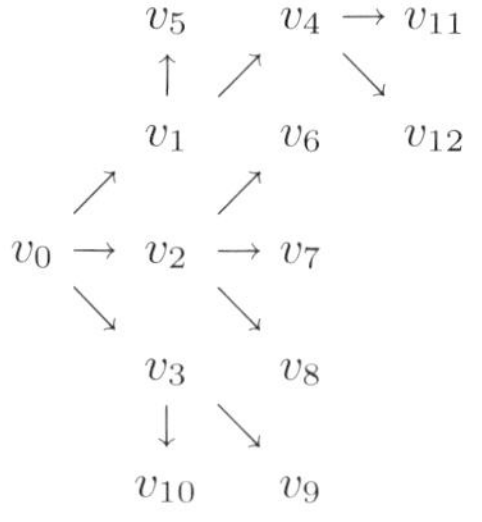

Fig. 6.1. A simple probability tree

$$\begin{aligned} &p_5 = \pi_1\pi_5, p_6 = \pi_2\pi_6, \quad p_7 = \pi_2\pi_7, \qquad p_8 = \pi_2\pi_8 \\ &p_9 = \pi_3\pi_9, p_{10} = \pi_3\pi_{10}, p_{11} = \pi_1\pi_4\pi_{11}, p_{12} = \pi_1\pi_4\pi_{12} \end{aligned} \tag{6.1}$$

under the linear constraints

$$\begin{aligned} \pi_1 + \pi_2 + \pi_3 &= 1 \\ \pi_4 + \pi_5 &= 1 \\ \pi_6 + \pi_7 + \pi_8 &= 1 \\ \pi_9 + \pi_{10} &= 1 \\ \pi_{11} + \pi_{12} &= 1 \end{aligned}$$

and the inequalities $0 \leq \pi_i \leq 1$, $i = 1, \ldots, 12$. □

If we ignore the inequality statements, a discrete probability model represented by a probability tree with N leaves can be thought of as N monomials in primitive probabilities that satisfy a collection of sum-to-one conditions.

Recall that a *discrete statistical model* is a collection of probability mass functions on a given support. We consider a collection of probability trees on a shared event tree $\mathcal{T}$ by allowing the primitive probabilities on $\mathcal{T}$ to be indeterminate but to satisfy a set of constraint equations. Algebraic types of constraints are common in the literature (see below). This provokes the following definition.

Definition 1. *An algebraically constrained tree (ACT) model is a triplet* $(\mathcal{T}, \Pi, C)$ *where*

1. $\mathcal{T} = (V(\mathcal{T}), E(\mathcal{T}))$ *is a tree;*
2. $\Pi = \{\pi(e) : e \in E(\mathcal{T})\}$ *is a set of (indeterminate) primitive probabilities; that is,* $\pi(e) \geq 0$ *and* $\sum_{e \in E(v)} \pi(e) = 1$ *for all* $v \in S(\mathcal{T})$ *and*
3. C *is a set of algebraic equations in the elements of* Π*.*

Example 2 (chain event graph models). These models, indicated with the acronym CEG, simply equate certain primitive probabilities. Specifically we say that two situations v_1 and v_2 in a probability tree T are in the same *stage* u if there is an invertible map $\Phi : E(v_1) \to E(v_2)$ such that, for all $e \in E(v_1)$

$$\pi(\Phi(e) = \pi(e)\,.$$

Such models where the probability edges emanating from a given situation are assumed to have the same probabilities are extremely common: see below. Here C is a set of linear equations equating certain primitive probabilities. For example in the tree in Fig. 6.1, we might have that all its situations lay in one of two stages $\{u_1, u_2\}$ where $u_1 = \{v_0, v_2\}$ and $u_2 = \{v_1, v_3, v_4\}$ and C is given by the following seven equations

$$\begin{aligned} &\pi_1 = \pi_6, \qquad \pi_2 = \pi_7, \qquad \pi_3 = \pi_8, \\ &\pi_4 = \pi_9 = \pi_{11}, \pi_5 = \pi_{10} = \pi_{12}\,. \end{aligned}$$

Substituting in (6.1) we obtain

$$p_5 = \pi_1\pi_5,\ p_6 = \pi_1\pi_2,\ \ p_7 = \pi_2^2, \qquad p_8 = \pi_2\pi_3,$$
$$p_9 = \pi_3\pi_4,\ p_{10} = \pi_3\pi_5,\ p_{11} = \pi_1\pi_4^2,\ p_{12} = \pi_1\pi_4\pi_5,$$

together with

$$\pi_1 + \pi_2 + \pi_3 = 1,$$
$$\pi_4 + \pi_5 = 1.$$

Example 3 (Bayesian networks). The geometry associated with the algebraic features of finite discrete Bayesian networks (BNs) has been vigorously studied (see for example Pistone et al. (2001), Garcia et al. (2005) and Geiger et al. (2006)). This class of models is a small subclass of CEG models. In the probability tree of a BN, root-to-leaf paths are all of the same length, valency of vertices the same distance from the root are all identical, and stages are only allowed to contain situations the same distance from the root vertex of the tree. The probability tree of the simplest, non-degenerate, Bayesian network $X_1 \to X_2 \to X_3$, where X_1, X_2, X_3 are all binary is given in Fig. 6.2.

In particular $X(v_0) = X_1$ and $X(v_1) = \{X_2|X_1 = 0\}$. If the atoms of the sample space for (X_1, X_2, X_3) are (x_1, x_2, x_3) where $x_i \in \{0, 1\}$ for $1 \le i \le 3$, then the eight atomic events of the tree in Fig. 6.2 are identified by

$$v_7 = (0,0,0),\ \ v_8 = (0,0,1),\ \ v_9 = (0,1,0),\ \ v_{10} = (0,1,1),$$
$$v_{11} = (1,0,0),\ v_{12} = (1,0,1),\ v_{13} = (1,1,0),\ v_{14} = (1,1,1).$$

Note that, for example, the edge (v_2, v_5) can be associated with the conditional event $\{X_2 = 0|X_1 = 1\}$. Since the Bayesian network is equivalent to specifying $\{X_3 \perp\!\!\!\perp X_1|X_2\}$ this model is equivalent to the saturated model given by the tree above together with the four linear constraints

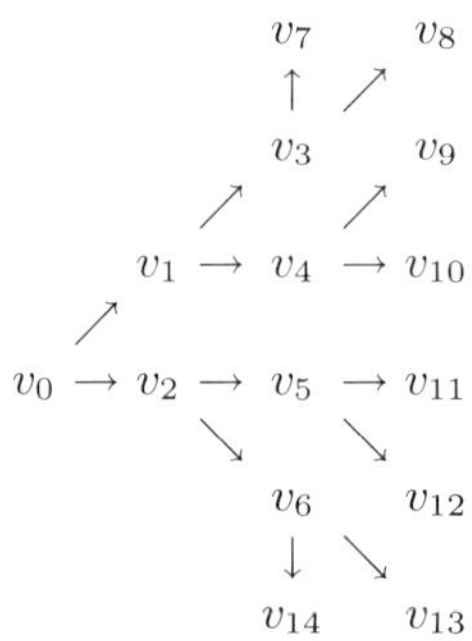

Fig. 6.2. Probability tree for the Bayesian network $X_1 \to X_2 \to X_3$

$$\text{Prob}(X_3 = x_3|X_1 = 0, X_2 = x_2) = \text{Prob}(X_3 = x_3|X_1 = 1, X_2 = x_2)$$

for $x_2, x_3 = 0, 1$ which in terms of primitive probabilities reduce to four linear equations

$$\pi_7 = \pi_{11}, \quad \pi_8 = \pi_{12}, \quad \pi_9 = \pi_{13}, \quad \pi_{10} = \pi_{14}.$$

Example 4 (generalizations of Bayesian networks). As the use of Bayesian networks increased it soon became apparent that there were circumstances when more symmetries could be demanded of a model than could be conveniently expressed as sets of conditional independence models. The stages of the corresponding CEGs where subsequently not only being determined by specific parent configurations but also unions of such classes: see for example Theisson et al. (1999). Furthermore the demand for the sample spaces associated with different configurations of parents to have the same dimension given their distance from the root vertex was also seen to be overly restrictive. Thus zeros in the primitives have been systematically modelled and structures built to efficiently process such information in these heterogeneous structures. Formally here we are simply adding further linear constraints $\{\pi(e) = 0 : e \in E^*(\mathcal{T}) \subset E(\mathcal{T})\}$ for some subset $E^*(\mathcal{T})$ of edges.

Example 5 (Bayesian linear constraint models). This class of models, first discussed in Riccomagno and Smith (2004), imposes on the saturated tree additional general linear constraints on the primitive probabilities. These arise quite naturally for example from constrained CEG models.

Example 6 (algebraic parametric models). These naturally arise in certain genetic models. They are CEG models such that the probability distributions indexed by the stages lie in a parametric family whose primitive probabilities can be expressed as an algebraic function of the hyper-parameters of the family. Examples of such families of distributions are the binomial and censored negative binomial. In the CEG model of Example 2 we might be able to assume that $X(v_0)$ has a binomial $Bi(2, \pi_0)$ distribution so that

$$\pi_1 = (1-\pi_0)^2, \pi_2 = 2\pi_0(1-\pi_0), \pi_3 - \pi_0^2$$

which, on substituting $\pi_5 = 1 - \pi_4$, gives

$$\begin{array}{lll} p_5 = (1-\pi_0)^2(1-\pi_4), & p_6 = 2\pi_0(1-\pi_0)^3, & p_7 = 4\pi_0^2(1-\pi_0)^2, \\ p_8 = 2\pi_0^3(1-\pi_0), & p_9 = \pi_0^2\pi_4, & p_{10} = \pi_0^2(1-\pi_4), \\ p_{11} = (1-\pi_0)^2\pi_4^2, & p_{12} = (1-\pi_0)^2\pi_4(1-\pi_4). & \end{array}$$

6.2 Manifest Probabilities and Solution Spaces

6.2.1 Introduction

A partition of $\Lambda(\mathcal{T}) = \{\Lambda_1, \ldots, \Lambda_M\}$ can be identified with a discrete random vector Y taking values $y_1, \ldots, y_M$ so that Λ_i corresponds to the event $\{Y = y_i\}$

for $i = 1, \ldots, M$. Assume that the probability mass function of Y is known, equivalently for $i = 1, \ldots, M$ the following values are known

$$q_i = \sum_{\lambda \in \Lambda_i} p(\lambda) = \sum_{\lambda \in \Lambda_i} \left(\prod_{j=1}^{n(\lambda)} \pi(e_j(\lambda)) \right) \in [0, 1],$$

which we can rewrite as

$$\sum_{\lambda \in \Lambda_i} \left(\prod_{j=1}^{n(\lambda)} \pi(e_j(\lambda)) \right) - q_j = 0. \tag{6.2}$$

The elements of the set $\boldsymbol{q} = \{q_1, \ldots, q_M\}$ are called *manifest probabilities.*

Definition 2. *1. The set $\boldsymbol{q}$ of manifest probabilities is* feasible *for the ACT model $(\mathcal{T}, \Pi, C)$ if there exist values of Π satisfying both (6.2) and the constraints in C. The set $I_{\boldsymbol{q}}$ of all such values is called the* inverse map *of $\boldsymbol{q}$.*

2. A function $\xi : \Pi \longrightarrow \mathbb{R}^n$ is said to be identified *by $\boldsymbol{q}$ in $(\mathcal{T}, \Pi, C)$ if, for all $\pi \in I_q$, $\xi(\pi)$ takes the same value, where $\mathbb{R}$ is the set of real number and n a positive integer number.*

Thus, $\boldsymbol{q}$ is feasible for $(\mathcal{T}, \Pi, C)$ if the systems of equations and inequalities obtained by adjoining those in Definition 1 defining the ACT model and those in (6.2) has a non-empty solution set. There are interesting problems connected with determining whether a set of manifest probabilities is feasible or a certain function is identified.

Example 7 (Cont. of Examples 3 and 6). Suppose we learn the probability $p_8 = \pi_1\pi_3\pi_8$ of the atomic event λ_8 in Example 6; that is, the set of manifest probabilities is $\boldsymbol{q} = \{p_8\}$. Then, since

$$p_8 = 2\pi_0^3(1 - \pi_0) \leq 27/128 \approx 0.21,$$

$\boldsymbol{q}$ is feasible only if $0 \leq p_8 \leq 0.21$.

Let $\widehat{\alpha}_1$ and $\widehat{\alpha}_2$ be the two solutions in π_0 of the equation

$$\pi_0^4 - \pi_0^3 + 0.5p_8 = 0.$$

Consider now the function $\xi = (\pi_0, \pi_4)$ with $\pi_0 \in \{\widehat{\alpha}_1, \widehat{\alpha}_2\}$ and $\pi_4 \in [0, 1]$. The model is therefore not identified. Now suppose we learn the values of p_8 and p_9. It is straightforward to check that the feasible region is $\{(p_1, \ldots, p_8) : 0 \leq p_9 \leq 4p_8, 0 \leq p_8 \leq 0.21\}$. The solution space is then

$$\{(\pi_0, \pi_4) : \pi_4 = 2p_9p_8^{-1}\pi_0(1 - \pi_0) \text{ where } \pi_0 = \widehat{\alpha}_1 \text{or } \widehat{\alpha}_2\}.$$

This is identifiable only when $\widehat{\alpha}_1 = \widehat{\alpha}_2$, that is only when $p_8 = 27/128$ and $\pi_0 = 3/4$. □

The solutions to the simple example above is typical and reflected in many common classes of models, see, for example, Settimi and Smith (2000) and Mond et al. (2003). When models have many arguments, usually it becomes necessary to use sophisticated elimination techniques or to derive characterization theorems to address identifiability and feasibility issues. Moreover, usually it is necessary in statistical models expressible as ACTs to address such question algebraically. For example, in a Bayesian network expressed as a chain event graph, the constraint equations associated with the known margins of a subset of the random variables on situations is usually not expressible as a single graphical constraint. To obtain a fuller description of the statistical model we need to append the Bayesian network with further algebraic equations. It is therefore more general and elegant to address issues of feasibility and identifiability within a model specification like the ACT which is explicitly algebraic.

6.3 Expressing Causal Effects Through Algebra

6.3.1 Causality Defined on Probability Trees

Suppose an uncontrolled (or *idle*) process respects a partially observed ACT model $(\mathcal{T}, \Pi, C^+(\boldsymbol{q}))$ with $C^+(\boldsymbol{q}) = C \cup C(\boldsymbol{q})$ where C denotes the original constraint equations on primitive probabilities as in Definition 1 and $C(\boldsymbol{q})$ is the set of equations generated by the observed marginals, our manifest probabilities as in (6.2). Suppose $\mathcal{T}$ is causal in the sense that the natural pre-ordering on its vertices is compatible with the order in which events occur in the modelled process.

A way to control (or *manipulate*) the system is by forcing a unit arriving at a situation v^* to pass along the edge $e^* = (v^*, v^{**})$. Let $V^* \subseteq S(\mathcal{T})$ be the set of such controlled situations and $E^* \subset E(\mathcal{T})$ the set of such edges. This gives rise to a new ACT model $(\mathcal{T}, \Pi^*, C^*)$ where $\mathcal{T}$ is like in the uncontrolled system and Π^* is defined in function of Π as follows. Let $e = (v^*, v) \in E(\mathcal{T})$ then

$$\pi^*(e) = \begin{cases} 1 & \text{if } e \in E^* \\ 0 & \text{if } v^* \in V^* \text{ and } e \notin E^* \\ \pi(e) & \text{otherwise.} \end{cases} \qquad (6.3)$$

That is, if v is not emanating from a controlled vertex then its probability remains unchanged, otherwise it takes value zero or one. This is equivalent to imposing a set of constraints to the system by setting some value in Π equal to zero or one. These constraints are part of C^*. Depending on what is more convenient we could work with a smaller set of primitives or a larger set of constrained equations. Note that the type of control described above can be seen as performing a sequence of projections of the simplex of probabilities $\{\pi(v^*, v) : v \in E(v^*)\}$, where v^* is a controlled situation, on the face $e^* = (v^*, v^{**}) \in E^*$.

On the basis of observing $\boldsymbol{q}$ in the idle system we are now interested in making predictions about the expectation of some quantity in the controlled system, after we perform the manipulation. In principle the set C^* could be different from the set $C^+(\boldsymbol{q})$ of original constraints and observational relationships. Some rules of consistency between C and C^* and the manipulations allowed on the system are formalized in Definitions 3 and 4. Definition 3 simply states that we consider manipulations of the type described above and that the manipulated system is still an ACT model. Definition 4 relates to the nature of the interaction between the idle system and the set of permitted manipulations. This interaction strongly depends on the underlying science and purpose of the modelling process.

Definition 3. *Let* $(\mathcal{T}, \Pi, C^+(\boldsymbol{q}))$ *be as above with* $C^+(\boldsymbol{q}) = C \cup C(\boldsymbol{q})$ *and* $(\mathcal{T}^*, \Pi^*, C^*)$ *be an ACT model. Then the map*

$$\kappa : \ (\mathcal{T}, \Pi, C^+(\boldsymbol{q})) \to (\mathcal{T}^*, \Pi^*, C^*)$$

is called causal *if* $\mathcal{T} = \mathcal{T}^*$ *and* Π^* *is defined in function of* Π *by (6.3).*

The context indicates what type of manipulations are permitted on the systems and which of the constraints in the idle system can be modified by the manipulation. For example, a physical law will hold whatever manipulation we attempt whilst model specific assumptions, like stages, might be chosen not to be respected by the enacted manipulation. Having decided on the form of the idle system, the manifest probabilities and the types of manipulation that may be enacted, the set of constraints C can be partitioned in *immutable* constraints C_i and *flexible* constraints C_f.

Definition 4. *A causal map is called* basic *if* $C_i \subseteq C^*$.

Clearly given a causal map κ we could have defined C_i^κ as the smallest polynomial ideal in the primitive probability which generates the same variety generated by $C \cap C^*$. This although certainly mathematically interesting will distract us from the focus of this chapter. We can summarize our original notion of flexible and immutable constraints by stating that flexible relationships are destroyed by a manipulation giving rise to a basic causal map whilst the immutable laws endure. Observational constraints are examples of flexible relationships. Pearl (1995) and others assume that causal maps are basic so that all relationships concerning probabilities associated with the manipulation no longer hold after the manipulation has taken place. We henceforth assume that all our causal maps are basic although the validity of this assumption is dependent on context. For further discussion of this and related issues see Riccomagno and Smith (2005). It is easily checked that when basic causal maps are always well defined.

To illustrate this construction consider a large random sample from a population whose evolution to an end point of, say, a disease is represented by one of the eight paths of the probability tree $\mathcal{T} = (V, E)$ in Fig. 6.1.

A treatment regime on a future sample from the same population is imposed. The treatment will take the following form. If a unit takes a path that arrives at a situation v^* in a set $V^* \subset V(\mathcal{T})$ then that individual will be forced to pass along just one edge $e^* \in E^*$ for some $E^* \subseteq E$. Let V^{**} be the set of children in edges in E^*.

Example 8. For the tree on Fig. 6.1 consider a manipulation operating on the vertex set $V^* = \{v_1, v_2\}$ with $E^* = \{(v_1, v_4), (v_2, v_6)\}$ and $V^{**} = \{v_3, v_6\}$. Let $[e_1, \ldots, e_{12}]$ be the order list of edge of $\mathcal{T}$ in Fig. 6.1. Under the stage assumptions made in Example 2 the edge probabilities are

$$\begin{aligned}\boldsymbol{\pi} &= [\pi_1, \pi_2, \pi_3, \pi_4, \pi_5, \pi_6, \pi_7, \pi_8, \pi_9, \pi_{10}, \pi_{11}, \pi_{12}] ,\\ &= [\pi_1, \pi_2, \pi_3, \pi_4, \pi_5, \pi_1, \pi_2, \pi_3, \pi_4, \pi_5, \pi_4, \pi_5] ,\end{aligned}$$

and under this manipulation they become

$$\begin{aligned}\boldsymbol{\pi}^* &= [\pi_1, \pi_2, \pi_3, 1, 0, 1, 0, 0, \pi_9, \pi_{10}, \pi_{11}, \pi_{12}] ,\\ &= [\pi_1, \pi_2, \pi_3, 1, 0, 1, 0, 0, \pi_4, \pi_5, \pi_4, \pi_5] .\end{aligned}$$

The new manipulated probabilities on the eight atomic events in (6.1) are therefore

$$\boldsymbol{p}^* = [0, \pi_2, 0, 0, \pi_3\pi_9, \pi_3\pi_{10}, \pi_1\pi_{11}, \pi_1\pi_{12}] . \tag{6.4}$$

□

The causal interpretation, given above, of the ACT of a probability tree signifies that we expect the treatment to have no influence on the processes that gave rise to the situations that happen before treatment (in our example for instance π_1 remains unchanged after manipulation). Furthermore it is assumed that the subsequent effect after having moved along an edge $e^* \in E^*$ will not be changed by the manipulation described above (for example π_{11} and π_{12} are unchanged). This latter assumption is substantive but often plausible. It demands that the manipulation is local in the sense that the only way the manipulation affects the system is where it is enacted and it has no further effect on later probability laws.

In general, the relationship between causal probabilities like $\boldsymbol{p}^*$ and conditional probabilities like $\boldsymbol{p}'$ is a complicated one as illustrated in Example 9. In a saturated model they are equal if and only if $\boldsymbol{V}^*$ is vertex set of a sub-tree of $\mathcal{T}$ rooted at v_0. Example 10 addresses two identification cases.

Example 9. Assume as in Example 6 $X(v_0)$ has a binomial $\mathrm{Bi}(2, \pi_0)$ distribution. Then substituting in (6.4) allows this to simplify further to

$$\boldsymbol{p}^* = [0, 2\pi_0(1-\pi_0), 0, 0, \pi_0^2\pi_4, \pi_0^2(1-\pi_4), (1-\pi_0)^2\pi_4, (1-\pi_0)^2(1-\pi_4)].$$

It is important to note that the manipulated probability vector $\boldsymbol{p}^*$ is *not* the same as the vector $\boldsymbol{p}'$ of probabilities conditional of the event composed by the set of atoms $\Lambda' = \{\lambda_6, \lambda_9, \lambda_{10}, \lambda_{11}, \lambda_{12}\}$ consistent with this manipulation. Indeed under the model of Example 6 we note that $\text{Prob}(\Lambda') = (1-\pi_0)^2(2\pi_0(1-\pi_0)+\pi_4)+\pi_0^2$ so that $\boldsymbol{p}' = (p_5, p_6, \ldots, p_{12})$ is given by

$$\boldsymbol{p}' = \frac{[0, 2\pi_0(1-\pi_0)^3, 0, 0, \pi_0^2\pi_4, \pi_0^2\pi_4^2(1-\pi_4), (1-\pi_0)^2, (1-\pi_0)^2\pi_4(1-\pi_4)]}{\text{Prob}(\Lambda')}$$

Example 10. Next, suppose the feature of interest is the probability $\xi_1(\Pi)$ of the event $\Lambda_I = \{\lambda_6, \lambda_9, \lambda_{11}\}$ – say, the event associated with the full recovery of the patient – after treatment. This probability is

$$\begin{aligned} p_6^* + p_9^* + p_{11}^* &= 2\pi_0(1-\pi_0) + \pi_0^2\pi_4 + (1-\pi_0)^2\pi_4 \\ &= \pi_4 + 2(1-\pi_4)\pi_0(1-\pi_0) \end{aligned}$$

whilst with no treatment it is

$$p_6 + p_9 + p_{11} = 2\pi_0(1-\pi_0)^3 + \pi_0^2\pi_4 + (1-\pi_0)^2\pi_4^2.$$

Now suppose we learn (p_8, p_9). Then substituting we see that whenever (p_8, p_9) is feasible

$$\xi_1(\Pi) = p_6^* + p_9^* + p_{11}^* = \pi_4^0(1 + (1-\pi_4^0)p_8p_9^{-1})$$

where $\pi_4^0 = 2p_9p_8^{-1}\pi_0(1-\pi_0)$ and where π_0 takes the values $\widehat{\alpha}_1$or $\widehat{\alpha}_2$. So although $\xi_1(\Pi)$ is not identified we know it can take only one of two possible values. So in particular this effect can be bounded above and below. Furthermore under a Bayesian analysis it will have an explicit posterior expectation depending on the weights on $\widehat{\alpha}_1$and $\widehat{\alpha}_2$ in the prior density of π_0.

Alternatively we might well be interested in the change in the probability of full recovery $\xi_2(\Pi)$ using the treatment rather than doing nothing. We have

$$\begin{aligned} \xi_2(\Pi) &= p_6^* + p_9^* + p_{11}^* - p_6 - p_9 - p_{11} \\ &= 2\pi_0(1-\pi_0)(1-(1-\pi_0)^2) + (1-\pi_0)^2(1-\pi_4^2) \\ &= (1-\pi_0)\{2\pi_0^2(2-\pi_0) + (1-\pi_0)(1-\pi_4^2)\}\,. \end{aligned}$$

Again we see that after learning (p_8, p_9) we still cannot identify $\xi(\Pi)$, but know it can take only one of two possible values

$$\{(\pi_0, \pi_4) : \pi_4 = 2p_9p_8^{-1}\pi_0(1-\pi_0) \text{ where } \pi_0 = \widehat{\alpha}_1 \text{ or } \widehat{\alpha}_2\}\,.$$

Note that in both these cases our object of interest is the solution space of an algebraic function. □

6.4 From Models to Causal ACTs to Analysis

6.4.1 The Unmanipulated System as a Probability Graph

As shown in Example 4 Bayesian networks are ACT models. To help compare these two model classes we develop in details an example which illustrates how certain causal statistical models can be analysed within these two frameworks. We are thus able to demonstrate within a practical framework why a Bayesian network approach is less expressive than its ACT counterpart whilst no easier to study algebraically.

Example 11 (lighting circuit). A domestic lighting system has a mains supply that can be tripped off or switched on or off. The mains supply is then cabled to room circuits which in turn can be tripped off or switched on or off. Consider the indicators

$$M = \begin{cases} 0 & \text{mains supply off} \\ 1 & \text{mains supply on} \end{cases}$$
$$H = \begin{cases} 0 & \text{hall circuit supply off} \\ 1 & \text{hall circuit supply on} \end{cases}$$
$$K = \begin{cases} 0 & \text{kitchen supply off} \\ 1 & \text{kitchen supply on} \end{cases}$$
$$L = \begin{cases} 0 & \text{light A in hall fails} \\ 1 & \text{light A in hall works} \end{cases}$$

There are two logical constraints in this system. Indeed if the mains are off then all power is off and if the hall circuit is off then A is off. Thus the atomic events of the event space are

$$\begin{aligned}
\lambda_1 &= \{M = 0\} = \{M = 0, H = 0, K = 0, L = 0\} \\
\lambda_2 &= \{M = 1, H = 0, K = 0\} = \{M = 1, H = 0, K = 0, L = 0\} \\
\lambda_3 &= \{M = 1, H = 0, K = 1\} = \{M = 1, H = 0, K = 1, L = 0\} \\
\lambda_4 &= \{M = 1, H = 1, K = 0, L = 0\} \\
\lambda_5 &= \{M = 1, H = 1, K = 0, L = 1\} \\
\lambda_6 &= \{M = 1, H = 1, K = 1, L = 0\} \\
\lambda_7 &= \{M = 1, H = 1, K = 1, L = 1\}
\end{aligned}$$

The sample space can be represented by the probability tree $\mathcal{T}_1$ in Fig. 6.3, where a leaf node has the same label as the corresponding root-to-leaf path.

The labelling of the edges on the paths corresponds to the sequence defining the path taken in the order (M, H, K, L). Thus, for example, $\lambda_6 = \{M = 1, H = 1, K = 1, L = 0\} = \{\rightarrow^1\rightarrow^1\rightarrow^1\rightarrow^0\}$. The situations are $S(\mathcal{T}_1) = \{v_i : 0 \leq 1 \leq 5\}$.

Six indeterminates suffice to describe the primitive probabilities. Indeed write $\pi_i = \text{Prob}(X(v_i) = 1)$, $\overline{\pi}_i = 1 - \pi_i$, $0 \leq i \leq 5$ and $\boldsymbol{\pi} = [\pi_0, \ldots, \pi_5]$. Let $\text{Prob}(\lambda_i) = p_i$, $1 \leq i \leq 7$ and let $\boldsymbol{p} = [p_1, p_2, \ldots, p_7]$. Then substituting the sum-to-one conditions gives

$$\boldsymbol{p} = [\overline{\pi}_0, \pi_0\overline{\pi}_1\overline{\pi}_2, \pi_0\overline{\pi}_1\pi_2, \pi_0\pi_1\overline{\pi}_3\overline{\pi}_4, \pi_0\pi_1\overline{\pi}_3\pi_4, \pi_0\pi_1\pi_3\overline{\pi}_5, \pi_0\pi_1\pi_3\pi_5]$$

where we also assume $\{0 < \pi_i < 1, 0 < \overline{\pi_i} < 1 : i = 0, \ldots, 5\}$. □

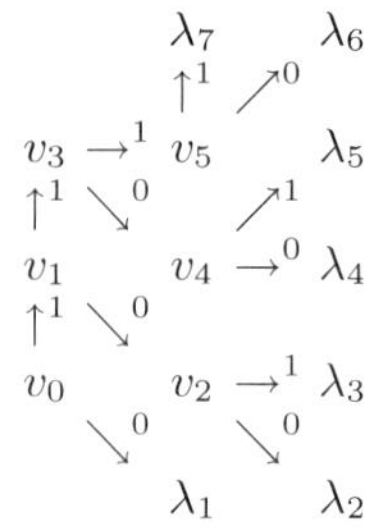

Fig. 6.3. A probability tree for the lighting circuit example

One simple way to restrict this model is to impose some conditional independence statements which can always be represented in terms of algebraic constraints as in Example 4. Thus suppose that we believe that

Item 1. Given the mains is on, the supplies in the kitchen and in the hall fail independently and

Item 2. Given the mains and hall lights are on, the kitchen supply and light in hall fail independently.

It is easily proved, by setting to zero suitable determinants of 2×2 matrices, that these two conditional independence assumptions are equivalent to the set of constraints $C = \{\pi_5 = \pi_4, \pi_3 = \pi_2\}$. Thus we can think of the observed probability vector $\boldsymbol{p}$ as a point in the variety described by setting to zero the components of $\boldsymbol{p}$ below

$$\boldsymbol{p} = [\overline{\pi}_0, \pi_0\overline{\pi}_1\overline{\pi}_2, \pi_0\overline{\pi}_1\pi_2, \pi_0\pi_1\overline{\pi}_2\overline{\pi}_4, \pi_0\pi_1\overline{\pi}_2\pi_4, \pi_0\pi_1\pi_2\overline{\pi}_4, \pi_0\pi_1\pi_2\pi_4] \qquad (6.5)$$

parametrized by the four parameters $\pi_0, \pi_1, \pi_2, \pi_4 \in (0, 1)$.

Computational commutative algebra provides a way to determine the corresponding variety in the $\boldsymbol{\pi}$ variable set. That is the set of polynomial relationships that the $\boldsymbol{\pi}$ indeterminates should satisfy to belong to the conditional model defined in (6.5). To do this, we compute the elimination ideal of the $\boldsymbol{\pi}$ indeterminates for the ideal generated by the seven polynomials associated with the system in (6.5), namely

$$\begin{array}{l} p_1 - (1 - \pi_0) \\ p_2 - \pi_0(1 - \pi_1)(1 - \pi_2) \\ p_3 - \pi_0(1 - \pi_1)\pi_2 \\ p_4 - \pi_0\pi_1(1 - \pi_2)(1 - \pi_4) \\ p_5 - \pi_0\pi_1(1 - \pi_2)\pi_4 \\ p_6 - \pi_0\pi_1\pi_2(1 - \pi_4) \\ p_7 - \pi_0\pi_1\pi_2\pi_4 \end{array} \qquad (6.6)$$

The actual computation of the elimination ideal is given in Appendix. The final result together its probabilistic interpretation is given by the four

Table 6.1. The light circuit system expressed in the π parameters

$p_4p_7 - p_5p_6$	gives the independence statement $K \perp\!\!\!\perp L \mid (M = 1, H = 1)$
$p_3p_4 - p_2p_6$	gives the independence statement $K \perp\!\!\!\perp H \mid (M = 1, L = 0)$
$p_2p_7 - p_3p_5$	gives the independence statement $K \perp\!\!\!\perp H \mid (M = 1, L = 1)$
$\sum_{i=1}^{7} p_i - 1$	the sum to one condition on the π parametrization is transferred onto the $\boldsymbol{p}$ parametrization

polynomials in Table 6.1. Note that the third polynomial is simply a consequence of one of our logical constraints.

6.4.2 A Bayesian Network for the Lighting Circuit Example

A more conventional representation of a system with conditional independence constraints is to use a Bayesian network. Usually the BN is constructed starting from the conditional independence hypotheses on the variables defining the process, here $\{M, H, K, L\}$. We might reasonably assert that

1. $H \sqcup K \mid M$. Thus if the mains is off then we ask that H must be off through our logical constraint whilst if $M = 1$ then this is the conditional independence statement in Item 1. of Sect. 6.4.1.
2. $L \perp\!\!\!\perp K \mid H, M$. Thus we know by our logical constraint that the lighting circuit is off with certainty (and so independently of K) unless both the mains and the hall supply are functioning. Furthermore when $M = 1 = H$ we are in the condition of Item 2. of the Sect. 6.4.1.
3. $L \sqcup M \mid H$. This is trivially satisfied since L can only be on if the hall supply is on.

These lead to the graph $\mathcal{G}$

$$\begin{array}{ccccc} & & K & & \\ & \nearrow & & & \\ M & \rightarrow & H & \rightarrow & L \end{array}$$

whose implied factorization, for $m, h, k, l \in \{0, 1\}$, is given by

$$\begin{aligned} &\text{Prob}(M = m, H = h, K = k, L = l) = \\ &\qquad \text{Prob}(M = m)\,\text{Prob}(H = h \mid M = m) \\ &\qquad \text{Prob}(K = k \mid M = m)\,\text{Prob}(L = l \mid H = h)\,. \end{aligned} \tag{6.7}$$

This implies that the system can be defined by the seven probabilities listed below and indicated by the θ letter. Some are equal to some π variables

$$\begin{aligned} \pi_1 &= \theta_M = \text{Prob}(M=1), \\ \pi_2 &= \theta_{K|M} = \text{Prob}(K=1|M=1), \\ &\theta_{K|\overline{M}} = \text{Prob}(K=1|M=0), \\ \pi_3 &= \theta_{H|M} = \text{Prob}(H=1|M=1), \\ &\theta_{H|\overline{M}} = \text{Prob}(H=1|M=0), \\ \pi_4 &= \theta_{L|H} = \text{Prob}(L=1|H=1), \\ &\theta_{L|\overline{H}} = \text{Prob}(L=1|H=0)\,. \end{aligned} \tag{6.8}$$

However the graph $\mathcal{G}$ is not a full specification of our problem since our logical constraints force further constraints on the space: namely that

$$\theta_{K|\overline{M}} = \theta_{H|\overline{M}} = \theta_{L|\overline{H}} = 0.$$

Adjoining this further constraints gives us the identical model space obtained more directly from the ACT. Note that because of the logical constraints the elicitation of this model structure was more involved for the Bayesian network. In particular, the known constraints were used not only directly at the end of the elicitation but also earlier when the conditional independence statements were elicited. It is common for causally induced logical constraints to seriously complicate Bayesian networks: so much so that Pearl (2000) chooses to preclude them. In contrast, the probability tree framework accommodates these constraints automatically. Note that in this example, because of this identification, the algebraic coding of the two models is also identical.

6.4.3 Expressiveness of Causal Trees Networks

Assume that the system is so insensitive it cannot trip. By this we mean that a light could not cause a circuit to fail, and a failure of the kitchen or hall subsystems could not cause the mains to fail. Call this Model 1. Then we could plausibly define a manipulation such that $V^* = \{v_1\}$ and $E^* = \{(v_1, v_2)\}$. The vector of manipulated probabilities becomes

$$\boldsymbol{p}^* = [\overline{\pi}_0, \pi_0\overline{\pi}_2, \pi_0\pi_2, 0, 0, 0, 0].$$

Under assumptions like those of Model 1 it might be argued that the Bayesian network $\mathcal{G}$ is also causal, see Pearl (2000). This would mean that the projection formula as applied to the Bayesian network would hold for any of its four variables when each is manipulated to either of its two levels. This of course implies that the formula in (6.7) of the Bayesian network $\mathcal{G}$ will hold in particular for the manipulation turning the hall system off. Under these assumptions we can compute the effect of this intervention by imposing

$$1 = \text{Prob}(E_H) = p_1 + p_2 + p_3$$

where $E_H = \lambda_1 + \lambda_2 + \lambda_3$ is the event "the hall circuit is off", and by adjoining the polynomial $p_1 + p_2 + p_3 - 1$ to the ideal generator set in (6.8). The reduced

Gröbner basis with respect to the lexicographic term ordering with initial ordering $p_7 < \cdots < p_1 < \theta_{L|H} < \theta_{K|M} < \theta_{H|M} < \theta_M$ is

$$\begin{aligned}
&p_4\,,\ p_5\,,\ p_6\,,\ p_7\,,\\
&p_1 + p_2 + p_3 - 1\,,\\
&\theta_M - p_2 - p_3\,,\\
&\sum_{i=1}^{7} p_i - 1\,.
\end{aligned}$$

The first four polynomials correctly set to zero the post intervention probability of events for which $H = 0$, that is in the complement event of E_H. The fifth polynomial is the rewriting of the sum to one condition with respect to the chosen `lex` ordering. The sixth and last polynomials retrieve the fact that the main and the kitchen circuit are not effected by the intervention on H. By the remaining polynomials we deduce that if $\theta_{H|M}$ is not zero then $p_2 = p_3 = 0$, which is possible only if $p_1 = 1$, that is the mains are off.

However note how in this context the causal assumption on the Bayesian network is ambiguous. In particular what does it mean to manipulate the hall circuit on, when the mains is down? This manipulation is logically precluded by the logical constraints on the model in Sect. 6.4.1! This illustrates just how restrictive the conditions for a causal Bayesian network can be and demonstrates the need for looser but analogous definitions based on trees of which ACT models are an example.

6.5 Equivalent Causal ACTs

Suppose we are solely interested in extending the probabilistic model in Model 1 in Sect. 6.4.3 so that the corresponding probability tree is causal with respect to the single manipulation of turning the hall circuit off i.e. setting $H = 0$. The probability tree $\mathcal{T}_2$ for this problem is at least as simple as $\mathcal{T}_1$ in the sense that all its root-to-leaf paths are no longer than those in $\mathcal{T}_1$ and the corresponding simplicial projection still gives us the required causal polynomials easily. This tree induces an equivalent but different parametrization of both the original probability space and its causal extension. It is given in Fig. 6.4.

Example 12. In $\mathcal{T}_2$ the variables are listed in the order (M, K) together then H and then L. For example, $(1, 1)$ stands for $M = K = 1$ and $(1, 0, 1)$ for $M = 1$, $K = 0$ and $H = 1$

The corresponding factorization is

$$\begin{aligned}
\mathrm{Prob}(M = m, H - h, K = k, L = l) &= \mathrm{Prob}(M = m, K = k)\\
\mathrm{Prob}(H = h|M = m, K = k)\,&\mathrm{Prob}(L = l|M = m, H = h, K = k)\,,
\end{aligned}$$

for $m, h, k, l \in \{0, 1\}$. □

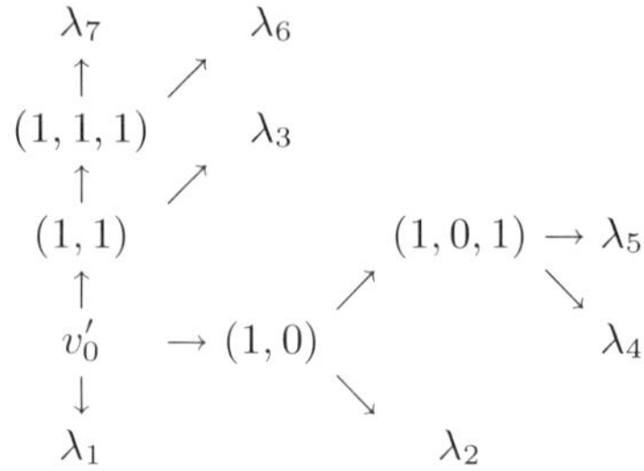

Fig. 6.4. An alternative ordering for the light circuit example

Seven parameters suffice to describe the primitive probabilities. They fall into three simplices: parameters upstream of the proposed manipulation, those associated with the manipulation itself (the reference) and the parameters associated with events downstream of the manipulation

$$\begin{aligned} &\text{Upstream parameters} \\ \psi^u_{00} &= \text{Prob}(M = 0 = K) \\ \psi^u_{10} &= \text{Prob}(M = 1, K = 0) \\ \psi^u_{11} &= \text{Prob}(M = 1 = K) \end{aligned}$$

$$\begin{aligned} &\text{Reference set parameters} \\ \psi^h_{0|10} &= \text{Prob}(H = 0|M = 1, K = 0) \\ \psi^h_{0|11} &= \text{Prob}(H = 0|M = 1 = K) \end{aligned}$$

$$\begin{aligned} &\text{Downstream parameters} \\ \psi^d_{0|110} &= \text{Prob}(L = 0|M = 1 = H, K = 0) \\ \psi^d_{0|111} &= \text{Prob}(L = 0|M = H = L = 1) \end{aligned}$$

Thus the root-to-leaf paths' probability become

$$\begin{aligned} p_1 &= \psi^u_{00}\psi^h_{0|00}\,, \\ p_2 &= \psi^u_{10}\psi^h_{0|10}\psi^d_{0|100}\,, \\ p_3 &= (1 - \psi^u_{00} - \psi^u_{10})\psi^h_{0|11}\psi^d_{0|101}\,, \\ p_4 &= \psi^u_{10}(1 - \psi^h_{0|10})\psi^d_{0|110}\,, \\ p_5 &= \psi^u_{10}(1 - \psi^h_{0|10})(1 - \psi^d_{0|110})\,, \\ p_6 &= (1 - \psi^u_{00} - \psi^u_{10})(1 - \psi^h_{0|11})\psi^d_{0|111}\,, \\ p_7 &= (1 - \psi^u_{00} - \psi^u_{10})(1 - \psi^h_{0|11})(1 - \psi^d_{0|111})\,. \end{aligned} \tag{6.9}$$

The ψ indeterminates can be written explicitly as functions of $\boldsymbol{p} = [p_1, \dots, p_7]$ by simple application of the definition of conditioning or by computing the

Gröbner basis of the ideal corresponding to the polynomial system in (6.9) with respect to the lexicographic ordering with initial ordering $p_7 \prec \cdots \prec p_1 \prec \psi_7 \prec \cdots \prec \psi_1$. This computation returns the set of polynomials

$$\begin{aligned}
&p_6 + p_7 - 1 + p_1 + p_5 + p_4 + p_2 + p_3\,,\\
&\psi^d_{0|111} p_6 + \psi^d_{0|111} p_7 - p_6\,,\\
&-p_4 + \psi^d_{0|110} p_5 + \psi^d_{0|110} p_4\,,\\
&-p_3 + \psi^h_{0|11} p_6 + \psi^h_{0|11} p_7 + h_{1,1}\, p_3\,,\\
&-p_2 + \psi^h_{0|10} p_5 + \psi^h_{0|10} p_4 + \psi^h_{0|10} p_2\,,\\
&-p_5 + \psi^u_{10} - p_4 - p_2\,,\\
&\psi^u_{00} + p_2 + p_3 + p_4 + p_5 + p_6 + p_7 - 1\,.
\end{aligned}$$

The first polynomial is the sum-to-one condition, the second and third ones return the downstream parameters, the next two polynomials return the reference parameters and finally the downstream parameters.

Example 13. Intervention forcing $H = 0$ again corresponds to setting two conditions $\psi^h_{0|10} = 1$ and $\psi^h_{0|11} = 1$. A Gröbner basis of the ideal obtained by adjoining the polynomials $\psi^h_{0|10} - 1$ and $\psi^h_{0|11} - 1$ to the ideal corresponding to the polynomial system in (6.9) is

$$\begin{aligned}
&p_4, p_5, p_6, p_7\,,\\
&p_1 + p_2 + p_3 - 1\,,\\
&\psi^h_{0|10} - 1, \psi^h_{0|11} - 1\,,\\
&\psi^u_{10} - p_2 \text{ and } \psi^u_{00} - 1 + p_2 + p_3\,.
\end{aligned}$$

The point made in this section is that the representation of a probability space and its causal extension with respect to a class of manipulations is not usually unique. However they will always agree at least on the interior of their definition space. Which tree is most appropriate will be largely determined by the context. The causal extension of deeper trees, like $\mathcal{T}_1$, will permit expression of more causal hypotheses but be less compact algebraically. Thus $\mathcal{T}_2$ has paths of length no greater than 3 but is incapable of expressing the manipulation of the kitchen circuit. In contrast, $\mathcal{T}_1$ can express this manipulation as a simplicial projection but has paths of length 4 and so it has a more complex structure. As a general principle we would advocate the encoding of a problem in terms of probability trees with minimal maximum path length sufficient to express all proposed manipulations.

6.5.1 Trees Causation and Causal Orders

Much of the early discussion in the literature on graphical expressions of causality was based around Bayesian networks. Since in a Bayesian network any order of how situations happen is subsumed under collections of conditional independence statements over variables, causal structures have largely been defined through these cross-sectional features. However one of the few transparent properties one might demand about a causal structure is that a cause should happen *before* an effect. If causal statistical models are based on a single Bayesian network over measurement variables then the contexts to which such models apply are severely restricted: precluding for example causal mechanisms that have the potential to be expressed in both directions. For example in our context we might know that the light failing may trip the hall circuit. So the light failing can cause the hall circuit to trip and the hall circuit failing definitely causes the light to fail.

A solution to this conundrum is simply to focus on describing the system in terms of a probability tree with an associated path event space that is detailed enough to separate different ways in which things can happen as it might pertain to the potential manipulations. The difficulty with the probability tree and its associated Bayesian network, we have discussed above, is that it implicitly models only the values of the pair of events (H, L), not their timing. This will automatically confound "H causing L" with "L causing H" if we are in a world where both are possibilities. But it may well be possible to gather information after observing that both failed to help determine whether on observing $(H, L) = (1, 1)$ the hall system went first or the light caused the hall system to fail. For example if the light bulb is still intact then it must be off because of a system failure whilst if it has blown then it may well have caused the hall system to trip. The semantics of the tree can be rich enough to accommodate this type of information when and if it arrives.

It is not hard to draw a new probability tree that expresses not only what happened but also in which order situations happened. The atomic events of the original tree form a partition of the atomic events of the new richer tree. Call the associated ACT model Model 2. The study of the effects of a given manipulation, as it relates to the behaviour of an unmanipulated system, can then be addressed even for a model where causality can be expressed in both directions.

Example 14. Let $\overline{HK}$ denote the event that "if the hall system fails then the kitchen system fails and the mains doesn't fail". The fact that the light fails is implicit through our logical constraint. Let $\emptyset$ denote no failures. Then, taking account of the logical constraints in the system, the paths of the probability tree giving the corresponding probability tree are

$$\begin{array}{llll}
\varrho_1 = \overline{M}, & \varrho_2 = \overline{HM}, & \varrho_3 = \overline{HKM}, & \varrho_4 = \overline{HK}, \\
\varrho_5 = \overline{H}, & \varrho_6 = \overline{L}, & \varrho_7 = \overline{LHM}, & \varrho_8 = \overline{LHKM}, \\
\varrho_9 = \overline{LHK}, & \varrho_{10} = \overline{LH}, & \varrho_{11} = \overline{LKM}, & \varrho_{12} = \overline{LKHM}, \\
\varrho_{13} = \overline{LKH}, & \varrho_{14} = \overline{LK}, & \varrho_{15} = \overline{KM}, & \varrho_{16} = \overline{KHM}, \\
\varrho_{17} = \overline{KH}, & \varrho_{18} = \overline{KLHM}, & \varrho_{19} = \overline{KLH}, & \varrho_{20} = \overline{KL}, \\
\varrho_{21} = \overline{K}, & \varrho_{22} = \overline{KLM}, & \varrho_{23} = \emptyset\,. &
\end{array}$$

Note that

$$\begin{aligned}
&\lambda_1 = \{\varrho_1, \varrho_2, \varrho_3, \varrho_7, \varrho_8, \varrho_{11}, \varrho_{12}, \varrho_{15}, \varrho_{16}, \varrho_{18}, \varrho_{22}\}\,,\\
&\lambda_2 = \{\varrho_4, \varrho_9, \varrho_{13}, \varrho_{17}, \varrho_{19}\},\ \lambda_3 = \{\varrho_5, \varrho_{10}\},\ \lambda_4 = \{\varrho_{14}, \varrho_{20}\}\,,\\
&\lambda_5 = \varrho_{21},\ \lambda_6 = \varrho_6,\ \lambda_7 = \varrho_{23}\,.
\end{aligned}$$

Write $\boldsymbol{p} = [p_1, p_2, \ldots, p_{24}]$ where $p_i = \text{Prob}(\varrho_i)$, $1 \le i \le 24$. Let for example $\pi^{\overline{L},\overline{K}}_{\overline{H}}$ denote the probability that the hall system fails given the light and then the kitchen circuit failed. Also write $\overline{\pi} = 1 - \pi$, and let $\emptyset$ subscript no subsequent failure if such a failure is logically possible. Then the new paths on the probability tree can be given in terms of its edge probabilities as below:

$$\begin{array}{lll}
p_1 = \pi_{\overline{M}} & p_2 = \pi_{\overline{H}}\pi^{\overline{H}}_{\overline{M}} & p_3 = \pi_{\overline{H}}\pi^{\overline{H}}_{\overline{K}}\pi^{\overline{HK}}_{\overline{M}}\\
p_4 = \pi_{\overline{H}}\pi^{\overline{H}}_{\overline{K}}\pi^{\overline{HK}}_{\emptyset} & p_5 = \pi_{\overline{H}}\pi^{\overline{H}}_{\emptyset} & p_6 = \pi_{\overline{L}}\pi^{\overline{L}}_{\emptyset}\\
p_7 = \pi_{\overline{L}}\pi^{\overline{L}}_{\overline{H}}\pi^{\overline{LH}}_{\overline{M}} & p_8 = \pi_{\overline{L}}\pi^{\overline{L}}_{\overline{H}}\pi^{\overline{LH}}_{\overline{H}}\pi^{\overline{LHK}}_{\overline{M}} & p_9 = \pi_{\overline{L}}\pi^{\overline{L}}_{\overline{H}}\pi^{\overline{LHK}}_{\emptyset}\\
p_{10} = \pi_{\overline{L}}\pi^{\overline{L}}_{\overline{H}}\pi^{\overline{LH}}_{\emptyset} & p_{11} = \pi_{\overline{L}}\pi^{\overline{L}}_{\overline{K}}\pi^{\overline{LK}}_{\overline{M}} & p_{12} = \pi_{\overline{L}}\pi^{\overline{L}}_{\overline{K}}\pi^{\overline{LK}}_{\overline{H}}\pi^{\overline{LKH}}_{\overline{M}}\\
p_{13} = \pi_{\overline{L}}\pi^{\overline{L}}_{\overline{K}}\pi^{\overline{LK}}_{\overline{H}}\pi^{\overline{LKH}}_{\emptyset} & p_{14} = \pi_{\overline{L}}\pi^{\overline{L}}_{\overline{K}}\pi^{\overline{LK}}_{\emptyset} & p_{15} = \pi_{\overline{K}}\pi^{\overline{K}}_{\overline{M}}\\
p_{16} = \pi_{\overline{K}}\pi^{\overline{K}}_{\overline{H}}\pi^{\overline{KH}}_{\overline{M}} & p_{17} = \pi_{\overline{K}}\pi^{\overline{K}}_{\overline{H}}\pi^{\overline{KH}}_{\emptyset} & p_{18} = \pi_{\overline{K}}\pi^{\overline{K}}_{\overline{L}}\pi^{\overline{KL}}_{\overline{H}}\pi^{\overline{KLH}}_{\overline{M}}\\
p_{19} = \pi_{\overline{K}}\pi^{\overline{K}}_{\overline{L}}\pi^{\overline{KL}}_{\overline{H}}\pi^{\overline{KLH}}_{\emptyset} & p_{20} = \pi_{\overline{K}}\pi^{\overline{K}}_{\overline{L}}\pi^{\overline{KL}}_{\emptyset} & p_{21} = \pi_{\overline{K}}\pi^{\overline{K}}_{\emptyset}\\
p_{22} = \pi_{\overline{K}}\pi^{\overline{K}}_{\overline{L}}\pi^{\overline{KL}}_{\overline{M}} & p_{23} = \pi_{\emptyset} &
\end{array}$$

The semantics of the tree in Example 14 are much richer than in the previous models of the lighting example. Various algebraic identities can be introduced to represent a more specific statistical model. The new tree also helps us to unpick exactly what we mean by manipulating the hall light off. A simple interpretation of this is that we are interested in the manipulation of the hall light off first. This would correspond to manipulating the root vertex, v_0, of this tree, with emanating edges $(\overline{M}, \overline{H}, \overline{K}, \overline{L}, \emptyset)$ with probabilities $(\pi_{\overline{M}}, \pi_{\overline{H}}, \pi_{\overline{K}}, \pi_{\overline{L}}, \pi_{\emptyset})$ to $(0, 1, 0, 0, 0)$. This projection allows us to construct the function associated with the total score and examine which functions on this richer tree allow this to be identified.

This shows that a tree rather than a Bayesian network is a natural and flexible framework with which to express the more complicated bidirectional causal structures whilst still coding the problem in a way amenable to algebraic analysis.

6.6 Conclusions

We believe that the natural framework within which causal manipulations of finite discrete models can be studied is the class of ACTs on a probability tree. These encompass a much wider class of models than can be expressed through competing graphical frameworks. They are closed to the type of information assumed to be present to embellish a statistical model. Furthermore their parametrization in terms of conditional probabilities is uniquely suited to this domain, where functions of interest, like the total cause, can be expressed as functions of projections into this space.

Of course many problems still need to be solved. First, the nature of the definition of causal functions of the type discussed in this chapter is contentious and their domain of appropriateness not yet fully understood. Second, the use of techniques in computer algebra to solve problems of significant size in this domain is a serious challenge because the space of polynomials of interest can be huge. There appears to be a strong need both for more specificity in the class of models considered, and a better understanding of the underlying geometry of these manipulation functions with specific important subclasses of ACTs. However, we believe that these issues can be addressed within this model class and that an understanding of the geometry of these algebraic systems could contribute greatly to the study of more general forms of manipulated discrete statistical models.

Appendix: Maple Code

Here we show how to perform the computation in Sect. 6.4.1 using the comprehensive computer system for advanced mathematics, Maple (Char et al., 1991). Maple, version 6, has two packages to perform Gröbner basis computations, `Groebner` and `grobner`, with `Groebner` being the newest. A package is called from the Maple command line with the command `with`, e.g.

```
> with(Groebner);
```

The list of indeterminates for Example 11 is entered below using the Maple command `seq`

```
> AtomsProb :=seq(p[i], i=1..7); AT:=AtomsProb :
```

$$AtomsProb := p_1, p_2, p_3, p_4, p_5, p_6, p_7$$

```
> PrimitiveProb :=seq(pi[i], i=0..5);
> PrimitiveProb ;
```

$$PrimitiveProb := \pi_0, \pi_1, \pi_2, \pi_3, \pi_4, \pi_5$$

The model we consider is the CEG model described in Example 11 with the two conditions expressed in Items 1. and 2. of Sect. 6.4.1. To input the polynomial ideal it is sufficient to enter the list of its generators in (6.6)

```
> IdealTree:=p[1]-(1-pi[0]),  p[2]-pi[0]*(1-pi[1])*(1-pi[2]),
p[3]-pi[0]*(1-pi[1])*pi[2],
p[4]-pi[0]*pi[1]*(1-pi[2])*(1-pi[4]),
p[5]-pi[0]*pi[1]*(1-pi[2])*pi[4],
p[6]-pi[0]*pi[1]*pi[2]*(1-pi[4]), p[7]-pi[0]*pi[1]*pi[2]*pi[4];
```

To determine the restrictions imposed on the $\boldsymbol{p}$ by the assumed model we run command to compute the relevant Gröbner basis

```
> GB:=gbasis( {IdealTree}, lexdeg([PP],[AP]));
```

and obtain

$$p_6 + p_4 + p_2 + p_3 + p_1 - 1 + p_5 + p_7$$
$$-p_7\, p_4 + p_5\, p_6$$
$$-p_7\, p_2 + p_3\, p_5$$
$$p_3\, p_4 - p_2\, p_6$$
$$\pi_4\, p_6 + \pi_4\, p_7 - p_7$$
$$-p_5 + \pi_4\, p_4 + \pi_4\, p_5$$
$$p_7\, \pi_2 - p_7 + \pi_2\, p_5$$
$$\pi_2\, p_4 + \pi_2\, p_6 - p_6$$
$$\pi_2\, p_2 + \pi_2\, p_3 - p_3$$
$$p_3\, \pi_1 + \pi_1\, p_6 + \pi_1\, p_7 - p_6 - p_7$$
$$\pi_1\, p_2 + \pi_1\, p_4 + \pi_1\, p_5 - p_4 - p_5$$
$$\pi_0 - p_2 - p_3 - p_4 - p_5 - p_6 - p_7$$

To study the effect of the intervention $H = 0$, we adjoin the polynomial $p_1 + p_2 + p_3 - 1$ to the ideal `IdealTree`

```
> H:=p1+p2+p3-1;
```

and perform the same type of Gröbner basis computation

```
> HGB:=gbasis( {H,IdealTree}, lexdeg([PP],[AP]));
```

obtaining

$$p_7$$
$$p_6$$
$$p_5$$
$$p_4$$
$$p_1 + p_2 + p_3 - 1$$
$$\pi_0 - p_2 - p_3$$
$$\pi_1\, p_3$$
$$p_2\, \pi_1$$
$$\pi_2\, p_2 - p_3 + \pi_2\, p_3$$

References

Char, B., Geddes, K., Gonnet, G., Leong, B., and Monogan, M. (1991). *MAPLE V Library Reference Manual.* Springer-Verlag, New York.

Garcia, L.D., Stillman, M., and Sturmfels, B. (2005). Algebraic geometry of Bayesian networks. *Journal of Symbolic Computation*, **39**(3-4), 331–355.

Geiger, D., Meek, C., and Sturmfels, B. (2006). On the toric algebra of graphical models. *The Annals of Statistics*, **34**, 1463–1492.

Mond, D., Smith, J., and van Straten, D. (2003). Stochastic factorizations, sandwiched simplices and the topology of the space of explanations. *Royal Society of London. Proceedings Series. A Mathematical, Physical and Engineering Sciences*, **459**(2039), 2821–2845.

Pearl, J. (1995). Causal diagrams for empirical research. *Biometrika*, **82**, 669–710.

Pearl, J. (2000). *Causality. Models, Reasoning, and Inference.* Cambridge University Press, Cambridge.

Pistone, G., Riccomagno, E., and Wynn, H.P. (2001). *Algebraic Statistics.* Chapman & Hall/CRC, Boca Raton.

Riccomagno, E. and Smith, J. (2004). Identifying a cause in models which are not simple bayesian networks. In *Proceedings of the International Conference on "Information Processing and Management of Uncertainty in knowledge-based systems"*, pages 1345–1322. IPMU, Perugia.

Riccomagno, E. and Smith, J.Q. (2005). The causal manipulation and bayesian estimation of chain event graphs. Technical report, CRiSM Paper. Centre for Research in Statistical Methodology, Warwick.

Settimi, R. and Smith, J.Q. (2000). Geometry, moments and conditional independence trees with hidden variables. *The Annals of Statistics*, **28**(4), 1179–1205.

Theisson, B., Meek, C., Chickering, D., and Heckerman, D. (1999). Computarionally efficient methods for selecting among mixtures of graphical models (with discussion). In J.M. Bernardo, J.O. Berger, A.P. Dawid, and A. Smith, editors, *Bayesian Statistics 6*, pages 631–656. Oxford Science Publications, Oxford.

7

Bayes Nets of Time Series: Stochastic Realizations and Projections

P.E. Caines, R. Deardon and H.P. Wynn

Summary. Graphical models in which every node holds a time-series are developed using special conditions from static multivariate Gaussian processes, particularly the notion of lattice conditional independence (LCI), due to Anderson and Perlman (1993). Under certain "feedback free" conditions, LCI imposes a special zero structure on the state space representation of processes which have a stochastic realisation. This structure comes directly from the transitive directed acyclic graph (TDAG) which is in one-to-one correspondence with the Boolean Hilbert lattice of the LCO formulation. Simple $AR(1)$ examples are presented.

7.1 Bayes Nets and Projections

Consider an undirected graphical model in which every node i is associated to a univariate Gaussian random variable X_i. The graph "holds" the conditional independence structure in the following way (see Lauritzen (1996)). Let the graph be $\mathcal{G}(E, V)$ where E and V represent, respectively, the edges and vertices and $|V| = d$, which we call the dimension of $\mathcal{G}$. Thus, associate an X_i with node i, $(i = 1, \ldots, d)$ and assume, for simplicity that $X = (X_1, \ldots, X_d)^{\mathrm{T}}$ (considered as a column vector) follows a multivariate normal distribution: $X \sim N(\mu, \Sigma)$. We shall assume that the $d \times d$ covariance matrix Σ has full rank d. For the algebraic results we shall be discussing it is enough to consider first and second moments and we can replace conditional independence by conditional orthogonality, but we shall use independence for ease of expression. Dahlhaus (2000) was one of the first to introduce graphical models to time series, and neuroscience is an important area of application (see Eichler (2005)).

The graph gives a number of conditional independence structure as follows. Let $\{E_1, E_2, E_3\}$ with $E_1 \cup E_2 \cup E_3 = E$, $E_i \cap E_j = \emptyset; i, j, k = 1, 2, 3,\ i \neq j$, be a partition of the edge set such that E_1 is disconnected (with respect to $\mathcal{G}$) from E_2. For an index set $J \subseteq \{1, \ldots, n\}$ we shall sometimes use the notation X_J to denote the set $\{X_j, j \in J\}$. We typically assume, in the undirected graph case, that the distribution has the strong Markov property, namely

L. Pronzato, A. Zhigljavsky (eds.), *Optimal Design and Related Areas in Optimization and Statistics*, Springer Optimization and its Applications 28,
DOI 10.1007/978-0-387-79936-0_7,

that for any such partition X_{E_1} is conditionally independent of X_{E_2} given X_{E_3}, written $X_{E_1} \perp\!\!\!\perp X_{E_2} \mid X_{E_3}$. In such cases we say that E_3 separates E_1 and E_2.

The case of directed graphs will be important here. For directed graphs the connection with conditional independence is the same except that the separation is with respect to the breaking of the common ancestry of E_1 and E_2 in the directed graph $\mathcal{G}$ in which every edge has a direction (orientation). The graph forces considerable structure on the covariance matrix Σ. We can represent this in various ways and we state them without proof.

1. Conditional covariance matrices. Clearly for Gaussian random variables conditional independence is equivalent to zero conditional covariance (or correlation). For a particular separating partition $\{E_1, E_2, E_3\}$, let $X_{E_i} = X^{(i)}$, $(i = 1, 2, 3)$ be the corresponding set of random variables and let the covariance matrix be full rank and after suitable rearrangement be partitioned as:

$$\Sigma = \begin{bmatrix} \Sigma_{11} & \Sigma_{12} & \Sigma_{13} \\ \Sigma_{21} & \Sigma_{22} & \Sigma_{23} \\ \Sigma_{31} & \Sigma_{32} & \Sigma_{33} \end{bmatrix},$$

where $\Sigma_{21} = \Sigma_{12}^{\mathrm{T}}$, etc.

Then conditional independence covariance of $X^{(1)}$ and $X^{(2)}$ given $X^{(3)}$ is equivalent to the conditional covariance being zero:

$$\Sigma_{12} - \Sigma_{13}\Sigma_{33}^{-1}\Sigma_{32} = 0\,.$$

2. Influence matrices (inverse covariance matrix). If we partition the inverse covariance matrix, or influence matrix, in the same way as Σ,

$$\Sigma^{-1} = \begin{bmatrix} \Sigma^{11} & \Sigma^{12} & \Sigma^{13} \\ \Sigma^{21} & \Sigma^{22} & \Sigma^{23} \\ \Sigma^{31} & \Sigma^{32} & \Sigma^{33} \end{bmatrix},$$

then the condition is $\Sigma^{12} = \Sigma^{21} = 0$.

3. Conditional expectations. Independence of random variables can be stated via conditional independence. Here we restrict ourselves to Gaussian random variables. Thus $E(\cdot|U)$ represents the formal conditional expectation operator, conditioning on U, which can be considered as a linear mapping on the space of all linear functions of the form $Z = \sum_{i=1}^{d} a_i X_i$ in which $Z \to E(Z|U)$. Denote I as the identity operator on this space: $I : Z \to Z$ for any such Z. Then conditional independence can be stated in a number of equivalent ways. Here are two:

(i) $E(\cdot|\{X^{(1)}, X^{(3)}\}) + E(\cdot|\{X^{(2)}, X^{(3)}\}) - E(\cdot|X^{(3)}) = I\,,$

(ii) $E(E(\cdot|\{X^{(1)}, X^{(3)}\})|\{X^{(2)}, X^{(3)}\}) = E(E(\cdot|\{X^{(2)}, X^{(3)}\})|\{X^{(1)}, X^{(3)}\})\,.$

Condition (ii) is a commutativity condition and can be written more transparently as

$$E_{13}E_{23} = E_{23}E_{13},$$

where $E_{ij} = E(\cdot|\{X^{(i)}, X^{(j)}\})$ and we interpreted the equation in terms of composition of operators.

4. Orthogonal projections. There is another representation in terms of orthogonal projections on an underlying space. This will play an important role here. Consider the R^d case, namely when each X_i is univariate. Consider an index set J and let X_J be a vector of zero mean random variables $\{X_i,\ i \in J\}$. Let $\varepsilon = (\varepsilon_1, \ldots, \varepsilon_d)^{\mathrm{T}}$ be a set of uncorrelated random variables with zero mean and unit variance. Then we can express every X_J as a linear combination of the ε_i: $X_J = A_J\varepsilon$, where A_J is a full rank $|J| \times n$ matrix. Then associate the orthogonal projector

$$P_J = A_J^{\mathrm{T}}(A_J A_J^{\mathrm{T}})^{-1}A_J$$

with X_J. Then the conditional orthogonality (independence in the Gaussian case) for disjoint index sets I, J, K is equivalent to the commutativity of the underlying orthogonal projections:

$$X_I \perp\!\!\!\perp X_J | X_K \Leftrightarrow P_{I\cup K}P_{J\cup K} = P_{J\cup K}P_{I\cup K}\,.$$

As simple example take three random variables given by

$$\begin{bmatrix} X_1 \\ X_2 \\ X_3 \end{bmatrix} = \begin{bmatrix} 1 & 2 & 3 \\ 2 & -1 & 2 \\ 1 & 1 & 1 \end{bmatrix} \begin{bmatrix} \varepsilon_1 \\ \varepsilon_2 \\ \varepsilon_3 \end{bmatrix}.$$

Then we have

$$P_{13} = \begin{bmatrix} \frac{5}{6} & \frac{1}{3} & -\frac{1}{6} \\ \frac{1}{3} & \frac{1}{3} & \frac{1}{3} \\ -\frac{1}{6} & \frac{1}{3} & \frac{5}{6} \end{bmatrix}, \quad P_{23} = \begin{bmatrix} \frac{1}{2} & 0 & \frac{1}{2} \\ 0 & 1 & \frac{1}{2} \\ \frac{1}{2} & 0 & \frac{1}{2} \end{bmatrix},$$

which commute, giving $X_1 \perp\!\!\!\perp X_2 \mid X_3$. As a check, assume the ε_i have unit variance then

$$\Sigma = \begin{bmatrix} 14 & 6 & 6 \\ 6 & 9 & 3 \\ 6 & 3 & 3 \end{bmatrix}, \quad \Sigma^{-1} = \begin{bmatrix} \frac{1}{2} & 0 & -1 \\ 0 & \frac{1}{6} & -\frac{1}{6} \\ -1 & -\frac{1}{6} & \frac{5}{2} \end{bmatrix},$$

with the structure indicated in condition (2.) above.

7.1.1 Subspace Lattices

There is an important relationship between orthogonal projections and the corresponding subspace lattice in R^d. Only after this discussion will we specialize to the issue of conditional independence. We list some basic facts about projections which can be used to derive consequences of factorizations arising for the Gaussian case from Theorem 2 below. For a basic text see Kadison and Ringrose (1983). It should also be stated that the use of linear operations for a second-order Bayes theory and projections is inherent in the work of Goldstein (1988).

(i) Every subspace of R^d has a unique projector on itself and the subspace lattice under inclusion induces a lattice on the projections: for subspaces V_1 and V_2 of R^d, $V_1 \subset V_2$ if and only if, for the corresponding projections into V_1 and V_2, $P_1 \leq P_2$, in the sense that $P_2 P_1 = P_1 P_2 = P_1$. In the infinite dimensional case, which we can take as $d \to \infty$, the lattice is sometimes called the *Hilbert lattice*.
(ii) The subspaces V_1 and V_2 are orthogonal if and only if $P_1 P_2 = 0$.
(iii) Projections do not necessarily commute but $P_1 \leq P_2$ implies $P_1 P_2 = P_2 P_1 = P_1$.
(iv) The projection corresponding to $V_1 \cap V_2$ is written $P_1 \wedge P_2$. If P_1 and P_2 commute then $P_1 \wedge P_2 = P_1 P_2 = P_2 P_1$. In general, $P_1 \wedge P_2 = \lim_{r \to \infty} (P_1 P_2)^r$.
(v) Define $V_1 \cup V_2$ as the smallest subspace containing V_1 and V_2 with corresponding projector written $P_1 \vee P_2$. If the projectors P_1 and P_2 commute then this is equal to the direct sum

$$V_1 \cup V_2 = V_1 + V_2$$

and

$$P_1 \vee P_2 = P_1 + P_2 - P_1 \wedge P_2 = P_1 + P_2 - P_1 P_2 \, .$$

(vi) If the identity projector I on R^d is partitioned as in $I = P + Q$ where P is a projector, then $Q = P^{\perp} = I - P$ is the projector orthogonal to P. More generally, if $I = \sum_{i=1}^{m} P_j$ where each P_j is a non-zero projector then all the P_j are mutually orthogonal. Such a decomposition is referred to as a partition of unity.
(vii) Full distributivity of $\wedge$ over $\vee$ and $\vee$ over $\wedge$ does not in general hold for projectors but the following *modularity* laws hold:

$$P_1 \vee (P_2 \wedge P_3) = (P_1 \vee P_2) \wedge P_3$$

$$P_1 \wedge (P_2 \vee P_3) = (P_1 \wedge P_2) \vee P_3$$

as does *orthomodularity*:

$$P_1 \wedge (P_1^{\perp} \vee P_2) = P_2 \, .$$

We see from this summary that where projectors commute, as when we have conditional independence, we can define a lattice structure which is Boolean and the projectors behave like Boolean variables. The Boolean logic is the logic of subspace inclusion. Thus if $U \subset V$ then we could write U *implies* V, in the sense of Boolean logic. Or we can write equivalently for projections $P_1 \leq P_2$, as in (i) above.

In the rest of this subsection we draw on the work of Andersson, Perlman and co-workers: Andersson and Perlman (1993), Andersson and Madsen (1998) and Andersson et al. (1995). The following example shows how the theory applies. The graph structure maps to a Boolean lattice in which all the relevant projectors commute.

Example 1. Consider a graph with four nodes and edges

$$1 \to 2, \;\; 1 \to 3, \;\; 1 \to 4, \;\; 2 \to 4, \;\; 3 \to 4\,.$$

This gives the conditional independence: $X_2 \perp\!\!\!\perp X_3 \mid X_1$. The equivalent condition for commutativity is that $P_{12}P_{13} = P_{13}P_{12}$. The subspace lattice has elements V_{1234}, the entire space, and the subspaces $V_{123}, V_{12}, V_{13}, V_1$ and $\emptyset$. The commutativity means that $V_1 = V_{12} \wedge V_{13}$ and that $P_1 = P_{12}P_{13}$.

In general, graphs can be written down where the relevant projectors do *not* all commute. An example is the well-known four-cycle where there are two conditional independence relations: $X_3 \perp\!\!\!\perp X_4 \mid (X_1, X_2)$ and $X_1 \perp\!\!\!\perp X_2 \mid (X_3, X_4)$, which give the commutativity relations

$$P_{123}P_{124} = P_{124}P_{123}, \quad P_{134}P_{234} = P_{234}P_{134}\,.$$

From this we deduce two maximal sets of commuting projectors:

$$\{P_{1234}, P_{123}, P_{124}, P_{12}\}, \quad \{P_{1234}, P_{134}, P_{234}, P_{34}\},$$

but these cannot in general be combined into a Boolean lattice.

Following this discussion we see that there is considerable simplification in the case where for some collection, $\mathcal{J}$, of index sets $J \subseteq \{1, \ldots, n\}$ all the corresponding projectors hold and equivalently all the corresponding conditional independence statements hold. The following definition and Theorem 1 are due to Andersson and Perlman (1993) and are basic to this chapter.

Definition 1. *Let $N_d = \{1, \ldots, d\}$ be a set of indices and $\mathcal{J}$ be a subset lattice of index sets (closed under union and intersection). Let $X_1, \ldots, X_d$ be a collection of (possibly multivariate) zero mean, finite variables, jointly distributed random variance such that for any $I, J \in \mathcal{J}$ we have*

$$X_I \perp\!\!\!\perp X_J \mid X_{I \cap J}\,.$$

Then we call the collection $X_{\mathcal{J}}$, lattice conditionally orthogonal (LCO) or lattice conditionally independent (LCI) in the Gaussian case.

In this case the lattice is in one-to-one correspondence with lattice of subspaces $V_J, J \in \mathcal{J}$, and all the projections $P_J, J \in \mathcal{J}$, form a Boolean algebra.

Example 1 can be generalized to a certain type of directed graphs namely acyclic directed graphs which are transitive: if there are edges $i \to j$ and $j \to k$ then the edge $i \to k$ is in the graph.

Theorem 1. *Every LCO (LCI) has a conditional independence structure derived from a transitive directed acyclic graph (TDAG), $\mathcal{G}$. Conversely, every LCO gives rise to a TDAG. In this sense the two structures are isomorphic.*

The isomorphism follows that in Example 1. The index sets J are chains in $\mathcal{G}$, that is, connected paths. We listed in Example 1 all the chains. Here is a more complex example.

Example 2. Consider a graph with 5 nodes and edges

$$3 \to 2,\ \ 3 \to 1,\ \ 2 \to 1,\ \ 3 \to 4,\ \ 3 \to 5,\ \ 4 \to 5\,.$$

The equivalent lattice has elements

$$12345,\ 1234,\ 2345,\ 123,\ 234,\ 345,\ 23,\ 34,\ 3$$

In this lattice, all intersections give independence statements: e.g. $123 \cap 345 = 3$ gives

$$\{X_1, X_2, X_3\} \perp\!\!\!\perp \{X_3, X_4, X_5\} \mid X_3$$

and all the corresponding projectors commute, e.g. $P_{123}P_{345} = P_{345}P_{123}$.

7.2 Time Series: Stochastic Realization and Conditional Independence

In previous papers, Caines et al. (2003), we give conditions on time series for conditional independence in what we call the non-anticipatory sense. There are two parts to the theory: (i) general condition under which the time series up *to time* t can be considered conditionally independent, (ii) when there is a finite dimensional stochastic realisation, or state-space representation, there are explicit forms of the observation and state equations for the conditions to hold. It is of importance, but in a sense no surprise, that in the latter case the structure of the realization perfectly reflects the conditional independence structure. The following example illustrates the structure.

Example 3. We motivate the discussion by considering a trivariate $AR(1)$ process $\{X_t, Y_t, Z_t\}$ written in the innovation form:

$$\begin{aligned}
X_t - a_{11}X_{t-1} - a_{12}Y_{t-1} - a_{13}Z_{t-1} &= \varepsilon_{1,t} \\
Y_t - a_{21}X_{t-1} - a_{22}Y_{t-1} - a_{23}Z_{t-1} &= \varepsilon_{2,t} \\
Z_t - a_{31}X_{t-1} - a_{32}Y_{t-1} - a_{33}Z_{t-1} &= \varepsilon_{3,t}
\end{aligned}$$

where $\{\varepsilon_t\} = \{\varepsilon_{1,t}, \varepsilon_{2,t}, \varepsilon_{3,t}\}$ is a white noise process; that is to say a Gaussian, zero mean, unit variance process independent over time and index. It is straightforward to check that if $a_{21} = a_{31} = a_{32} = a_{12} = 0$, and stationarity conditions hold, the processes up to time t are conditionally independent.

Thus, defining the X-process up to time t as $X^{(t)} = \{\ldots, X_{t-1}, X_t\}$, and similarly for $Y^{(t)}$ and $Z^{(t)}$, it holds that, under these conditions,

$$X^{(t)} \perp\!\!\!\perp Y^{(t)} \mid Z^{(t-1)} .$$

There are several ways to show this (i) via conditional expectations, (ii) via the covariance generating functions and (iii) via explicit construction of the semi-infinite projections, which are the analogues of the projections P_J of the last section.

Another method is to explicitly solve for X_t and Y_t using the shift operator z: $zV_t = V_{t-1}$:

$$X_t = \tfrac{\varepsilon_{1,t} + a_{13} z Z_t}{1 - a_{11} z} ,$$

$$Y_t = \tfrac{\varepsilon_{2,t} + a_{13} z Z_t}{1 - a_{11} z} .$$

Now if Z_s, $t \leq s$, is fixed we see that X_t and Y_t only depend on $\varepsilon_{2,s}$, $s \leq t$, and $\varepsilon_{2,s}$, $s \leq t$, respectively, which are independent.

To generalize these results we assume first that $\{X_{1,t}, \ldots, X_{d,t}\}$ are a collection of wide sense stationary (time invariant mean and covariance) stochastic processes each one of which may be multivariate. (We shall sometimes drop the "t" suffix.) For a process X_i up to time t we write $\{\ldots, X_{i,t-1}, X_{i,t}\} = X_i^{(t)}$. A pair of wide sense stationary processes (X_i, X_j), $i \neq j$, are said to be (strongly) feedback free if and only if the joint spectral density has a factorization where the first matrix factor is stable and inverse stable:

$$G_{X_i,X_j}(z) = \begin{bmatrix} A & B \\ 0 & C \end{bmatrix} \begin{bmatrix} A^* & 0 \\ B^* & C^* \end{bmatrix} (z),$$

where A^* is the conjugate of A, etc. This is equivalent (see Caines et al. (2003)) to $X_{t,i}|X_j = X_{t,i}|X_j^{(t-1)}$, where "|" represents orthogonal projection, which is the conditional expectation in the zero mean Gaussian case. Henceforth, we shall use this family of equalities as the definition of the (strong) feedback free property in the more general non-stationary Gaussian process case. We shall also require that wherever there is a conditional independence hypothesis it comes with an appropriate feedback free condition. In this section, we handle a single conditional independence.

Conditional orthogonality (independence) for three different stochastic processes $(X_i, X_j, X_k) = \{\ldots, (X_{i,t}, X_{j,t}, X_{k,t}), \ldots\}$ can be defined in one basic way, but it has two equivalent expressions: for the whole process $t = -\infty \ldots \infty$, which we may call *global* or for the processes up to t, $\{X_i^{(t)}, X_j^{(t)}, X_k^{(t)}\}$, which we call *local.* The following result links the *strongly feedback free* condition to conditional orthogonality. In the following, X_i refers to the whole process $\{X_{i,t}\}_{-\infty}^{\infty}$. The proofs of the next two theorems are given in Caines et al. (2003).

Theorem 2. *Under the assumption that each pair $\{X_i, X_k\}$ and $\{X_j, X_k\}$ is feedback free each of the following types of conditional independence is equivalent*

1. $X_i \perp\!\!\!\perp X_j \mid X_k$,
2. $X_i^{(t)} \perp\!\!\!\perp X_j^{(t)} \mid X_k^{(t)}, \quad t = \ldots, -1, 0, 1, \ldots$
3. $X_i^{(t)} \perp\!\!\!\perp X_j^{(t)} \mid X_k^{(t+s-1)}, \quad t = \ldots, -1, 0, 1, \ldots; s = 0, 1, \ldots$

The next result shows that there is a state-space or stochastic realization which, under a natural condition, is equivalent to the conditional independence in Theorem 1. The extra condition is simply to guarantee a finite dimensional state-space representation.

Theorem 3. *Let $\{X_{i,t}, X_{j,t}, X_{k,t}\}$ be a collection of three jointly distributed stochastic processes such that the joint process possesses a finite dimensional stochastic realisation, then the two conditions (i) each pair $\{X_i, X_k\}$ and $\{X_j, X_k\}$ is feedback free and (ii) $X_i^{(t)} \perp\!\!\!\perp X_j^{(t)} \mid X_k^{t+s-1}$, $s = 0, 1, \ldots; t = \ldots, -1, 0, 1, \ldots$, are (jointly) equivalent to the statement that $\{X_{i,t}, X_{j,t}, X_{k,t}\}$ have either of the following stochastic state-space realizations:*

Form 1:

$$\begin{bmatrix} s_{i,r+1} \\ s_{j,r+1} \\ s_{k,r+1} \end{bmatrix} = \begin{bmatrix} F_{ii} & 0 & 0 \\ 0 & F_{jj} & 0 \\ 0 & 0 & F_{kk} \end{bmatrix} \begin{bmatrix} s_{i,r} \\ s_{j,r} \\ s_{k,r} \end{bmatrix} + \begin{bmatrix} M_{ii} & 0 & M_{ik} \\ 0 & M_{jj} & M_{jk} \\ 0 & 0 & M_{kk} \end{bmatrix} \begin{bmatrix} u_{i,r} \\ u_{j,r} \\ u_{k,r} \end{bmatrix}$$

$$\begin{bmatrix} X_{i,r} \\ X_{j,r} \\ X_{k,r} \end{bmatrix} = \begin{bmatrix} H_{ii} & 0 & 0 \\ 0 & H_{jj} & 0 \\ 0 & 0 & H_{kk} \end{bmatrix} \begin{bmatrix} s_{i,r} \\ s_{j,r} \\ s_{k,r} \end{bmatrix} + \begin{bmatrix} N_{ii} & 0 & N_{ik} \\ 0 & N_{jj} & N_{jk} \\ 0 & 0 & N_{kk} \end{bmatrix} \begin{bmatrix} u_{i,r} \\ u_{j,r} \\ u_{k,r} \end{bmatrix}$$

Form 2:

$$\begin{bmatrix} \tilde{s}_{i,r+1} \\ \tilde{s}_{j,r+1} \\ \tilde{s}_{k,r+1} \end{bmatrix} = \begin{bmatrix} \tilde{F}_{ii} & 0 & \tilde{F}_{ik} \\ 0 & \tilde{F}_{jj} & \tilde{F}_{ik} \\ 0 & 0 & \tilde{F}_{kk} \end{bmatrix} \begin{bmatrix} \tilde{s}_{i,r} \\ \tilde{s}_{j,r} \\ \tilde{s}_{k,r} \end{bmatrix} + \begin{bmatrix} \tilde{M}_{ii} & 0 & \tilde{M}_{ik} \\ 0 & \tilde{M}_{jj} & \tilde{M}_{jk} \\ 0 & 0 & \tilde{M}_{kk} \end{bmatrix} \begin{bmatrix} u_{i,r} \\ u_{j,r} \\ u_{k,r} \end{bmatrix}$$

$$\begin{bmatrix} X_{i,r} \\ X_{j,r} \\ X_{k,r} \end{bmatrix} = \begin{bmatrix} \tilde{H}_{ii} & 0 & \tilde{H}_{ik} \\ 0 & \tilde{H}_{jj} & \tilde{H}_{jk} \\ 0 & 0 & \tilde{H}_{kk} \end{bmatrix} \begin{bmatrix} s_{i,r} \\ s_{j,r} \\ s_{k,r} \end{bmatrix} + \begin{bmatrix} N_{ii} & 0 & N_{ik} \\ 0 & N_{jj} & N_{jk} \\ 0 & 0 & N_{kk} \end{bmatrix} \begin{bmatrix} u_{i,r} \\ u_{j,r} \\ u_{k,r} \end{bmatrix}$$

where the joint system input and observation noise process, $\begin{bmatrix} u_{i,r} \\ u_{j,r} \\ u_{k,r} \end{bmatrix}$, *is a zero mean orthogonal process with block diagonal (instantaneous) covariance.*

7.3 LCO/LCI Time Series

It is our purpose in this section to lift the LCO/LCI condition of Sect. 7.1 into the framework just developed. Our development is a little informal. Here are the initial steps, which we express in a constructive style.

1. Start with a TDAG, $\mathcal{G}(E, V)$, or the equivalent distributive Boolean lattice $\mathcal{L}$.
2. To every vertex V_i associate a stochastic process X_i, require that the entire joint process $\{X_1, \ldots, X_d\}$ is wide-sense stationary.
3. For every conditional independence statement that can written down from $\mathcal{G}$, or the corresponding Boolean lattice $\mathcal{L}$, require that the corresponding feedback free condition holds. Note that we need to group processes together according to special index sets. Thus, for an index set J we say $X_J = \{X_{j,t},\ j \in J\}$. If our conditional independence is to be
$$X_J \amalg X_K | X_{J \cap K}, \quad J, K \in \mathcal{L},$$
then we require that pairs $(X_J, X_{J\cap K})$ and $(X_J, X_{J\cap K})$ are each feedback free.
4. Finally, assume that each of the up-to-t version of conditional independence statement given by the graph $\mathcal{G}$ holds (or either of the two equivalent conditions in Theorem 3). With obvious notation for every $J, K \in \mathcal{L}$
$$X_J^{(t)} \amalg X_K^{(t)} | X_{J\cap K}^{(t)} .$$

If the above conditions hold we say that $X_1, \ldots, X_d$ is $\mathcal{G}$-compatible. To summarize: all those pairs of feedback free conditions and conditional independence conditions must hold which can be listed from the original TDAG, $\mathcal{G}$, or equivalently the Boolean lattice $\mathcal{L}$.

To complete the construction we need to generalize Theorem 3 to the LCO/LCI case. We do this by extending the zero structures of the representations in Theorem 3.

Definition 2. *A square matrix A is said to be of block form if it can be partitioned into blocks B_{ij}, $(i, j = 1, \ldots, d)$, such that B_{ij} is $n_i \times n_j$, $(i, j = 1, \ldots, d)$ and the blocks are stacked in standard index order: larger i means lower position, larger j means further to the right.*

Definition 3. *We say that a square matrix A has a zero structure compatible with a transitive directed graph $\mathcal{G}(E, V)$ with d vertices if the following hold. (i) A is of block form as in Definition 2 and (ii) $B_{ij} = 0$ whenever $i \to j$ is NOT in the edge set E of $\mathcal{G}$.*

It is clear that since $\mathcal{G}$ is directed and transitive the $\mathcal{G}$-compatible matrix A must be an upper triangular matrix. The reader may refer to Examples 4 and 5, below, to see two compatible structures in which the blocks are scalar.

Looking at the structure of the stochastic realization in Theorem 5, we see that the upper triangular form with a zero block in the requisite condition is compatible with the graph $1 \leftarrow 3 \rightarrow 2$, because the zero blocks are the blocks $21, 31, 32, 12$, corresponding to the missing directed edges. It should be clear, then, how to extend this structure to at collection of processes $X_1, \ldots, X_d$, in a way which is $\mathcal{G}$-compatible.

We state the main result in a brief form to avoid cumbersome notation and omit the proof, which is an extension of that for Theorem 3.

Theorem 4. *Let $\{X_1, \ldots, X_d\}$ be a set of jointly wide-sense stationary stochastic processes with a (joint) rational spectral density. Let $\mathcal{G}$ be a TDAG, with d vertices. Then $X_1, \ldots, X_d$ is $\mathcal{G}$-compatible if and only if $\{X_1, \ldots, X_d\}$ has a stochastic realization of one of the two alternative forms given in Theorem 3, but with a zero structure compatible with $\mathcal{G}$, in which n_i is the dimension of $X_i, i = 1, \ldots, d$.*

If the conditions of Theorem 4 hold we call the process an LCO/LCI process compatible with $\mathcal{G}$.

It is important to note that under the conditions of Theorem 4 the zero structure, which is inherited from $\mathcal{G}$, is also induced by the structure of the spectral resolution for the whole processes:

$$G(z) = \Psi\Psi^*(z) \tag{7.1}$$

where Ψ has a zero structure compatible with $\mathcal{G}$.

In both the examples below each node represents a univariate $AR(1)$:

$$X^{(t)} = \Phi X^{(t-1)} + \varepsilon^{(t)}$$

and we use Theorem 4 to claim a particular form for Φ.

Example 4. We give the time series version of Example 1. We have changed the labelling: the graph is $\{4 \rightarrow 1,\ 4 \rightarrow 2,\ 4 \rightarrow 3,\ 2 \rightarrow 1, 3 \rightarrow 1\}$.

$$\Phi = \begin{bmatrix} a_{11} & a_{12} & a_{13} & a_{14} \\ 0 & a_{22} & 0 & a_{24} \\ 0 & 0 & a_{33} & a_{34} \\ 0 & 0 & 0 & a_{44} \end{bmatrix}.$$

Example 5. This is the analogue of Example 2. In this case we take the graph:

$$5 \rightarrow 3, 5 \rightarrow 1, 3 \rightarrow 1, 5 \rightarrow 4, 5 \rightarrow 2, 4 \rightarrow 2\,.$$

The form of Φ is

$$\Phi = \begin{bmatrix} a_{11} & 0 & a_{13} & 0 & a_{15} \\ 0 & a_{22} & 0 & a_{24} & a_{25} \\ 0 & 0 & a_{33} & 0 & a_{35} \\ 0 & 0 & 0 & a_{44} & a_{45} \\ 0 & 0 & 0 & 0 & a_{55} \end{bmatrix}.$$

We can confirm the conditional independence of the up-to-t series by elimination. For example to confirm

$$X_1 \perp\!\!\!\perp X_2 \mid (X_2, X_3, X_5)$$

we eliminate $\varepsilon_3, \varepsilon_4, \varepsilon_5$ from the equation

$$X = (I - z\Phi)^{-1}\varepsilon \tag{7.2}$$

to obtain equations for $X_{1,t}$ and $X_{2,t}$ in terms of $X_{3,t}, X_{4,t}, X_{5,t}, \varepsilon_{1,t}, \varepsilon_{2,t}$,

$$X_{1,t} = \frac{a_{13}zX_{3,t} + a_{15}zX_{5,t} + \varepsilon_{1,t}}{1 - za_{11}},$$
$$X_{2,t} = \frac{a_{24}zX_{4,t} + a_{25}zX_{5,t} + \varepsilon_{2,t}}{1 - a_{22}z}.$$

We can easily read off the conditional independence from these generating functions.

The spectral resolution can be written down, noting that $\Phi = (I - z\Phi)^{-1}$ and using (7.1). The moving average representation in the $AR(1)$ case is given by (7.2), and it is useful to understand the structure. This arises from the fact that the zero structure of Ψ is the same as that of Φ. This is a generic fact: the $\mathcal{G}$ compatible matrices (with non-zero diagonals and prescribed dimensions) form a group in that the product of any two such matrices is of the same form and so are the inverses. In the simple examples given here we see that the infinite moving average, which derives from an $AR(1)$ in the LCI class, has a structure in which $X^{(i)}$ is a function only of present and past innovations which can be traced backwards in the TDAG.

As an example we compute the moving average representation for $X_{3,t}$ in Example 5 as

$$X_{3,t} = \frac{e_3 + (-e_{3,t}a_{55} + a_{35}e_{5,t})z}{(1 - za_{33})(1 - za_{55})},$$

confirming the dependence only on $e_{3,t}$ and $e_{5,t}$: the innovations reachable by reverse arrows.

7.4 TDAG as Generalized Time

It should be clear that combining the "spatial" dependence of the TDAG with the infinite TDAGs for each separate process based time itself gives a "super" TDAG. This leads to the notion that the TDAG/LCI itself is an appropriate setting to carry out a generalization of time which suites the development of a stochastic realisation or state-space theory of which the present results are a special, but informative, case. This is the subject of current research but many of the ingredients are given here: the basic use of the LCI/LCO set up, the feedback free conditions, the idea of TDAG compatible transfer matrices and the underlying group structure and invertibility.

References

Andersson, S. and Madsen, J. (1998). Symmetry and lattice conditional independence in a multivariate normal distribution. *The Annals of Statistics,* **26**, 525–572.

Andersson, S. and Perlman, M. (1993). Lattice models for conditional independence in a multivariate normal distribution. *The Annals of Statistics,* **21**, 1318–1358.

Andersson, S., Madigan, D., Perlman, M., and Trigg, C. (1995). On the relationship between conditional independence models determined by finite distributive lattices and directed acyclic graph. *Journal of Statistical Planning and Inference,* **48**, 25–46.

Caines, P., Deardon, R., and Wynn, H. (2003). Conditional orthogonality and conditional stochastic realization. *Directions in Mathematical Systems Theory and Optimization, Lecture Notes in Control and Information Sciences,* A. Rantzer and C.I. Byrnes (Eds.), **286**, 71–84.

Dahlhaus, R. (2000). Graphical interaction models for time series. *Metrika,* **51**, 151–72.

Eichler, M. (2005). A graphical approach for evaluating effective connectivity in neural systems. *Philosophical Transactions of the Royal Society of London B,* **360**, 953–967.

Goldstein, M. (1988). Adjusting belief structures. *Journal of the Royal Statistical Society B,* **50**, 133–154.

Kadison, R. and Ringrose, J. (1983). *Fundamentals of the Theory of Operator Algebras, volume 1.* Volume 15 of *Graduate Studies in Mathematics.* American Mathematical Society, San Diego.

Lauritzen, L. (1996). *Graphical Models.* Clarendon Press, Oxford.

8

Asymptotic Normality of Nonlinear Least Squares under Singular Experimental Designs

A. Pázman and L. Pronzato

Summary. We study the consistency and asymptotic normality of thc LS cstimator of a function $h(\theta)$ of the parameters θ in a nonlinear regression model with observations $y_i = \eta(x_i, \theta) + \varepsilon_i$, $i = 1, 2 \ldots$ and independent errors ε_i. Optimum experimental design for the estimation of $h(\theta)$ frequently yields singular information matrices, which corresponds to the situation considered here. The difficulties caused by such singular designs are illustrated by a simple example: depending on the true value of the model parameters and on the type of convergence of the sequence of design points $x_1, x_2 \ldots$ to the limiting singular design measure ξ, the convergence of the estimator of $h(\theta)$ may be slower than $1/\sqrt{n}$, and, when convergence is at a rate of $1/\sqrt{n}$ and the estimator is asymptotically normal, its asymptotic variance may differ from that obtained for the limiting design ξ (which we call *irregular asymptotic normality* of the estimator). For that reason we focuss our attention on two types of design sequences: those that converge strongly to a discrete measure and those that correspond to sampling randomly from ξ. We then give assumptions on the limiting expectation surface of the model and on the estimated function h which, for the designs considered, are sufficient to ensure the regular asymptotic normality of the LS estimator of $h(\theta)$.

8.1 Introduction

Although the singularity of regression models has been somewhat neglected in the literature on parameter estimation and experimental design, the difficulties it induces for statistical inference have been noticed and investigated in several domains of applications, see, e.g. Stoica (2001) for signal processing, Hero et al. (1996) for image processing (emission tomography) or Sjöberg et al. (1995) for general black-box modeling of dynamical systems. There, the singularity comes from an over-parametrization and the attention is directed to the Cramèr–Rao bound. The motivation of this chapter is different and comes from optimum design theory. In nonsingular nonlinear models, a standard way to design an optimum experiment for parameter estimation is to optimize a criterion based on the asymptotic normality of the estimator. This

L. Pronzato, A. Zhigljavsky (eds.), *Optimal Design and Related Areas in Optimization and Statistics*, Springer Optimization and its Applications 28,
DOI 10.1007/978-0-387-79936-0_8,

is justified by the fact that the asymptotic variance of the estimator depends on the limiting design measure (it is the inverse of the associated information matrix), but not on the way this measure is approached by the sequence of design points. Can we use a similar approach also for singular models? It is the purpose of this chapter to present easily interpreted conditions which allow this. It is important since optimum designs may produce singular models. Indeed, when the interest is in a function $h(\theta)$ of the parameters θ of the model, or more simply in a subset of θ, the information matrix at the optimum design is often singular, see, e.g., Silvey (1980, p. 58) and Example 1 below for the case of c-optimality. We then say that *the design is singular.*

Singular designs cause no special difficulty in linear regression (which partly explains why singularity issues have been disregarded in the design literature), as briefly shown hereafter: a linear combination $\mathbf{c}^\top\theta$ of the parameters is either estimable or not, depending on the direction of $\mathbf{c}$.

8.1.1 Singular Designs in Linear Models

The standard set-up for an optimum experimental design problem in a linear model is as follows. Given a set $\mathcal{X} \subset \mathbb{R}^k$, the design space, a design (an *exact design*, or a *design of size* N) is a choice $x_1, \ldots, x_N$ of points from $\mathcal{X}$. According to this design, we observe N random variables $y(x_1), \ldots, y(x_N)$ modelled by

$$y\left(x_i\right) = \mathbf{f}^\top(x_i)\bar{\theta} + \varepsilon_i\,,\; i = 1, \ldots, N \tag{8.1}$$

where the errors ε_i are independent and $\mathbb{E}(\varepsilon_i) = 0$, $\mathrm{Var}(\varepsilon_i) = \sigma^2$ for all i. Here, the vectors $\mathbf{f}(x_i) \in \mathbb{R}^p$ are known, $\sigma^2 \in \mathbb{R}^+$ is unknown and $\bar{\theta} \in \mathbb{R}^p$ is the unknown true value of the model parameters θ. We emphasize that throughout the chapter the choice of design points $x_1, \ldots, x_N$ is independent of the observed variables $y(x_i)$ (that is, the design is *not sequential*). If the information matrix

$$\mathbf{M}(x_1, \ldots, x_N) = \sum_{i=1}^{N} \mathbf{f}(x_i)\mathbf{f}^\top(x_i)$$

is nonsingular, then the least squares estimator (LSE) of θ,

$$\hat{\theta}^N \in \arg\min_{\theta} \sum_{i=1}^{N} \left[y(x_i) - \mathbf{f}^\top(x_i)\theta\right]^2 \tag{8.2}$$

is unique and its variance is

$$\mathrm{Var}(\hat{\theta}^N) = \sigma^2\mathbf{M}^{-1}(x_1, \ldots, x_N)\,.$$

On the other hand, if $\mathbf{M}(x_1, \ldots, x_N)$ is singular then $\hat{\theta}^N$ is not defined uniquely. However, $\mathbf{c}^\top\hat{\theta}^N$ does not depend on the choice of the solution $\hat{\theta}^N$ of (8.2) if and only if $\mathbf{c} \in \mathrm{Range}[\mathbf{M}(x_1, \ldots, x_N)]$. Then

$$\mathrm{Var}(\mathbf{c}^\top\hat{\theta}^N) = \sigma^2\mathbf{c}^\top\mathbf{M}^-(x_1,\ldots,x_N)\mathbf{c} \tag{8.3}$$

where the choice of the g-inverse matrix $\mathbf{M}^-$ is arbitrary. This last expression can be used as a criterion (the criterion of c-optimality) for an optimal choice of the N-point design $x_1,\ldots,x_N$, and the design minimizing this criterion may be singular, see Silvey (1980) and Pázman (1980) for some properties and Wu (1980, 1983) for a detailed investigation of the consistency of $\mathbf{c}^\top\hat{\theta}^N$ when $N\to\infty$.

8.1.2 Designs in Nonlinear Models

Consider now the same set-up with, however, one noticeable difference, namely that the observations are modelled by

$$y(x_i) = \eta(x_i,\bar{\theta}) + \varepsilon_i\,,\ i=1,\ldots,N$$

where the function $\eta(\cdot,\cdot):\mathscr{X}\times\Theta\to\mathbb{R}$ is nonlinear in the parameters $\theta\in\Theta\subset\mathbb{R}^p$ with unknown true value $\bar{\theta}$. One can often suppose, at least approximately, that the parameter space Θ is bounded and closed, hence compact. Also, it is standard in experimental design to take $\mathscr{X}$ as a compact subset of $\mathbb{R}^k$. Besides these classical assumptions we shall also assume that Θ has no isolated points, i.e. that $\Theta\subset\overline{\mathrm{int}(\Theta)}$ (the closure of the interior of Θ), and that $\Theta\subset\Theta_0$, with Θ_0 and open subset of $\mathbb{R}^p$ such that for any $x\in\mathscr{X}$ the function $\eta(\theta,x)$ is defined and two times continuously differentiable on Θ_0.

The LSE of θ, defined by

$$\hat{\theta}^N \in \arg\min_{\theta\in\Theta}\sum_{i=1}^N [y(x_i)-\eta(x_i,\theta)]^2\,,$$

is a random vector with rather complicated statistical properties even in seemingly simple situations. In the case of normal errors, optimum design can be based on fairly accurate approximations of the distribution of $\hat{\theta}^N$ for small N, see Pázman and Pronzato (1992) and Pronzato and Pázman (1994). However, this is technically difficult and the standard approach (for large N) is to use the much simpler asymptotic normal approximation of the distribution of $\hat{\theta}^N$. Under suitable conditions on the sequence of design points $x_1,x_2,\ldots$ and on the model response $\eta(x,\theta)$, see Jennrich (1969), Gallant (1987) and Ivanov (1997), $\hat{\theta}^N$ is proved to be strongly consistent ($\lim_{N\to\infty}\hat{\theta}^N=\bar{\theta}$ a.s.) and to converge in distribution to a normal random vector

$$\sqrt{N}(\hat{\theta}^N-\bar{\theta}) \xrightarrow{\mathrm{d}} \nu \sim \mathscr{N}\left(0,\sigma^2\mathbf{M}_\infty^{-1}(\bar{\theta})\right),\ N\to\infty\,. \tag{8.4}$$

Here, the limit information matrix $\mathbf{M}_\infty(\theta)$, defined by

$$\mathbf{M}_\infty(\theta) = \lim_{N\to\infty}\frac{1}{N}\sum_{i=1}^N \mathbf{f}_\theta(x_i)\mathbf{f}_\theta^\mathrm{T}(x_i)\,,$$

with the notation

$$\mathbf{f}_\theta(x) = \frac{\partial \eta(x,\theta)}{\partial \theta}\,, \tag{8.5}$$

is supposed to exist and to be nonsingular at $\bar{\theta}$. We denote by ξ the probability measure (called the *design measure*, or simply the design, a concept introduced by Kiefer and Wolfowitz (1959)), that corresponds to the limit of relative frequencies of the sequence $x_1, x_2 \ldots$ when it exists. In that case, we can write

$$\mathbf{M}_\infty(\theta) = \mathbf{M}(\xi,\theta) = \int_{\mathcal{X}} \mathbf{f}_\theta(x)\mathbf{f}_\theta^\top(x)\,\xi(\mathrm{d}x) \tag{8.6}$$

and $\sigma^2\mathbf{M}^{-1}(\xi,\bar{\theta})$ forms an approximation of $N\,\mathrm{Var}_{\bar{\theta}}(\hat{\theta}^N)$ for large N. This is basic in experimental design theory; for instance D-optimum design corresponds to maximizing $\det \mathbf{M}(\xi,\bar{\theta})$ with respect to ξ, see Fedorov (1972); Silvey (1980); Pázman (1986); Atkinson and Donev (1992).

On the other hand, when $\mathbf{M}(\xi,\bar{\theta})$ is singular the estimator $\hat{\theta}^N$ may not be asymptotically normal, or even not consistent and not uniquely defined. Still, it might seem possible to base the design on a generalization of the c-optimality criterion used for linear models, of the form

$$\Phi[\mathbf{M}(x_1,\ldots,x_N;\,\theta)] = \mathrm{Var}_\theta[h(\hat{\theta}^N)] + \mathcal{O}(1/N^2)$$

for a suitable choice of the function $h(\cdot)$, where Var_θ denotes the variance conditional to θ being the true value of the model parameters. In the regular case, this is justified by the delta-method, see Lehmann and Casella (1998, p. 61): under some regularity conditions on $h(\theta)$, from the asymptotic normality (8.4) of $\hat{\theta}^N$ one obtains the approximation

$$N\,\mathrm{Var}_\theta[h(\hat{\theta}^N)] = \sigma^2\,\frac{\partial h(\theta)}{\partial \theta^\top}\mathbf{M}^{-1}(\xi,\theta)\frac{\partial h(\theta)}{\partial \theta} + \mathcal{O}(1/N)\,, \tag{8.7}$$

an expression similar to that used for c-optimality in linear models, see (8.3). Also, the estimator $h(\hat{\theta}^N)$ is asymptotically normal,

$$\sqrt{N}[h(\hat{\theta}^N) - h(\bar{\theta})] \xrightarrow{\mathrm{d}} \nu \sim \mathcal{N}\left(0, \sigma^2\left[\frac{\partial h(\theta)}{\partial \theta^\top}\mathbf{M}^{-1}(\xi,\theta)\frac{\partial h(\theta)}{\partial \theta}\right]_{\bar{\theta}}\right),\ N \to \infty\,,$$

which we shall call *regular asymptotic normality* (with $\mathbf{M}^{-1}$ replaced by a g-inverse when $\mathbf{M}$ is singular, see Sect. 8.5).

The question is for which functions $h(\cdot)$, and under which conditions on the model $\eta(\cdot,\cdot)$ and design sequence $x_1, x_2 \ldots$ a formula similar to (8.7) is justified and regular asymptotic normality also holds in the singular case? We shall show that essentially three types of conditions must be fulfilled:

(i) Conditions on the convergence of the design sequence $x_1, x_2 \ldots$ to the design measure ξ, as discussed in Sect. 8.2;

(ii) Conditions on $\bar{\theta}$, the true value of the model parameters θ, in relation to the geometry of the model (Sect. 8.4);

(iii) Conditions on the function $h(\cdot)$.

The consistency of $\hat{\theta}^N$ and $h(\hat{\theta}^N)$ is considered in Sect. 8.3, based on the conditions of Sect. 8.2 on the design sequence. The regular asymptotic normality of $h(\hat{\theta}^N)$ is investigated in Sect. 8.5. The extension to a multidimensional function of interest $H(\theta)$ is considered in Sect. 8.6.

8.2 The Convergence of the Design Sequence to a Design Measure

The investigation of the asymptotic properties of the estimator in a regression model requires the specification of the asymptotic behavior of the design sequence $(x_i)_i$. Rather than formulating conditions in terms of finite tail products, as in the seminal paper of Jennrich (1969), we shall state conditions in terms of limiting design measures. To each truncated subsequence $x_1, \ldots, x_N$, we associate the empirical design measure ξ_N and its cumulative distribution function (c.d.f.)

$$\mathbb{F}_{\xi_N}(x) = \sum_{i=1,\ x_i \leq x}^{N} \frac{1}{N}$$

(where the inequality $x_i \leq x$ must be understood componentwise). The sequence $(\xi_N)_N$ is said to converge weakly to a limit design measure ξ with the c.d.f. $\mathbb{F}_\xi$ if $\lim_{N\to\infty} \mathbb{F}_{\xi_N}(x) = \mathbb{F}_\xi(x)$ at every continuity point of $\mathbb{F}_\xi$ (which corresponds to the weak convergence of probability measures, see, e.g. Billingsley (1971) and Shiryaev (1996, p. 314)). However, even in linear models, weak convergence is not enough to ensure regular asymptotic normality of the estimator when the limiting design is singular. This is illustrated by the example below.

Example 1. Consider the linear regression model (8.1) with $p = 2$ and $\mathbf{f}(x) = (x \;\; x^2)^\top$. The true value $\bar{\theta}$ of the model parameters $\theta = (\theta_1 \;\; \theta_2)^\top$ is assumed to satisfy $\bar{\theta}_1 \geq 0$, $\bar{\theta}_2 < 0$. The errors ε_i are i.i.d., with zero mean and variance 1. We are interested in the estimation of the point x where $\eta(x,\theta) = \mathbf{f}^\top(x)\theta$ is maximum, that is, $h = h(\theta) = -\theta_1/(2\theta_2)$, with $h \geq 0$ and

$$\frac{\partial h(\theta)}{\partial \theta} = -\frac{1}{2\theta_2}\begin{pmatrix} 1 \\ 2h \end{pmatrix}.$$

Let θ^* be a prior guess for θ with $\theta_1^* \geq 0$, $\theta_2^* < 0$, $h_* = -\theta_1^*/(2\theta_2^*)$ denote the corresponding prior guess for h and define $x_* = 2h_*$. The c-optimum design ξ_* supported in $\mathcal{X} = [0,1]$ that minimizes $[\partial h(\theta)/\partial\theta^\top \mathbf{M}^-(\xi)\, \partial h(\theta)/\partial\theta]_{\theta^*}$ is easily computed from Elfving's Theorem (1952), and is given by

$$\xi_* = \begin{cases} \gamma_*\delta_{\sqrt{2}-1} + (1-\gamma_*)\delta_1 & \text{if } 0 \leq x_* \leq \sqrt{2}-1 \text{ or } 1 \leq x_* \\ \delta_{x_*} & \text{otherwise} \end{cases} \tag{8.8}$$

with δ_x the delta measure that puts weight 1 at x and

$$\gamma_* = \frac{\sqrt{2}}{2} \frac{1 - x_*}{2(\sqrt{2}-1) - x_*} .$$

Here we suppose that the prior guess θ^* is such that $\sqrt{2}-1 < x_* \leq 1$ so that the c-optimum design ξ_* puts mass 1 at x_*. When ξ_* is used, $\theta_1 x_* + \theta_2 x_*^2$ is estimable since $\mathbf{u}_* = (x_* \quad x_*^2)^\top$ is in the range of

$$\mathbf{M}(\xi_*) = \begin{pmatrix} x_*^2 & x_*^3 \\ x_*^3 & x_*^4 \end{pmatrix} .$$

The variance of $\mathbf{u}_*^\top \hat{\theta}^N$ for ξ_*, which we denote $\mathrm{Var}(\mathbf{u}_*^\top \hat{\theta}^N | \xi_*)$, satisfies

$$N \,\mathrm{Var}(\mathbf{u}_*^\top \hat{\theta}^N | \xi_*) = \mathbf{u}_*^\top \mathbf{M}^-(\xi_*) \mathbf{u}_* = 1 ,$$

with $\mathbf{M}^-$ any g-inverse of $\mathbf{M}$.

We consider design sequences that converge to ξ_* and investigate the case when convergence is weak. Suppose that the design points satisfy

$$x_i = \begin{cases} x_* & \text{if } i = 2k-1 \\ x_* + (1/k)^\alpha & \text{if } i = 2k \end{cases} \tag{8.9}$$

for some $\alpha \geq 0$, $i = 1, 2, \ldots$, with $\sqrt{2}-1 < x_* \leq 1$. From Corollary 1 of Wu (1980), one can show that the LSE $\hat{\theta}^N$ is strongly consistent when $\alpha \leq 1/2$, which we suppose in the rest of the example (see Pázman and Pronzato (2006) for details when $\alpha = 1/4$).

The variance of $\mathbf{u}_*^\top \hat{\theta}^N$, with $\mathbf{u}_* = (x_* \quad x_*^2)^\top$, for the design ξ_N satisfies $N \,\mathrm{Var}(\mathbf{u}_*^\top \hat{\theta}^N | \xi_N) = \mathbf{u}_*^\top \mathbf{M}^{-1}(\xi_N) \mathbf{u}_*$ with,

$$\mathbf{M}(\xi_N) = \begin{pmatrix} \mu_2(N) & \mu_3(N) \\ \mu_3(N) & \mu_4(N) \end{pmatrix}$$

and, for $N = 2M$, $\mu_i(N) = x_*^i/2 + (1/N) \sum_{k=1}^{M} [x_* + (1/k)^\alpha]^i$, $i = 2, 3, 4$. We then obtain

$$\lim_{N\to\infty} N \,\mathrm{Var}(\mathbf{u}_*^\top \hat{\theta}^N | \xi_N) = V(\alpha) = \frac{2(1-\alpha)^2}{\alpha^2 + (1-\alpha)^2} , \tag{8.10}$$

which is monotonically decreasing in α for α varying between 0 and $1/2$, with $V(0) = 2$ and $V(1/2) = 1$. For any $\alpha \in [0, 1/2)$ we thus have

$$\lim_{N\to\infty} N \,\mathrm{Var}(\mathbf{u}_*^\top \hat{\theta}^N | \xi_N) = V(\alpha) > N \,\mathrm{Var}(\mathbf{u}_*^\top \hat{\theta}^N | \xi_*) = 1 ,$$

that is, the limiting variance for ξ_N is always larger than the variance for the limiting design ξ_* (this is due to the discontinuity of the function $\mathbf{M}(\xi) \mapsto N \,\mathrm{Var}(\mathbf{u}_*^\top \hat{\theta}^N | \xi)$ at $\mathbf{M}(\xi_*)$, see Pázman (1980, 1986 p. 67)).

Moreover, we can easily show that Lindeberg's condition is satisfied for any linear combination of θ, see, e.g., Shiryaev (1996), and, for any $\mathbf{u} \neq 0$

$$\sqrt{N}\frac{\mathbf{u}^\top(\hat{\theta}^N - \bar{\theta})}{(\mathbf{u}^\top \mathbf{M}^{-1}(\xi_N)\mathbf{u})^{1/2}} \xrightarrow{\mathrm{d}} \zeta \sim \mathcal{N}(0,1)\,,\ N \to \infty\,.$$

Since $\mathbf{u}_*^\top \mathbf{M}^{-1}(\xi_N)\mathbf{u}_*$ tends to $V(\alpha)$,

$$\sqrt{N}\mathbf{u}_*^\top(\hat{\theta}^N - \bar{\theta}) \xrightarrow{\mathrm{d}} \zeta_* \sim \mathcal{N}(0, V(\alpha))\,,\ N \to \infty\,.$$

On the other hand, $\mathbf{u}^\top \mathbf{M}^{-1}(\xi_N)\mathbf{u}$ grows as $N^{2\alpha}$ for $\mathbf{u}$ not parallel to $\mathbf{u}_*$. For instance, for $\mathbf{u} = \mathbf{u}_0 = (1\ \ 0)^\top$ we obtain

$$N^{1/2-\alpha}\mathbf{u}_0^\top(\hat{\theta}^N - \bar{\theta}) \xrightarrow{\mathrm{d}} \zeta_0 \sim \mathcal{N}(0, W(\alpha))\,,\ N \to \infty\,,$$

with

$$W(\alpha) = 2^{2(1-\alpha)}\,\frac{(1-2\alpha)(1-\alpha)^2}{\alpha^2 + (1-\alpha)^2}\,. \tag{8.11}$$

Come back now to the estimation of $h(\theta) = -\theta_1/(2\theta_2)$.

When $\bar{\theta}_1 + x_*\bar{\theta}_2 \neq 0$, that is when $h(\bar{\theta}) \neq h_*$, which corresponds to the typical situation, we have

$$h(\hat{\theta}^N) = h(\bar{\theta}) + (\hat{\theta}^N - \bar{\theta})^\top \left[\frac{\partial h(\theta)}{\partial \theta}_{|\bar{\theta}} + o_p(1)\right]$$

with $\partial h(\theta)/\partial\theta_{|\bar{\theta}} = -1/(2\bar{\theta}_2)[1\ \ 2h(\bar{\theta})]^\top$ not parallel to $\mathbf{u}_*$, and

$$N^{1/2-\alpha}\left[h(\hat{\theta}^N) - h(\bar{\theta})\right] \xrightarrow{\mathrm{d}} \zeta \sim \mathcal{N}(0, v_{\bar{\theta}})\,,\ N \to \infty\,,$$

with $v_{\bar{\theta}} = W(\alpha)[x_* - 2h(\bar{\theta})]^2/(4\bar{\theta}_2^2 x_*^2)$ where $W(\alpha)$ is given by (8.11); $h(\hat{\theta}^N)$ is thus asymptotically normal but converges as $N^{\alpha-1/2}$.

In the particular situation where the prior guess h_* coincides with the true value $h(\bar{\theta})$, $\bar{\theta}_1 + x_*\bar{\theta}_2 = 0$ and we write

$$\begin{aligned} h(\hat{\theta}^N) &= h(\bar{\theta}) + (\hat{\theta}^N - \bar{\theta})^\top \frac{\partial h(\theta)}{\partial \theta}_{|\bar{\theta}} \\ &\quad + \frac{1}{2}(\hat{\theta}^N - \bar{\theta})^\top \left[\frac{\partial^2 h(\theta)}{\partial\theta\partial\theta^\top}_{|\bar{\theta}} + o_p(1)\right](\hat{\theta}^N - \bar{\theta}) \end{aligned} \tag{8.12}$$

with

$$\frac{\partial h(\theta)}{\partial \theta}_{|\bar{\theta}} = -\frac{1}{2\bar{\theta}_2 x_*}\mathbf{u}_* \ \text{ and } \ \frac{\partial^2 h(\theta)}{\partial\theta\partial\theta^\top}_{|\bar{\theta}} = \frac{1}{2\bar{\theta}_2^2}\begin{pmatrix} 0 & 1 \\ 1 & 2x_* \end{pmatrix}.$$

Define $\delta_N = \hat{\theta}^N - \bar{\theta}$ and $E_N = 2\bar{\theta}_2^2\,\delta_N^\top\,\partial^2 h(\theta)/(\partial\theta\partial\theta^\top)_{|\bar{\theta}}\,\delta_N$. The eigenvector decomposition of $\partial^2 h(\theta)/(\partial\theta\partial\theta^\top)_{|\bar{\theta}}$ gives

$$E_N = \beta\left[(\mathbf{v}_1^\top \delta_N)^2 - (\mathbf{v}_2^\top \delta_N)^2\right]$$

with $\mathbf{v}_{1,2} = (1 \;\; x_* \pm \sqrt{1+x_*^2})$ and $\beta = (x_* + \sqrt{1+x_*^2})/[2(1+x_*^2+x_*\sqrt{1+x_*^2})]$. We then obtain

$$N^{1/2-\alpha}\mathbf{v}_{1,2}^\top\delta_N \xrightarrow{\text{d}} \zeta_{1,2} \sim \mathcal{N}(0, [1+1/x_*^2]W(\alpha))\,,\;\; N \to \infty\,.$$

From (8.12), the limiting distribution of $h(\hat{\theta}^N)$ is not normal when $\alpha \geq 1/4$. When $\alpha < 1/4$ we have $\sqrt{N}\,E_N = o_p(1)$, $N \to \infty$, and (8.12) implies

$$\sqrt{N}[h(\hat{\theta}^N) - h(\bar{\theta})] \xrightarrow{\text{d}} \zeta \sim \mathcal{N}(0, V(\alpha)/(4\bar{\theta}_2^2 x_*^2))\,,\;\; N \to \infty\,,$$

with $V(\alpha)$ given by (8.10). Note that the limiting variance is larger than

$$[\partial h(\theta)/\partial\theta^\top\,\mathbf{M}^-(\xi_*)\,\partial h(\theta)/\partial\theta]_{\bar{\theta}} = 1/(4\bar{\theta}_2^2 x_*^2)\,.$$

To summarize, the estimation of $h(\theta)$ requires $\alpha \leq 1/2$ in the design (8.9), $h(\hat{\theta}^N)$ is then generally asymptotically normal but converges as slowly as $N^{\alpha-1/2}$. In the special case where the design is optimum for the true value $h(\bar{\theta})$, $h(\hat{\theta}^N)$ is not asymptotically normal when $1/4 \leq \alpha \leq 1/2$; it is asymptotically normal for $\alpha < 1/4$ and converges as $1/\sqrt{N}$, but the limiting variance differs from that computed from the limiting optimum design ξ_*. □

Regular asymptotic normality may thus fail to hold when the design sequence converges weakly to a singular design. Stronger types of convergence are required and we shall consider two situations that arise quite naturally.

The first one concerns the case where *the limiting design ξ is discrete.* Since, from Caratheodory's Theorem, optimum designs can be written as discrete probability measures (see, e.g., Fedorov (1972) and Silvey (1980)), sequences of design points that converge to discrete measures are of special interest. In that case, we shall require *strong convergence* (or convergence in variation, see Shiryaev (1996, p. 360)) of the empirical measure ξ_N.

Definition 1. *Let ξ be a discrete probability measure on $\mathcal{X}$, with finite support*

$$\mathcal{S}_\xi = \{x \in \mathcal{X} : \xi(\{x\}) > 0\}\,.$$

We say that the design sequence $(x_i)_i$ converges strongly to ξ when

$$\lim_{N\to\infty} \xi_N(\{x\}) = \xi(\{x\}) \;\; \textit{for any} \;\; x \in \mathcal{X}\,.$$

In the second situation the limiting design measure ξ is not necessary discrete and ξ_N converges weakly to ξ, but we require that *the design sequence is a random sample from ξ.*

Definition 2. *Let ξ be a probability measure on $\mathcal{X}$. We say that the design sequence $(x_i)_i$ is a randomized design with measure ξ if the points $x_i \in \mathcal{X}$ are independently sampled according to the probability measure ξ.*

The following example shows, however, that strong convergence of the design sequence is not enough to ensure the regular asymptotic normality of the estimator and further conditions on the model are required. They will be presented in Sects. 8.4 and 8.5.

Example 2. This is a slight extension of the example considered in Pázman and Pronzato (2006). We consider the same linear regression model as in Example 1, but now the design is such that $N - m$ observations are taken at $x = x_* = 2h_* \in (0, 1]$, with $h_* = -\theta_1^*/(2\theta_2^*)$ a prior guess for the location of the maximum of the function $\theta_1 x + \theta_2 x^2$, and m observations are taken at $x = z \in (0, 1]$, $z \neq x_*$. We shall suppose that either m is fixed[1] or $m \to \infty$ with $m/N \to 0$ as N tends to infinity. In both cases the sequence $(x_i)_i$ converges strongly to δ_{x_*} as $N \to \infty$, in the sense of Definition 1. Note that $\delta_{x_*} = \xi_*$, the c-optimum design measure for $h(\theta_*)$, when $\sqrt{2} - 1 < x_* \leq 1$, see (8.8).

The LSE $\hat{\theta}^N$ is given by

$$\hat{\theta}^N = \bar{\theta} + \frac{1}{x_* z(x_* - z)} \left[\frac{\beta_m}{\sqrt{m}} \begin{pmatrix} x_*^2 \\ -x_* \end{pmatrix} + \frac{\gamma_{N-m}}{\sqrt{N-m}} \begin{pmatrix} -z^2 \\ z \end{pmatrix} \right] \tag{8.13}$$

where $\beta_m = (1/\sqrt{m}) \sum_{x_i = z} \varepsilon_i$ and $\gamma_{N-m} = (1/\sqrt{N-m}) \sum_{x_i = x_*} \varepsilon_i$ are independent random variables that tend to be distributed $\mathcal{N}(0, 1)$ as $m \to \infty$ and $N - m \to \infty$. Obviously, $\hat{\theta}^N$ is consistent if and only if $m \to \infty$. However, $h(\hat{\theta}^N)$ is also consistent when m is finite provided that $\bar{\theta}_1 + x_* \bar{\theta}_2 = 0$. Indeed, for m finite we have

$$\hat{\theta}^N \stackrel{\text{a.s.}}{\to} \hat{\theta}^\# = \bar{\theta} + \frac{1}{z(x_* - z)} \frac{\beta_m}{\sqrt{m}} \begin{pmatrix} x_* \\ -1 \end{pmatrix}, \quad N \to \infty,$$

and $h(\hat{\theta}^\#) = -\hat{\theta}_1^\# / (2\hat{\theta}_2^\#) = x_*/2 = h(\bar{\theta})$. Also,

$$\sqrt{N}[h(\hat{\theta}^N) - h(\bar{\theta})] = \sqrt{N}[h(\hat{\theta}^N) - h(\hat{\theta}^\#)] = \sqrt{N}(\hat{\theta}^N - \hat{\theta}^\#)^\top \left[\frac{\partial h(\theta)}{\partial \theta}_{|\hat{\theta}^\#} + o_p(1) \right]$$

with $\partial h(\theta)/\partial \theta_{|\hat{\theta}^\#} = -1/(2\hat{\theta}_2^\#)[1 \quad x_*]^\top$ and

$$\sqrt{N}(\hat{\theta}^N - \hat{\theta}^\#) = \frac{\sqrt{N}}{x_*(x_* - z)} \frac{\gamma_{N-m}}{\sqrt{N-m}} \begin{pmatrix} -z \\ 1 \end{pmatrix}.$$

Therefore, $\sqrt{N}[h(\hat{\theta}^N) - h(\bar{\theta})] \stackrel{\text{d}}{\to} \nu/(2\zeta)$ with $\nu \sim \mathcal{N}(0, 1/x_*^2)$ and $\zeta \sim \mathcal{N}(\bar{\theta}_2, 1/[mz^2(x_* - z)^2])$, and $h(\hat{\theta}^N)$ is not asymptotically normal.

[1] Taking only a finite number of observations at another place than x_* might seem an odd strategy; note, however, that the algorithm of Wynn (1972) for the minimization of $[\partial h(\theta)/\partial \theta^\top \mathbf{M}^-(\xi) \, \partial h(\theta)/\partial \theta]_{\theta^*}$ generates such a sequence of design points when the design space is $\mathcal{X} = [-1, 1]$, see Pázman and Pronzato (2006), or when $\mathcal{X}$ is a finite set containing x_*.

Suppose now that $m = m(N) \to \infty$ with $m/N \to 0$ as $N \to \infty$. If $\bar{\theta}_1 + x_*\bar{\theta}_2 \neq 0$ we can write

$$\sqrt{m}[h(\hat{\theta}^N) - h(\bar{\theta})] = \sqrt{m}(\hat{\theta}^N - \bar{\theta})^\top \left[\frac{\partial h(\theta)}{\partial \theta}_{|\bar{\theta}} + o_p(1)\right]$$

and using (8.13) we get

$$\sqrt{m}[h(\hat{\theta}^N) - h(\bar{\theta})] \xrightarrow{\text{d}} \zeta \sim \mathcal{N}\left(0, \frac{(\bar{\theta}_1 + x_*\bar{\theta}_2)^2}{4\bar{\theta}_2^4 z^2 (x_* - z)^2}\right), \quad N \to \infty .$$

$h(\hat{\theta}^N)$ thus converges as $1/\sqrt{m}$ and is asymptotically normal with a limiting variance depending on z. If $\bar{\theta}_1 + x_*\bar{\theta}_2 = 0$,

$$\sqrt{N}[h(\hat{\theta}^N) - h(\bar{\theta})] = \sqrt{N}\left(-\frac{\hat{\theta}_1^N}{2\hat{\theta}_2^N} - \frac{x_*}{2}\right) = -\frac{\sqrt{N}}{2}\frac{\gamma_{N-m}}{\sqrt{N-m}}\frac{1}{x_*\hat{\theta}_2^N}$$

and

$$\sqrt{N}[h(\hat{\theta}^N) - h(\bar{\theta})] \xrightarrow{\text{d}} \zeta \sim \mathcal{N}(0, 1/(4\bar{\theta}_2^2 x_*^2)) .$$

This is the only situation within Examples 1 and 2 where regular asymptotic normality holds: $h(\hat{\theta}^N)$ converges as $1/\sqrt{N}$, is asymptotically normal and has a limiting variance that can be computed from the limiting design ξ_*, that is, which coincides with $[\partial h(\theta)/\partial\theta^\top \mathbf{M}^-(\xi_*)\, \partial h(\theta)/\partial\theta]_{\bar{\theta}}$. Note that assuming that $\bar{\theta}_1 + x_*\bar{\theta}_2 = 0$ amounts to assuming that the prior guess $h_* = x_*/2$ coincides with the true location of the maximum of the model response, which is rather unrealistic.

8.3 Consistency of Estimators

The (strong) consistency results presented below are based on the following two lemmas which respectively concern designs satisfying the conditions of Definitions 1 and 2. The proofs are given in Appendix.

Lemma 1. *Let the sequence $(x_i)_i$ converge strongly to a discrete design ξ in the sense of Definition 1. Assume that $a(x, \theta)$ is a bounded function on $\mathcal{X} \times \Theta$ and that $(\alpha_i)_i$ is an i.i.d. sequence of random variables having finite mean and variance. Then*

$$\lim_{N\to\infty} \frac{1}{N} \sum_{k=1}^{N} a(x_k, \theta)\, \alpha_k = \mathbb{E}\{\alpha_1\} \sum_{x \in \mathcal{S}_\xi} a(x, \theta)\xi(\{x\})$$

a.s. with respect to $\alpha_1, \alpha_2 \ldots$ and uniformly on Θ.

Lemma 2. *Let $(z_i)_i$ be a sequence of i.i.d. random vectors from $\mathbb{R}^r$ and $a(z,\theta)$ be a Borel measurable real function on $\mathbb{R}^r \times \Theta$, continuous in $\theta \in \Theta$ for any z, with Θ a compact subset of $\mathbb{R}^p$. Assume that*

$$\mathbb{E}\left\{\max_{\theta\in\Theta} |a(z_1,\theta)|\right\} < \infty\,,$$

then $\mathbb{E}\{a(z_1,\theta)\}$ is continuous in $\theta \in \Theta$ and

$$\lim_{N\to\infty} \frac{1}{N}\sum_{i=1}^{N} a(z_i,\theta) = \mathbb{E}\{a(z_1,\theta)\}$$

a.s. and uniformly on Θ.

Next result (Theorem 1) is quite standard and concerns the set of possible limiting points of the sequence of LSE of θ. We shall denote $y^N = (y(x_1),\ldots,y(x_N))$, $y = (y(x_i))_i$ the sequence of observations $y(x_i) = \eta(x_i,\bar\theta) + \varepsilon_i$, where $\bar\theta$ is the unknown true value of the model parameters, and

$$\begin{aligned} J_N(\theta,y^N) &= \frac{1}{N}\sum_{i=1}^{N} [y(x_i) - \eta(x_i,\theta)]^2\,, \\ J(\theta) &= \int_{\mathcal{X}} \left[\eta(x,\bar\theta) - \eta(x,\theta)\right]^2 \xi(\mathrm{d}x) + \sigma^2\,. \end{aligned} \tag{8.14}$$

We shall assume the following.

Assumption A1: *Θ is a compact subset of $\mathbb{R}^p$ and $\eta(x,\theta)$ is bounded on $\mathcal{X}\times\Theta$ and continuous in θ for any $x \in \mathcal{X}$, with $\mathcal{X} \subset \mathbb{R}^k$.*

Theorem 1. *Let the sequence $(x_i)_i$ either converge to ξ in the sense of Definition 1, or be generated by ξ according to Definition 2. Then, under A1, with probability one all limit points of the sequence $(\hat\theta^N(y^N))_N$ of LSE,*

$$\hat\theta^N(y^N) \in \arg\min_{\theta\in\Theta} J_N(\theta,y^N)\,, \tag{8.15}$$

are elements of the set

$$\Theta^{\#} = \arg\min_{\theta\in\Theta} J(\theta)\,.$$

Proof. Lemmas 1 and 2 allow to proceed essentially in a standard way. We can write

$$\begin{aligned} \frac{1}{N}\sum_{k=1}^{N} [y(x_k) - \eta(x_k,\theta)]^2 = \frac{1}{N}\sum_{k=1}^{N} \varepsilon_k^2 + \frac{2}{N}\sum_{k=1}^{N} \left[\eta(x_k,\bar\theta) - \eta(x_k,\theta)\right]\varepsilon_k \\ + \frac{1}{N}\sum_{k=1}^{N} \left[\eta(x_k,\bar\theta) - \eta(x_k,\theta)\right]^2\,, \end{aligned}$$

with $\varepsilon_k = y(x_k) - \eta(x_k, \bar{\theta})$. The first term on the right-hand side converges a.s. to σ^2 by the strong law of large numbers. As required in Lemma 1, $\eta(x, \bar{\theta}) - \eta(x, \theta)$ is bounded on $\mathscr{X} \times \Theta$, hence

$$\lim_{N\to\infty} \frac{1}{N} \sum_{k=1}^{N} \left[\eta(x_k, \bar{\theta}) - \eta(x_k, \theta)\right] \varepsilon_k = \sum_{x\in\mathcal{S}_\xi} \left[\eta(x, \bar{\theta}) - \eta(x, \theta)\right] \xi(\{x\}) \mathbb{E}\{\varepsilon_1\} = 0$$

a.s. and uniformly on Θ. Similarly, taking $z = (x, \varepsilon)$ and $a(z, \theta) = [\eta(x, \bar{\theta}) - \eta(x, \theta)]\varepsilon$ in Lemma 2, we obtain

$$\lim_{N\to\infty} \frac{1}{N} \sum_{k=1}^{N} \left[\eta(x_k, \bar{\theta}) - \eta(x_k, \theta)\right] \varepsilon_k = \mathbb{E}_\xi \left\{\eta(x, \bar{\theta}) - \eta(x, \theta)\right\} \mathbb{E}\{\varepsilon_1\} = 0$$

a.s. and uniformly on Θ. By the same arguments we also obtain

$$\lim_{N\to\infty} \frac{1}{N} \sum_{k=1}^{N} \left[\eta(x_k, \bar{\theta}) - \eta(x_k, \theta)\right]^2 = \int_{\mathscr{X}} \left[\eta(x, \bar{\theta}) - \eta(x, \theta)\right]^2 \xi(\mathrm{d}x)$$

a.s. and uniformly on Θ. We have thus proved that the sequence $(J_N(\theta, y^N))_N$ converges a.s. and uniformly on Θ to $J(\theta)$.

Let $\theta^\# = \theta^\#(y)$ be a limit point of $(\hat{\theta}^N(y^N))_N$ (which exists since Θ is compact). There exists a subsequence $(\hat{\theta}^{N_t})_t$ of this sequence that converges to $\theta^\#$. From the definitions of $J_N(\theta, y^N)$ and $\hat{\theta}^N(y^N)$, we can write

$$J_{N_t}(\bar{\theta}, y^{N_t}) = \frac{1}{N_t} \sum_{k=1}^{N_t} \varepsilon_k^2 \geq J_{N_t}(\hat{\theta}^{N_t}, y^{N_t}) .$$

The left-hand side converges to σ^2 a.s. as $t \to \infty$ while the right-hand side converge a.s. to $J(\theta^\#) = \int_{\mathscr{X}} \left[\eta(x, \bar{\theta}) - \eta(x, \theta^\#)\right]^2 \xi(\mathrm{d}x) + \sigma^2$ (since $J_N(\theta, y^N)$ converges a.s. and uniformly in θ to $J(\theta)$). This implies that $\int_{\mathscr{X}} \left[\eta(x, \bar{\theta}) - \eta(x, \theta^\#)\right]^2 \xi(\mathrm{d}x) = 0$, i.e., that $\theta^\# \in \Theta^\#$. □

One may notice that Theorem 1 does not mean that $\hat{\theta}^N$ can take any value in $\Theta^\#$. For instance, in Example 2 we have $\Theta^\# = \{\theta : \theta_1 + x_*\theta_2 = \bar{\theta}_1 + x_*\bar{\theta}_2\}$ although $\hat{\theta}^N$ converges a.s. to $\bar{\theta}$ when the number m of observations at $z \neq x_*$ tends to infinity, see also Wu (1981).[2]

Next result concerns the consistency of $h(\hat{\theta}^N)$. We shall require the following.

[2] This reference presents sufficient conditions for consistency that are weaker than those considered here, in the sense that they do not require the design sequence $(x_i)_i$ to obey Definitions 1 or 2. However, this may lead to irregular asymptotic normality, as shown in Examples 1 and 2, and is not considered in this chapter.

ASSUMPTION A2: *The function* $h(\theta)$ *is continuous in* $\theta \in \Theta$ *and such that*

$$\int_{\mathcal{X}} \left[\eta(x,\theta) - \eta(x,\bar{\theta})\right]^2 \xi(\mathrm{d}x) = 0 \Rightarrow h(\theta) = h(\bar{\theta}) .$$

Theorem 2 (Consistency of $h(\hat{\theta}^N)$). *Under the assumptions of Theorem 1 we suppose that* $h(\cdot)$ *satisfies A2. Let* $(\hat{\theta}^N(y^N))_N$ *be any sequence defined by (8.15). Then*

$$\lim_{N\to\infty} h[\hat{\theta}^N(y^N)] = h(\bar{\theta}) \quad a.s.$$

Proof. As in the proof of Theorem 1 we extract from $(\hat{\theta}^N(y^N))_N$ a subsequence $(\hat{\theta}^{N_t})_t$ converging to $\theta^\# = \theta^\#(y) \in \Theta^\#$. From the continuity of $h(\cdot)$, $\lim_{t\to\infty} h(\hat{\theta}^{N_t}) = h(\theta^\#)$, and, from Theorem 1 and A2, $h(\theta^\#) = h(\bar{\theta})$ a.s. Therefore, every converging subsequence of $(h[\hat{\theta}^N(y^N)])_N$ has a.s. the same limit $h(\bar{\theta})$, which is the limit of the whole sequence $(h[\hat{\theta}^N(y^N)])_N$. □

8.4 On the Geometry of the Model Under the Design Measure ξ

We shall assume the following in the rest of the chapter.

ASSUMPTION A3: $\bar{\theta} \in \mathrm{int}(\Theta)$ *and, for any* $x \in \mathcal{X}$, $\eta(x,\theta)$ *is two times continuously differentiable with respect to* $\theta \in \Theta_0$*; these first two derivatives are bounded on* $\mathcal{X} \times \Theta$.

Let ξ be a design measure, and $\mathcal{L}_2(\xi)$ be the Hilbert space of real valued functions ϕ on $\mathcal{X}$ which are square integrable, i.e.,

$$\int_{\mathcal{X}} \phi^2(x)\,\xi(\mathrm{d}x) < \infty .$$

Two functions ϕ and ϕ^* are equivalent in $\mathcal{L}_2(\xi)$ if $\int_{\mathcal{X}} \left[\phi(x) - \phi^*(x)\right]^2 \xi(\mathrm{d}x) = 0$, that is, $\phi(x) = \phi^*(x)$ ξ-a.s., which we shall denote

$$\phi \overset{\xi}{=} \phi^* .$$

Note that for a discrete measure ξ the equivalence $\phi \overset{\xi}{=} \phi^*$ means that $\phi(x) = \phi^*(x)$ on the support of ξ. The elements of $\mathcal{L}_2(\xi)$ are thus classes of equivalent functions rather than functions; by the sentence "the function ϕ belongs to a subset $\mathcal{A}$ of $\mathcal{L}_2(\xi)$" we mean that the whole class containing ϕ belongs to $\mathcal{A}$, which we shall denote $\phi \overset{\xi}{\in} \mathcal{A}$. The notation $\mathcal{A} \overset{\xi}{\subset} \mathcal{B}$ is defined similarly. The inner product and the norm in $\mathcal{L}_2(\xi)$ are, respectively, defined by

$$\langle \phi, \phi^* \rangle_\xi = \int_{\mathcal{X}} \phi(x)\, \phi^*(x)\, \xi(\mathrm{d}x) \quad \text{and} \quad \|\phi\|_\xi = \left[\int_{\mathcal{X}} \phi^2(x)\, \xi(\mathrm{d}x) \right]^{1/2} .$$

Under A3, the functions $\eta(\cdot, \theta)$ and

$$\{\mathbf{f}_\theta\}_i(\cdot) = \partial\eta(\cdot, \theta)/\partial\theta_i \,, \; i = 1, \ldots, p \,, \tag{8.16}$$

belong to $\mathcal{L}_2(\xi)$. We shall denote

$$\theta \overset{\xi}{\sim} \theta^*$$

when the values θ and θ^* satisfy $\eta(\cdot, \theta) \overset{\xi}{=} \eta(\cdot, \theta^*)$. Notice that $\theta \overset{\xi}{\sim} \theta^*$ does not imply that $\{\mathbf{f}_\theta\}_i \overset{\xi}{=} \{\mathbf{f}_{\theta_*}\}_i$ (this is the object of Lemma 3 below). Also note that $\theta \overset{\xi}{\sim} \bar{\theta}$ is equivalent to $\theta \in \Theta^\#$ as defined in Theorem 1.

For any $\theta \in \Theta_0$ we define an operator P_θ which acts upon any $\phi \in \mathcal{L}_2(\xi)$ as follows,

$$(P_\theta \phi)(x') = \mathbf{f}_\theta^\top(x') \mathbf{M}^+(\xi, \theta) \int_{\mathcal{X}} \mathbf{f}_\theta(x)\, \phi(x)\, \xi(\mathrm{d}x)$$

where $\mathbf{M}^+$ denotes the Moore–Penrose g-inverse of $\mathbf{M}$, $\mathbf{M}(\xi, \theta)$ is defined in (8.6) and $\mathbf{f}_\theta(x)$ in (8.5). We keep the same notation $P_\theta \phi$ when ϕ is vector valued with components in $\mathcal{L}_2(\xi)$. We denote

$$\mathcal{L}_\theta = \left\{ \alpha^\top \mathbf{f}_\theta(\cdot) : \alpha \in \mathbb{R}^p \right\} .$$

From $\mathbf{M}\mathbf{M}^+\mathbf{M} = \mathbf{M}$ and $\mathbf{M}^+\mathbf{M}\mathbf{M}^+ = \mathbf{M}^+$ we obtain

$$\begin{aligned}
&\phi \in \mathcal{L}_2(\xi) \Rightarrow P_\theta \phi \overset{\xi}{\in} \mathcal{L}_\theta \,, \\
&\phi \in \mathcal{L}_\theta \Rightarrow P_\theta \phi \overset{\xi}{=} \phi \,, \\
&\phi ,\; \phi^* \in \mathcal{L}_2(\xi) \Rightarrow \langle \phi, P_\theta \phi^* \rangle_\xi = \langle P_\theta \phi, \phi^* \rangle_\xi \,,
\end{aligned}$$

hence P_θ is the orthogonal projector onto $\mathcal{L}_\theta$.

We shall need the following technical assumptions on the geometry of the model.

Assumption A4: *For any point $\theta^* \overset{\xi}{\sim} \bar{\theta}$ there exists a neighborhood $\mathcal{V}(\theta^*)$ such that*

$$\forall \theta \in \mathcal{V}(\theta^*) \,, \; \mathrm{rank}[\mathbf{M}(\xi, \theta)] = \mathrm{rank}[\mathbf{M}(\xi, \theta^*)] \,.$$

Assumption A5: *Define*

$$\mathcal{S}_\epsilon = \left\{ \theta \in \mathrm{int}(\Theta) : \left\| \eta(\cdot, \theta) - \eta(\cdot, \bar{\theta}) \right\|_\xi^2 < \epsilon \right\} .$$

There exists $\epsilon > 0$ *such that for every* $\theta^{\#}, \theta^* \in \mathcal{S}_\epsilon$ *we have*

$$\left[\frac{\partial}{\partial\theta}\left\|\eta(\cdot,\theta)-\eta(\cdot,\theta^{\#})\right\|_\xi^2\right]_{\theta=\theta^*} = 0 \;\Rightarrow\; \theta^{\#} \overset{\xi}{\sim} \theta^* .$$

The assumptions A4 and A5 admit a straightforward geometrical and statistical interpretation in the case where the measure ξ is discrete. Let $\{x^{(1)}, \ldots, x^{(k)}\}$ denote the support of ξ and define

$$\eta(\theta) = (\eta(x^{(1)},\theta), \ldots, \eta(x^{(k)},\theta))^\top .$$

The set $\mathcal{E} = \{\eta(\theta) : \theta \in \Theta\}$ is then the expectation surface of the model under the design ξ and $\theta^* \overset{\xi}{\sim} \bar{\theta}$ is equivalent to $\eta(\theta^*) = \eta(\bar{\theta})$. When A4 is not satisfied, it means that the surface $\mathcal{E}$ possesses edges, and the point $\eta(\bar{\theta})$ belongs to such an edge (although $\bar{\theta} \in \mathrm{int}(\Theta)$). When A5 is not satisfied, it means that the surface $\mathcal{E}$ intersects itself at the point $\eta(\bar{\theta})$ and therefore there are points $\eta(\theta)$ arbitrarily close to $\eta(\bar{\theta})$ with θ far from $\bar{\theta}$. In any of such circumstances asymptotic normality of the least squares estimator does not hold (from the geometrical interpretation of least squares estimation as the projection of the vector of observations onto the expectation surface $\mathcal{E}$).

The geometrical assumptions above yield the following.

Lemma 3. *Under A4 and A5,* $\bar{\alpha} \overset{\xi}{\sim} \bar{\theta}$ *implies*

$$\forall \phi, \phi^* \in \mathcal{L}_2(\xi)\,,\;\; \langle \phi, P_{\bar{\alpha}}\phi^*\rangle_\xi = \langle \phi, P_{\bar{\theta}}\phi^*\rangle_\xi \,.$$

The proof is given in Appendix.

One may notice that Assumptions A4 and A5 are only used to prove

$$\bar{\alpha} \overset{\xi}{\sim} \bar{\theta} \Rightarrow \mathcal{L}_{\bar{\alpha}} \overset{\xi}{=} \mathcal{L}_{\bar{\theta}} \,,$$

to be used in Theorem 3, see the proof of Lemma 3. This result could also be obtained in a different way. Indeed, from the properties of the projector $P_{\bar{\theta}}$ we have

$$\begin{aligned}
\mathcal{L}_{\bar{\alpha}} \overset{\xi}{\subset} \mathcal{L}_{\bar{\theta}} &\Leftrightarrow P_{\bar{\theta}}\{\mathbf{f}_{\bar{\alpha}}\}_i \overset{\xi}{=} \{\mathbf{f}_{\bar{\alpha}}\}_i\,,\; i=1,\ldots,p\,,\\
&\Leftrightarrow \|P_{\bar{\theta}}\{\mathbf{f}_{\bar{\alpha}}\}_i - \{\mathbf{f}_{\bar{\alpha}}\}_i\|_\xi^2 = 0\,,\; i=1,\ldots,p\,,\\
&\Leftrightarrow \{\mathbf{M}(\xi,\bar{\alpha})\}_{ii} - \left\{\mathbf{N}(\xi,\bar{\alpha},\bar{\theta})\mathbf{M}^+(\xi,\bar{\theta})\mathbf{N}(\xi,\bar{\theta},\bar{\alpha})\right\}_{ii} = 0\,,\; i=1,\ldots,p\,,
\end{aligned}$$

where we used the notation

$$\mathbf{N}(\xi,\bar{\alpha},\bar{\theta}) = \int_{\mathcal{X}} \mathbf{f}_{\bar{\alpha}}(x)\,\mathbf{f}_{\bar{\theta}}^\top(x)\,\xi(\mathrm{d}x)\,.$$

Therefore, in Theorem 3 below instead of Assumptions A4 and A5 we can use equivalently the following, which may be easier to check although it does not have a geometrical interpretation.

Assumption B1: $\bar{\alpha} \stackrel{\xi}{\sim} \bar{\theta}$ *implies*

$$\{\mathbf{M}(\xi,\bar{\alpha})\}_{ii} = \{\mathbf{N}(\xi,\bar{\alpha},\bar{\theta})\mathbf{M}^+(\xi,\bar{\theta})\mathbf{N}(\xi,\bar{\theta},\bar{\alpha})\}_{ii} ,$$
$$\{\mathbf{M}(\xi,\bar{\theta})\}_{ii} = \{\mathbf{N}(\xi,\bar{\theta},\bar{\alpha})\mathbf{M}^+(\xi,\bar{\alpha})\mathbf{N}(\xi,\bar{\alpha},\bar{\theta})\}_{ii} ,$$

for every $i = 1, \ldots, p$.

8.5 The Regular Asymptotic Normality of $h(\hat{\theta}^N)$

Definition 3. *We say that* $h(\hat{\theta}^N)$ *satisfies the property of* regular asymptotic normality *when* $\sqrt{N}[h(\hat{\theta}^N) - h(\bar{\theta})]$ *converges in distribution to a variable distributed* $\mathcal{N}(0, [\partial h(\theta)/\partial\theta^\top \mathbf{M}^-(\xi,\theta)\, \partial h(\theta)/\partial\theta]_{\bar{\theta}})$.

Examples 1 and 2 demonstrate that for a singular ξ regular asymptotic normality does not hold in general when $h(\cdot)$ is a nonlinear function of θ. We thus introduce an assumption on $h(\cdot)$ in addition to A2.

Assumption A6: *The function* $h(\cdot)$ *is defined and has a continuous nonzero vector of derivatives* $\partial h(\theta)/\partial\theta$ *on* Θ_0. *Moreover, for any* $\theta \stackrel{\xi}{\sim} \bar{\theta}$ *there exists a linear mapping* A_θ *from* $\mathcal{L}_2(\xi)$ *to* $\mathbb{R}$ *(a continuous linear functional on* $\mathcal{L}_2(\xi)$*), such that* $A_\theta = A_{\bar{\theta}}$ *and that*

$$\frac{\partial h(\theta)}{\partial\theta_i} = A_\theta\left[\{\mathbf{f}_\theta\}_i\right] , \ i = 1, \ldots, p ,$$

where $\{\mathbf{f}_\theta\}_i$ *is defined by (8.16).*

This receives a simple interpretation when ξ is a discrete design measure with support $\{x^{(1)}, \ldots, x^{(k)}\}$. Suppose that Assumption A2 holds for every $\bar{\theta} \in \Theta$. Then A6 is equivalent to the assumption that there exists a function Ψ, with continuous gradient, such that $h(\theta) = \Psi[\eta(\theta)]$, with $\eta(\theta) = (\eta(x^{(1)},\theta), \ldots, \eta(x^{(k)},\theta))^\top$, and we obtain

$$\frac{\partial h(\theta)}{\partial\theta^\top} = \frac{\partial\Psi(t)}{\partial t^\top}_{|t=\eta(\theta)} \frac{\partial\eta(\theta)}{\partial\theta^\top} .$$

A6 thus holds for every $\bar{\theta} \in \mathrm{int}(\Theta)$ with $A_\theta = \partial\Psi(t)/\partial t^\top_{|t=\eta(\theta)}$.

It is useful to discuss A2 and A6 in the context of Example 2. There the limiting design is $\xi_* = \delta(x_*)$, the measure that puts mass one at x_*. Therefore, $\theta \stackrel{\xi_*}{\sim} \bar{\theta} \Leftrightarrow \theta_1 + x_*\theta_2 = \bar{\theta}_1 + x_*\bar{\theta}_2$. It follows that $\theta \stackrel{\xi_*}{\sim} \bar{\theta} \Rightarrow h(\theta) = h(\bar{\theta})$ only if $\bar{\theta}_1 + x_*\bar{\theta}_2 = 0$, and this is the only case where A2 holds. We have seen in Example 2 that regular asymptotic normality does not hold when $\bar{\theta}_1 + x_*\bar{\theta}_2 \neq 0$, hence the importance of A2. Consider now the derivative of $h(\theta)$. We have $\partial h(\theta)/\partial\theta = -1/(2\theta_2)[1 \quad -\theta_1/\theta_2]^\top$ and $\partial\eta(x_*,\theta)/\partial\theta = (x_* \quad x_*^2)^\top$. Therefore, even if $\bar{\theta}_1 + x_* \bar{\theta}_2 = 0$, we obtain $\theta \overset{\xi_*}{\sim} \bar{\theta} \Rightarrow \partial h(\theta)/\partial\theta = -1/(2\theta_2)[1 \;\; x_*]^\top = -1/(2x_*\theta_2)\partial\eta(x_*, \theta)/\partial\theta$, and A6 does not hold if $\theta \neq \bar{\theta}$. When m is fixed in Example 2, $\Theta^\# = \arg\min J(\theta)$, with $J(\theta)$ given by (8.14), contains other points than $\bar{\theta}$; then A6 does not hold and there is no regular asymptotic normality for $h(\hat{\theta}^N)$. On the contrary, when $m \to \infty$ $\Theta^\# = \{\bar{\theta}\}$ and A6 holds trivially: this is the only situation in the example where regular asymptotic normality holds.

When ξ is a continuous design measure, an example where A6 holds is when

$$h(\theta) = \Psi[h_1(\theta), \ldots, h_k(\theta)]$$

with Ψ a continuously differentiable function of k variables, and with

$$h_i(\theta) = \int_{\mathcal{X}} g_i[\eta(x, \theta), x]\, \xi(\mathrm{d}x)\,, \;\; i = 1, \ldots, k\,,$$

for some functions $g_i(t, x)$ differentiable with respect to t for any x in the support of ξ. Then, supposing that we can interchange the order of derivatives and integrals, we obtain

$$\frac{\partial h(\theta)}{\partial \theta_i} = \sum_{j=1}^{k} \left[\frac{\partial \Psi(v)}{\partial v_j}\right]_{v_j = h_j(\theta)} \int_{\mathcal{X}} \left[\frac{\partial g_j(t, x)}{\partial t}\right]_{t = \eta(x,\theta)} \{\mathbf{f}_\theta\}_i(x)\, \xi(\mathrm{d}x)\,,$$

and for any $\phi \in \mathcal{L}_2(\xi)$

$$A_\theta(\phi) = \sum_{j=1}^{k} \left[\frac{\partial \Psi(v)}{\partial v_j}\right]_{v_j = h_j(\theta)} \int_{\mathcal{X}} \left[\frac{\partial g_j(t, x)}{\partial t}\right]_{t = \eta(x,\theta)} \phi(x)\, \xi(\mathrm{d}x)$$

so that A6 holds when A2 is satisfied.

We can now formulate the main result of the chapter, concerning regular asymptotic normality.

Theorem 3. *Let the sequence $(x_i)_i$ either converge to ξ in the sense of Definition 1, or be generated by ξ according to Definition 2. Suppose that the model response $\eta(x, \theta)$ satisfies assumptions A1, A3 and that the function of interest $h(\theta)$ satisfies A2, A6. Let $(\hat{\theta}^N(y^N))_N$ be a sequence of solutions of (8.15). Then, under A4, A5, or under B1, regular asymptotic normality holds: the sequence $\sqrt{N}\left\{h[\hat{\theta}^N(y^N)] - h(\bar{\theta})\right\}$ converges in distribution as $N \to \infty$ to a random variable distributed*

$$\mathcal{N}\left(0, \sigma^2 \left[\frac{\partial h(\theta)}{\partial \theta^\top} \mathbf{M}^-(\xi, \theta) \frac{\partial h(\theta)}{\partial \theta}\right]_{\theta = \bar{\theta}}\right)$$

where the choice of the g-inverse is arbitrary.

Proof. From the properties of P_θ we have for every $\theta \overset{\xi}{\sim} \bar{\theta}$

$$\frac{\partial h(\theta)}{\partial \theta_i} = A_\theta \left[P_\theta \{\mathbf{f}_\theta\}_i \right] = \mathbf{u}^\top(\theta) \{\mathbf{M}(\xi, \theta)\}_{.i} \,, \tag{8.17}$$

where

$$u_j(\theta) = A_\theta \left[\mathbf{f}_\theta^\top \{\mathbf{M}^+(\xi, \theta)\}_{.j} \right] .$$

From Lemma 3, we have

$$\theta^* \overset{\xi}{\sim} \bar{\theta} \Rightarrow \mathbf{u}^\top(\theta^*) \mathbf{f}_{\theta^*} \overset{\xi}{=} \mathbf{u}^\top(\bar{\theta}) \mathbf{f}_{\bar{\theta}} \,. \tag{8.18}$$

Indeed, for any ϕ in $\mathcal{L}_2(\xi)$ we can write

$$\langle \mathbf{u}^\top(\theta^*) \mathbf{f}_{\theta^*}, \phi \rangle_\xi = \sum_{j=1}^{p} A_{\theta^*} \left[\mathbf{f}_{\theta^*}^\top \{\mathbf{M}^+(\xi, \theta^*)\}_{.j} \right] \int_{\mathcal{X}} \{\mathbf{f}_{\theta^*}\}_j(x) \, \phi(x) \, \xi(\mathrm{d}x) \,.$$

From the linearity of A_{θ^*} this is equal to

$$\begin{aligned} A_{\theta^*} \left[\sum_{j=1}^{p} \mathbf{f}_{\theta^*}^\top \{\mathbf{M}^+(\xi, \theta^*)\}_{.j} \int_{\mathcal{X}} \{\mathbf{f}_{\theta^*}\}_j(x) \, \phi(x) \, \xi(\mathrm{d}x) \right] &= A_{\theta^*}[P_{\theta^*} \phi] \\ &= A_{\bar{\theta}}[P_{\bar{\theta}} \phi] = \langle \mathbf{u}^\top(\bar{\theta}) \mathbf{f}_{\bar{\theta}}, \phi \rangle_\xi \end{aligned}$$

so that $\mathbf{u}^\top(\theta^*) \mathbf{f}_{\theta^*} \overset{\xi}{=} \mathbf{u}^\top(\bar{\theta}) \mathbf{f}_{\bar{\theta}}$.

Let $(\hat{\theta}^N(y^N))_N$ be a sequence of solutions of (8.15) and $(\hat{\theta}^{N_t})_t$ be a subsequence converging to a limit point $\theta^\# = \theta^\#(y) \overset{\xi}{\sim} \bar{\theta}$, see Theorem 1. By the Taylor formula we have

$$\begin{aligned} 0 &= \frac{\partial J_{N_t}(\theta, y^{N_t})}{\partial \theta}_{|\hat{\theta}^{N_t}} \\ &= \frac{\partial J_{N_t}(\theta, y^{N_t})}{\partial \theta}_{|\theta^\#} + \left[\frac{\partial^2 J_{N_t}(\theta, y^{N_t})}{\partial \theta \partial \theta^\top} \right]_{\beta_t} (\hat{\theta}^{N_t} - \theta^\#) \,, \end{aligned} \tag{8.19}$$

where $\beta_t = \beta_t(y)$ lies on the segment joining $\hat{\theta}^{N_t}$ with $\theta^\#(y)$. (Notice that $\lim_{t \to \infty} \beta_t(y) = \theta^\#(y)$ a.s.) Now,

$$\begin{aligned} -\sqrt{N_t}\, \mathbf{u}^\top(\theta^\#) \left[\frac{\partial J_{N_t}(\theta, y^{N_t})}{\partial \theta} \right]_{\theta^\#} &= \\ \frac{2}{\sqrt{N_t}} \sum_{i=1}^{N_t} \mathbf{u}^\top(\theta^\#) \left[\frac{\partial \eta(x_i, \theta)}{\partial \theta} \right]_{\theta^\#} &\left[y(x_i) - \eta(x_i, \theta^\#) \right]. \end{aligned} \tag{8.20}$$

Consider first the case where $(x_i)_i$ converges strongly to a discrete design in the sense of Definition 1. We decompose the sum on the right-hand side of (8.20) into two sums: the first one corresponds to indices i such that $x_i \notin S_\xi$, the support of ξ, the second one is for $x_i \in S_\xi$. The first sum then tends to zero in probability. For the second, we use (8.18) and the fact that $\eta[x_i, \theta^\#(y)] = \eta(x_i, \bar{\theta})$ for $x_i \in S_\xi$ to obtain

$$\frac{2}{\sqrt{N_t}} \sum_{i=1,\, x_i \in S_\xi}^{N_t} \mathbf{u}^\top(\bar{\theta})\, \mathbf{f}_{\bar{\theta}}(x_i)\, \varepsilon_i \,,$$

which, by the central limit theorem, converges in distribution to $\mathcal{N}(0, 4D)$ with

$$D = \sigma^2\, \mathbf{u}^\top(\bar{\theta}) \mathbf{M}(\xi, \bar{\theta}) \mathbf{u}(\bar{\theta}) \,. \tag{8.21}$$

Consider alternatively the case of a randomized design in the sense of Definition 2. For almost every sequence of errors $\varepsilon_1, \varepsilon_2 \ldots$ we have

$$\xi(\{x : \eta(x, \bar{\theta}) = \eta[x, \theta^\#(y)]\}) = 1 \,,$$

see Theorem 1, and

$$\xi\left(\{x : \mathbf{u}^\top[\theta^\#(y)] \mathbf{f}_{\theta^\#(y)}(x) = \mathbf{u}^\top(\bar{\theta}) \mathbf{f}_{\bar{\theta}}(x)\}\right) = 1 \,,$$

see (8.18). Hence (8.20) implies that for each t

$$-\sqrt{N_t}\, \mathbf{u}^\top(\theta^\#) \left[\frac{\partial J_{N_t}(\theta, y^{N_t})}{\partial \theta} \right]_{\theta^\#} = \frac{2}{\sqrt{N_t}} \sum_{i=1}^{N_t} \mathbf{u}^\top(\bar{\theta})\, \mathbf{f}_{\bar{\theta}}(x_i)\, \varepsilon_i \,, \ \xi\text{-a.s.}$$

From the independence of x_i and ε_i, $\mathbb{E}\{\varepsilon_i\} = 0$ and $\mathrm{Var}\{\varepsilon_i\} = \sigma^2$ we obtain, again by the central limit theorem, that the last sum converges in distribution to a variable distributed $\mathcal{N}(0, 4D)$.

Therefore, for both types of designs we have

$$-\sqrt{N_t}\, \mathbf{u}^\top(\theta^\#) \left[\frac{\partial J_{N_t}(\theta, y^{N_t})}{\partial \theta} \right]_{\theta^\#} \xrightarrow{\mathrm{d}} \nu \sim \mathcal{N}(0, 4D) \,, \ t \to \infty \,, \tag{8.22}$$

with D given by (8.21).

Moreover, Lemmas 1 and 2 imply that $\partial^2 J_{N_t}(\theta, y^{N_t}) / \partial\theta \partial\theta^\top$ converges a.s. and uniformly in θ to

$$-2 \int_{\mathcal{X}} [\eta(x, \bar{\theta}) - \eta(x, \theta)]\, \frac{\partial^2 \eta(x, \theta)}{\partial\theta \partial\theta^\top}\, \xi(\mathrm{d}x) + 2\mathbf{M}(\xi, \theta)$$

as $t \to \infty$. Therefore, for both types of designs we have

$$\begin{aligned} \mathbf{u}^\top(\theta^\#) \left[\frac{\partial^2 J_{N_t}(\theta, y^{N_t})}{\partial\theta \partial\theta^\top} \right]_{\beta_t(y)} &\xrightarrow{\text{a.s.}} 2\mathbf{u}^\top(\theta^\#)\, \mathbf{M}(\xi, \theta^\#) \\ &= 2 \left[\frac{\partial h(\theta)}{\partial \theta^\top} \right]_{\theta^\#} \,, \ t \to \infty \,. \end{aligned}$$

Hence from (8.19) and (8.22) we obtain

$$2\sqrt{N_t}\left[\frac{\partial h(\theta)}{\partial \theta^\top}\right]_{\theta^\#}(\hat{\theta}^{N_t}-\theta^\#)\stackrel{\mathrm{d}}{\to}\nu\sim\mathcal{N}(0,4D)\,,\ t\to\infty\,. \tag{8.23}$$

Applying the Taylor formula again, we obtain

$$\begin{aligned}\sqrt{N_t}\left[h(\hat{\theta}^{N_t})-h(\bar{\theta})\right] &= \sqrt{N_t}\left[h(\hat{\theta}^{N_t})-h(\theta^\#)\right]\\ &= \sqrt{N_t}\left[\frac{\partial h(\theta)}{\partial \theta^\top}\right]_{\delta_t(y)}(\hat{\theta}^{N_t}-\theta^\#)\,,\ t\to\infty\,,\end{aligned}$$

where $\delta_t(y)$ is on the segment connecting $\hat{\theta}^{N_t}$ with $\theta^\#$, hence $\delta_t(y)\stackrel{\mathrm{a.s.}}{\to}\theta^\#$ and $\left[\partial h(\theta)/\partial\theta^\top\right]_{\delta_t(y)}\stackrel{\mathrm{a.s.}}{\to}\left[\partial h(\theta)/\partial\theta^\top\right]_{\theta^\#}$, $t\to\infty$. Finally, we obtain from (8.23)

$$\sqrt{N_t}\left[h(\hat{\theta}^{N_t})-h(\bar{\theta})\right]\stackrel{\mathrm{d}}{\to}\frac{\nu}{2}\sim\mathcal{N}(0,D)\,,\ t\to\infty\,,$$

and, according to (8.17), $[\partial h(\theta)/\partial\theta]_{\bar{\theta}}$ is in the range of $\mathbf{M}(\xi,\bar{\theta})$, which gives

$$D=\sigma^2\,\mathbf{u}^\top(\bar{\theta})\mathbf{M}(\xi,\bar{\theta})\mathbf{u}(\bar{\theta})=\sigma^2\left[\frac{\partial h(\theta)}{\partial\theta^\top}\mathbf{M}^-(\xi,\theta)\frac{\partial h(\theta)}{\partial\theta}\right]_{\bar{\theta}}$$

for any choice of the g-inverse $\mathbf{M}^-(\xi,\bar{\theta})$. □

8.6 Estimation of a Multidimensional Function $H(\theta)$

Let $H(\theta)=[h_1(\theta),\ldots,h_q(\theta)]^\top$ be a q-dimensional function defined on Θ. We shall assume the following.

ASSUMPTION A2*: *The functions $h_i(\theta)$ are continuous in Θ, $i=1,\ldots,q$, and such that*

$$\theta\stackrel{\xi}{\sim}\bar{\theta}\Rightarrow H(\theta)=H(\bar{\theta})\,.$$

We then have the following straightforward extension of Theorem 2.

Theorem 4. *Under the assumptions of Theorem 2, but with A2* replacing A2, we have*

$$\lim_{N\to\infty}H[\hat{\theta}^N(y^N)]=H(\bar{\theta})\ \text{a.s.}$$

for $(\hat{\theta}^N(y^N))_N$ any sequence defined by (8.15).

Consider now the following assumption; its substitution for A6 in Theorem 3 gives Theorem 5 below.

ASSUMPTION A6*: *The vector function $H(\theta)$ has a continuous Jacobian $\partial H(\theta)/\partial\theta^\top$ on Θ_0. Moreover, for each $\theta\stackrel{\xi}{\sim}\bar{\theta}$ there exists a continuous linear mapping B_θ from $L_2(\xi)$ to $\mathbb{R}^q$ such that $B_\theta=B_{\bar{\theta}}$ and that*

$$\frac{\partial H(\theta)}{\partial \theta_i} = B_\theta \left[\{\mathbf{f}_\theta\}_i\right] \,, \; i = 1, \ldots, p \,,$$

where $\{\mathbf{f}_\theta\}_i$ *is given by (8.16).*

Theorem 5. *Under the assumptions of Theorem 3, but with A2* and A6* replacing A2 and A6, for* $(\hat{\theta}^N(y^N))_N$ *a sequence of solutions of (8.15),* $\sqrt{N}\left\{H[\hat{\theta}^N(y^N)] - H(\bar{\theta})\right\}$ *converges in distribution as* $N \to \infty$ *to a random vector distributed normally*

$$\mathcal{N}\left(0, \sigma^2 \left[\frac{\partial H(\theta)}{\partial \theta^\top} \mathbf{M}^-(\xi, \theta) \frac{\partial H^\top(\theta)}{\partial \theta}\right]_{\bar{\theta}}\right)$$

where the choice of the g-inverse is arbitrary.

Proof. Take any $\mathbf{c} \in \mathbb{R}^q$, and define $h_c(\theta) = \mathbf{c}^\top H(\theta)$. Evidently $h_c(\theta)$ satisfies the assumptions of Theorem 3, and $\sqrt{N}\left\{h_c[\hat{\theta}^N(y^N)] - h_c(\bar{\theta})\right\}$ converges in distribution as $N \to \infty$ to a random variable distributed

$$\mathcal{N}\left(0, \sigma^2 \, \mathbf{c}^\top \left[\frac{\partial H(\theta)}{\partial \theta^\top} \mathbf{M}^-(\xi, \theta) \frac{\partial H^\top(\theta)}{\partial \theta}\right]_{\bar{\theta}} \mathbf{c}\right) .$$

□

Appendix. Proofs of Lemmas 1–3

Proof of Lemma 1. We can write

$$\left| \frac{1}{N} \sum_{k=1}^{N} a(x_k, \theta)\alpha_k - \mathbb{E}\{\alpha_1\} \sum_{x \in S_\xi} a(x, \theta)\, \xi(\{x\}) \right|$$

$$\leq \left| \frac{1}{N} \sum_{k=1,\, x_k \notin S_\xi}^{N} a(x_k, \theta)\alpha_k \right|$$

$$+ \sum_{x \in S_\xi} \sup_{\theta \in \Theta} |a(x, \theta)| \left| \frac{N(x)}{N} \left(\frac{1}{N(x)} \sum_{k=1,\, x_k = x}^{N} \alpha_k \right) - \mathbb{E}\{\alpha_1\}\, \xi(\{x\}) \right| ,$$

where $N(x)/N$ is the relative frequency of the point x in the sequence $x_1, x_2, \ldots, x_N$. The last sum for $x \in S_\xi$ tends to zero a.s. and uniformly on Θ, since $N(x)/N$ tends to $\xi(\{x\})$, and $[1/N(x)] \sum_{k=1,\, x_k=x}^{N} \alpha_k$ converges a.s. to $\mathbb{E}\{\alpha_1\}$. The first sum on the right-hand side is bounded by

$$\sup_{x \in \mathcal{X},\, \theta \in \Theta} |a(x, \theta)| \, \frac{N(\mathcal{X} \setminus S_\xi)}{N} \, \frac{1}{N(\mathcal{X} \setminus S_\xi)} \sum_{k=1,\, x_k \in \mathcal{X} \setminus S_\xi} \alpha_k \,.$$

This expression tends a.s. to zero, since $N(\mathcal{X} \setminus S_\xi)/N$ tends to zero, and the law of large numbers applies for the remaining part in case $N(\mathcal{X} \setminus S_\xi) \to \infty$. □

Proof of Lemma 2. We use a construction similar to that in Bierens (1994, p. 43). Take some fixed $\theta^1 \in \Theta$ and consider the set

$$\mathcal{B}(\theta^1, \delta) = \left\{\theta \in \Theta : \left\|\theta - \theta^1\right\| \leq \delta\right\} .$$

Define $\bar{a}_\delta(z)$ and $\underline{a}_\delta(z)$ as the maximum and the minimum of $a(z, \theta)$ over the set $\mathcal{B}(\theta^1, \delta)$.

The expectations $\mathbb{E}\{|\underline{a}_\delta(z)|\}$ and $\mathbb{E}\{|\bar{a}_\delta(z)|\}$ are bounded by

$$\mathbb{E}\{\max_{\theta \in \Theta} |a(z, \theta)|\} < \infty .$$

Also, $\bar{a}_\delta(z) - \underline{a}_\delta(z)$ is an increasing function of δ. Hence, we can interchange the order of the limit and expectation in the following expression

$$\lim_{\delta \searrow 0} [\mathbb{E}\{\bar{a}_\delta(z)\} - \mathbb{E}\{\underline{a}_\delta(z)\}] = \mathbb{E}\left\{\lim_{\delta \searrow 0} [\bar{a}_\delta(z) - \underline{a}_\delta(z)]\right\} = 0 ,$$

which proves the continuity of $\mathbb{E}\{a(z, \theta)\}$ at θ^1 and implies

$$\forall \beta > 0 ,\ \exists \delta(\beta) > 0 \ \text{ such that } \ \left|\mathbb{E}\{\bar{a}_{\delta(\beta)}(z)\} - \mathbb{E}\{\underline{a}_{\delta(\beta)}(z)\}\right| < \frac{\beta}{2} .$$

Hence we can write for every $\theta \in \mathcal{B}(\theta^1, \delta(\beta))$

$$\begin{aligned}
\frac{1}{N}\sum_k \underline{a}_{\delta(\beta)}(z_k) - \mathbb{E}\{\underline{a}_{\delta(\beta)}(z)\} - \frac{\beta}{2} &\leq \frac{1}{N}\sum_k \underline{a}_{\delta(\beta)}(z_k) - \mathbb{E}\{\bar{a}_{\delta(\beta)}(z)\} \\
&\leq \frac{1}{N}\sum_k a(z_k, \theta) - \mathbb{E}\{a(z, \theta)\} \\
&\leq \frac{1}{N}\sum_k \bar{a}_{\delta(\beta)}(z_k) - \mathbb{E}\{\underline{a}_{\delta(\beta)}(z)\} \\
&\leq \frac{1}{N}\sum_k \bar{a}_{\delta(\beta)}(z_k) - \mathbb{E}\{\bar{a}_{\delta(\beta)}(z)\} + \frac{\beta}{2} .
\end{aligned}$$

From the strong law of large numbers, we have that $\forall \gamma > 0$, $\exists N_1(\beta, \gamma)$ such that

$$\text{Prob}\left\{\forall N > N_1(\beta, \gamma) ,\ \left|\frac{1}{N}\sum_k \bar{a}_{\delta(\beta)}(z_k) - \mathbb{E}\{\bar{a}_{\delta(\beta)}(z)\}\right| < \frac{\beta}{2}\right\} > 1 - \frac{\gamma}{2} ,$$

$$\text{Prob}\left\{\forall N > N_1(\beta, \gamma) ,\ \left|\frac{1}{N}\sum_k \underline{a}_{\delta(\beta)}(z_k) - \mathbb{E}\{\underline{a}_{\delta(\beta)}(z)\}\right| < \frac{\beta}{2}\right\} > 1 - \frac{\gamma}{2} .$$

Combining with previous inequalities, we obtain

$$\text{Prob}\left\{\forall N > N_1(\beta,\gamma)\,,\quad \max_{\theta\in\mathcal{B}(\theta^1,\delta(\beta))}\left|\frac{1}{N}\sum_k a(z_k,\theta) - \mathbb{E}\{a(z,\theta)\}\right| < \beta\right\} > 1-\gamma\,.$$

It only remains to cover Θ with a finite numbers of sets $\mathcal{B}(\theta^i,\delta(\beta))$, $i = 1,\ldots,n(\beta)$, which is always possible from the compactness assumption. For any $\alpha > 0, \beta > 0$, take $\gamma = \alpha/n(\beta)$, $N(\beta) = \max_i N_i(\beta,\gamma)$. We obtain

$$\text{Prob}\left\{\forall N > N(\beta)\,,\ \max_{\theta\in\Theta}\left|\frac{1}{N}\sum_k a(z_k,\theta) - \mathbb{E}\{a(z,\theta)\}\right| < \beta\right\} > 1-\alpha\,,$$

which completes the proof. □

Proof of Lemma 3. Since P_θ is the orthogonal projector onto $\mathcal{L}_\theta$ it is sufficient to prove that $\bar{\alpha} \overset{\xi}{\sim} \bar{\theta}$ implies that any element of $\mathcal{L}_{\bar{\alpha}}$ is in $\mathcal{L}_{\bar{\theta}}$.

From $\{\mathbf{f}_{\bar{\theta}}\}_1,\ldots,\{\mathbf{f}_{\bar{\theta}}\}_p$ we choose r functions that form a linear basis of $\mathcal{L}_{\bar{\theta}}$. Without any loss of generality we can suppose that they are the first r ones. Decompose θ into $\theta = (\beta,\gamma)$, where β corresponds to the first r components of θ and γ to the $p-r$ remaining ones. Define similarly $\bar{\theta} = (\bar{\beta},\bar{\gamma})$. From A4, the components of $\partial\eta[x,(\beta,\gamma)]/\partial\gamma$ are linear combinations of components of $\partial\eta[x,(\beta,\gamma)]/\partial\beta$ not only for $\theta = \bar{\theta}$ but also for θ in some neighborhood of $\bar{\theta}$.

Define the following mapping G from $\mathbb{R}^{r+p}$ to $\mathbb{R}^r$ by

$$G(\beta,\alpha) = \int_{\mathcal{X}} \frac{\partial\eta[x,(\beta,\bar{\gamma})]}{\partial\beta}\,\{\eta[x,(\beta,\bar{\gamma})] - \eta(x,\alpha)\}\,\xi(\mathrm{d}x)\,.$$

From $\bar{\alpha} \overset{\xi}{\sim} \bar{\theta}$ we obtain $G(\bar{\beta},\bar{\alpha}) = 0$. The matrix

$$\frac{\partial G(\beta,\alpha)}{\partial\beta^\top}_{|\bar{\beta},\bar{\alpha}} = \int_{\mathcal{X}} \frac{\partial\eta[x,(\beta,\bar{\gamma})]}{\partial\beta}_{|\bar{\beta}}\,\frac{\partial\eta[x,(\beta,\bar{\gamma})]}{\partial\beta^\top}_{|\bar{\beta}}\,\xi(\mathrm{d}x)$$

is a nonsingular $r\times r$ submatrix of $\mathbf{M}(\xi,\bar{\theta})$, with $\text{rank}[\mathbf{M}(\xi,\theta)] = r$ for θ in a neighborhood of $\bar{\theta}$. From the Implicit Function Theorem, see Spivak (1965, Th. 2–12, p. 41), there exist neighborhoods $\mathcal{V}(\bar{\alpha})$, $\mathcal{W}(\bar{\beta})$ and a differentiable mapping $\psi : \mathcal{V}(\bar{\alpha}) \to \mathcal{W}(\bar{\beta})$ such that $\psi(\bar{\alpha}) = \bar{\beta}$ and that $\alpha\in\mathcal{V}(\bar{\alpha})$ implies $G[\psi(\alpha),\alpha] = 0$. It follows that

$$\begin{aligned}&\frac{\partial}{\partial\beta}\int_{\mathcal{X}} \{\eta[x,(\beta,\bar{\gamma})] - \eta(x,\alpha)\}^2_{\beta=\psi(\alpha)}\,\xi(\mathrm{d}x) \qquad (8.24)\\ &= 2\int_{\mathcal{X}} \left[\frac{\partial\eta[x,(\beta,\bar{\gamma})]}{\partial\beta}\right]_{\beta=\psi(\alpha)} \{\eta[x,(\psi(\alpha),\bar{\gamma})] - \eta(x,\alpha)\}\,\xi(\mathrm{d}x) = 0\,.\end{aligned}$$

Since the components of $\partial\eta[x,(\beta,\gamma)]/\partial\gamma$ are linear combinations of the components of $\partial\eta[x,(\beta,\gamma)]/\partial\beta$ for any $\theta = (\beta,\gamma)$ in some neighborhood of $\bar{\theta}$, we obtain from (8.24)

$$\frac{\partial}{\partial \gamma} \int_{\mathcal{X}} \{\eta[x, (\beta, \gamma)] - \eta(x, \alpha)\}^2_{\beta=\psi(\alpha), \gamma=\bar{\gamma}} \, \xi(\mathrm{d}x) =$$
$$2 \int_{\mathcal{X}} \left[\frac{\partial \eta[x, (\beta, \gamma)]}{\partial \gamma} \right]_{\beta=\psi(\alpha), \gamma=\bar{\gamma}} \{\eta[x, (\psi(\alpha), \bar{\gamma})] - \eta(x, \alpha)\} \, \xi(\mathrm{d}x) = 0 \, .$$

Combining with (8.24) we obtain that

$$\left\{ \frac{\partial}{\partial \theta} \int_{\mathcal{X}} [\eta(x, \theta) - \eta(x, \alpha)]^2 \, \xi(\mathrm{d}x) \right\}_{\theta=[\psi(\alpha), \bar{\gamma}]} = 0$$

for all α belonging to some neighborhood $\mathcal{U}(\bar{\alpha})$. We can make $\mathcal{U}(\bar{\alpha})$ small enough to satisfy the inequality $\left\| \eta[x, (\psi(\alpha), \bar{\gamma})] - \eta(x, \bar{\theta}) \right\|_{\xi}^2 < \epsilon$ required in A5. It follows that $(\psi(\alpha), \bar{\gamma}) \overset{\xi}{\sim} \alpha$, that is, $\eta(\cdot, \alpha) \overset{\xi}{=} \eta[\cdot, (\psi(\alpha), \bar{\gamma})]$ for all α in a neighborhood of $\bar{\alpha}$. By taking derivatives we then obtain

$$\left[\frac{\partial \eta(\cdot, \alpha)}{\partial \alpha^\top} \right]_{\bar{\alpha}} \overset{\xi}{=} \left[\frac{\partial \eta[\cdot, (\psi(\alpha), \bar{\gamma})]}{\partial \alpha^\top} \right]_{\bar{\alpha}}$$
$$\overset{\xi}{=} \left[\frac{\partial \eta[\cdot, (\beta, \bar{\gamma})]}{\partial \beta^\top} \right]_{(\psi(\bar{\alpha}), \bar{\gamma})} \left[\frac{\partial \psi(\alpha)}{\partial \alpha^\top} \right]_{\bar{\alpha}} ,$$

that is, $\mathcal{L}_{\bar{\alpha}} \overset{\xi}{\subset} \mathcal{L}_{(\psi(\bar{\alpha}), \bar{\gamma})} = \mathcal{L}_{\bar{\theta}}$.

By interchanging $\bar{\alpha}$ with $\bar{\theta}$ we obtain $\mathcal{L}_{\bar{\theta}} \overset{\xi}{\subset} \mathcal{L}_{\bar{\alpha}}$. □

Acknowledgments

The research of the first author has been supported by the VEGA grant No. 1/3016/06. The work of the second author was partially supported by the IST Programme of the European Community, under the PASCAL Network of Excellence, IST-2002-506778. This publication only reflects the authors's view.

References

Atkinson, A. and Donev, A. (1992). *Optimum Experimental Design.* Oxford University Press, NY, USA.

Bierens, H. (1994). *Topics in Advanced Econometrics.* Cambridge University Press, Cambridge.

Billingsley, P. (1971). *Weak Convergence of Measures: Applications in Probability.* SIAM, Philadelphia.

Elfving, G. (1952). Optimum allocation in linear regression. *The Annals of Mathematical Statistics,* **23**, 255–262.

Fedorov, V. (1972). *Theory of Optimal Experiments.* Academic Press, New York.

Gallant, A. (1987). *Nonlinear Statistical Models.* Wiley, New York.

Hero, A., Fessler, J., and Usman, M. (1996). Exploring estimator bias-variance tradeoffs using the uniform CR bound. *IEEE Transactions on Signal Processing*, **44**, 2026–2041.

Ivanov, A. (1997). *Asymptotic Theory of Nonlinear Regression.* Kluwer, Dordrecht.

Jennrich, R. (1969). Asymptotic properties of nonlinear least squares estimation. *The Annals of Mathematical Statistics,* **40**, 633–643.

Kiefer, J. and Wolfowitz, J. (1959). Optimum designs in regression problems. *The Annals of Mathematical Statistics,* **30**, 271–294.

Lehmann, E. and Casella, G. (1998). *Theory of Point Estimation.* Springer, Heidelberg.

Pázman, A. (1980). Singular experimental designs. *Math. Operationsforsch. Statist., Ser. Statistics*, **16**, 137–149.

Pázman, A. (1986). *Foundations of Optimum Experimental Design.* Reidel (Kluwer group), Dordrecht (co-pub. VEDA, Bratislava).

Pázman, A. and Pronzato, L. (1992). Nonlinear experimental design based on the distribution of estimators. *Journal of Statistical Planning and Inference,* **33**, 385–402.

Pázman, A. and Pronzato, L. (2006). On the irregular behavior of LS estimators for asymptotically singular designs. *Statistics & Probability Letters,* **76**, 1089–1096.

Pronzato, L. and Pázman, A. (1994). Second-order approximation of the entropy in nonlinear least-squares estimation. *Kybernetika*, **30**(2), 187–198. *Erratum* 32(1):104, 1996.

Shiryaev, A. (1996). *Probability.* Springer, Berlin.

Silvey, S. (1980). *Optimal Design.* Chapman & Hall, London.

Sjöberg, J., Zhang, Q., Ljung, L., Benveniste, A., Delyon, B., Glorennec, P.-Y., Hjalmarsson, H., and Juditsky, A. (1995). Nonlinear black-box modeling in system identification: a unified overview. *Automatica*, **31**(12), 1691–1724.

Spivak, M. (1965). *Calculus on Manifolds. A Modern Approach to Classical Theorems of Advanced Calculus.* W. A. Benjamin, Inc., New York.

Stoica, P. (2001). Parameter estimation problems with singular information matrices. *IEEE Transactions on Signal Processing*, **49**, 87–90.

Wu, C.-F. (1980). Characterizing the consistent directions of least squares estimates. *The Annals of Statistics*, **8**(4), 789–801.

Wu, C.-F. (1981). Asymptotic theory of nonlinear least squares estimation. *The Annals of Statistics*, **9**(3), 501–513.

Wu, C.-F. (1983). Further results on the consistent directions of least squares estimators. *The Annals of Statistics*, **11**(4), 1257–1262.

Wynn, H. (1972). Results in the theory and construction of D-optimum experimental designs. *Journal of the Royal Statistical Society B*, **34**, 133–147.

9

Robust Estimators in Non-linear Regression Models with Long-Range Dependence

A. Ivanov and N. Leonenko

Summary. We present the asymptotic distribution theory for M-estimators in non-linear regression model with long-range dependence (LRD) for a general class of covariance functions in discrete and continuous time. Our limiting distributions are not always Gaussian and they have second moments. We present non-Gaussian distributions in terms of characteristic functions rather then the multiple Ito–Wiener integrals. These results are some variants of the non-central limit theorems of Taqqu (1979), however the normalizing factors and limiting distributions are of more general type. Beran (1991) observed, in the case of a Gaussian sample with LRD, that the M-estimators and the least-squares estimator of the location parameter have equal asymptotic variances. We present a similar phenomenon for a general non-linear regression model with LRD in discrete and continuous time (see Corollary 1), that is in the case of Gaussian non-linear regression models with LRD errors the M-estimates (for smooth score functions) of the regression parameters are asymptotically equivalent in the first order to the least-squares estimator.

9.1 Introduction

The study of random processes with correlations decaying at a hyperbolic rate, i.e. those with long-range dependence (LRD), presents interesting and challenging probabilistic as well as statistical problems. Progress has been made in the past two decades on theoretical aspects of the subject. On the other hand, recent applications have confirmed that data in a large number of fields (including hydrology, geophysics, turbulence, economics and finance) display LRD. Many stochastic and statistical models have been developed for the description and analysis of this phenomenon. For recent developments, see Beran (1994), Barndorff-Nielsen and Sheppard (2001), Barndorff-Nielsen and Leonenko (2005) and Heyde and Leonenko (2005). The volume of Doukhan et al. (2003) contains outstanding surveys of the field. In particular, the volume discusses different definitions of LRD of stationary processes in terms of the autocorrelation function (the integral of the correlation function diverges) and the spectrum (the spectral density has a singularity at zero).

L. Pronzato, A. Zhigljavsky (eds.), *Optimal Design and Related Areas in Optimization and Statistics*, Springer Optimization and its Applications 28,
DOI 10.1007/978-0-387-79936-0_9,

We will use these features as an indication of LRD of stationary processes with discrete and continuous time.

Statistical problems for stochastic processes with LRD were studied by many authors. The books of Beran (1994) and Daukhan et al. (2003) contain a reasonably complete bibliography on the subject. In particular, Yajima (1988, 1991), Künsch et al. (1993), Koul (1992), Koul and Mukherjee (1993, 1994), Dahlhaus (1995), Robinson and Hidalgo (1997), Deo (1997), Deo and Hurvich (1998), Beran and Ghosh (1998) and Choy and Taniguchi (2001) considered regression models with LRD in discrete time, while Ivanov and Leonenko (1989), Leonenko (1999), Leonenko et al. (2000) and Ivanov and Leonenko (2000, 2002, 2004) have studied regression models with LRD in continuous time. Note that Holevo (1976) considered regression estimates for processes with zeros in the spectrum. The mentioned results differ significantly from the corresponding results for regression models with weak dependent errors (see, Ibragimov and Rozanov (1970) or Grenander and Rosenblatt (1984), Ivanov (1997)).

The results of Adenstedt (1974), Samarov and Taqqu (1988), Yajima (1988, 1991) and Künsch et al. (1993) indicate that the efficiency of the least squares estimates (LSE) for the coefficients of linear regression with LRD errors may still be very good. The works of Yajima (1988, 1991) contain some central limit theorems (under conditions on cumulants of all orders) for LSE of the coefficients of linear regression with LRD errors. Yajima (1988, 1991) also obtained formulae for the asymptotic covariance matrix of LSE and generalized LSE, and conditions of the consistency and efficiency (see, also Adenstedt (1974) and Samarov and Taqqu (1988) for corresponding results for the sample mean). Dahlhaus (1995) proved asymptotic normality for the generalized LSE. Künsch et al. (1993) discussed the effect of LRD errors on standard independence-based inference rules in the context of experimental design. Koul (1992), Koul and Mukherjee (1993, 1994), Koul and Surgailis (2000a), Giraitis et al. (1996) and Mukherjee (2000) considered the asymptotic properties of various robust estimates of regression coefficients. Robinson and Hidalgo (1997) proved a central limit theorem for the time series regression estimates in the presence of LRD in both errors and stochastic regression. Deo (1997) and Deo and Hurvich (1998) established also the asymptotic normality of the LSE in regression model with LRD noise. Ivanov and Leonenko (1989) and Leonenko (1999) presented Gaussian and non-Gaussian limiting distributions of the LSE of linear regression coefficients of LRD random processes and fields.

Non-linear regression models with independent or weakly dependent errors have been considered by many authors (see, for example, Hannan (1973), Ivanov (1997), Ivanov and Leonenko (1989), Skouras (2000), Pollard and Radchenko (2006) and Pázman and Pronzato (2006) and the references therein). The first results on non-linear regression with LRD were obtained by Robinson and Hidalgo (1997). They presented conditions for the consistency of some estimates of a parameter of non-linear regression with LRD errors in discrete time models. Important results on the asymptotic distribution of

M-estimators in non-linear regression models with discrete time and LRD property of the noise process are presented in Koul (1996) both for smooth and more general score functions ψ. However, our assumptions on the regression function are of different nature and are suitable for continuous time models. Moreover our results on M-estimators for smooth functions ψ cannot be derived from the Koul (1996) paper (see also Koul (2002) and Koul and Baillie (2003)). For technical reasons the results for non-smooth functions ψ will be given in a separate paper. Robinson and Hidalgo (1997) presented conditions for the consistency of some estimates of a parameter of non-linear regression with LRD errors in discrete time models based on the limit theorems for martingale-differences.

One of the popular classes of estimators in linear models is the so-called class of M-estimators. See, e.g. Huber (1981), Hampel et al. (1986), Ronner (1984), Wu and Zen (1999), Bardadym and Ivanov (1999), Qian and Künsch (1998), Cantoni and Ronchetti (2001) and Arcones (2001) and the references therein. Beran (1991), Koul (1992), Koul and Mukherjee (1993, 1994), Giraitis et al. (1996), Koul and Surgailis (2000a,b, 2003) and Koul et al. (2004) considered the asymptotic properties of M-estimates of location parameter or regression coefficients in linear regression models with LRD. Beran (1991) observed, in the case of a Gaussian sample with LRD, that the M-estimators and LSE of the location parameter have equal asymptotic variances (see, also Koul (1992), Koul and Mukherjee (1993, 1994), Giraitis et al. (1996), Koul and Surgailis (1997) for more general models). We present a similar phenomenon for a general non-linear regression model with LRD in continuous time (see Corollary 1 below).

Note that we consider discrete and continuous time regression. This is of importance in view of the fact that the procedure of discretization leads sometimes to a loss of information on important parameters of the process observed such as the parameter of intermittency of the fractional Riesz–Bessel motion (see, for example, Leonenko (1999) p. 42 or Anh et al. (2001)).

The asymptotic theory of LSE in non-linear regression with LRD has been considered by Ivanov and Leonenko (2000, 2004) and Mukherjee (2000). The papers by Ivanov and Leonenko (2002) and Ivanov and Orlovsky (2001) discuss the asymptotic distributions of a class of M-estimates and L_p-estimates $(1 < p < 2)$ in the non-linear regression model with LRD. Our chapter is a continuation of these papers. We present the asymptotic distribution theory for M-estimators in non-linear regression model with LRD for a general class of covariance functions (c.f.'s) in discrete and continuous time. Our limiting distributions are not always Gaussian and they have second moments. In contrast to the paper of Ivanov and Leonenko (2002), we present non-Gaussian distributions in terms of characteristic functions (ch.f.'s) rather then the multiple Ito–Wiener integrals. Generally speaking, these results are some variants of the non-central limit theorems of Taqqu (1979) and Dobrushin and Major (1979) however the normalizing factors and limiting distributions are of more general type.

We do not consider here the problem of estimating the dependence index or Hurst's parameter (see, the book of Beran (1994) and recent papers of Robinson (1994a, b, 1995a, b), Lobato and Robinson (1996), Robinson and Hidalgo (1997), Giraitis and Koul (1997), Hall et al. (1997), Leonenko and Woyczynski (1999), Koul and Surgailis (2000b), Doukhan et al. (2003), Sibbersten, P. (2003), Kato and Masry (2003), Leonenko and Sakhno (2006) and Ivanov and Leonenko (2008), among others).

9.2 Main Results

9.2.1 Stationary Processes with LRD

This subsection reviews a number of known results, in particular from Ivanov and Leonenko (2004).

We wish to consider a stationary processes with discrete and continuous time simultaneously. That is why we introduce the following notation: $\mathbb{S} = \mathbb{Z}$ for discrete time ($t \in \mathbb{Z}$) and $\mathbb{S} = \mathbb{R}$ for continuous time ($t \in \mathbb{R}$).

Let $(\Omega, \mathbb{F}, P)$ be a complete probability space, and $\xi(t) = \xi(\omega, t) : \Omega \times \mathbb{S} \to \mathbb{R}$ be a random process. We assume that this process is measurable and mean-square continuous in the case of continuous time. We also denote $\mu(\mathrm{d}t)$ the counting measure in the case of discrete time (that is $\mu(\{t\}) = 1$, $t \in \mathbb{Z}$) and the Lebesgue measure $\mathrm{d}t$ for continuous time (that is $\mu(\mathrm{d}t) = \mathrm{d}t$ if $t \in \mathbb{R}$). According to this notation, all integrals $\int_0^T f(t)\xi(t)\mu(\mathrm{d}t)$ mean the sums $\sum_{t=0}^{T-1} \xi(t)f(t)$ for discrete time and the Lebesgue integrals $\int_0^T f(t)\xi(t)\mathrm{d}t$ for continuous time, where $f(t)$ is a non-random (measurable for continuous time) function.

Let $\xi(t), t \in \mathbb{S}$ be a stochastic process satisfying the assumption:
A. $\xi(t)$, $t \in \mathbb{S}$, is a stationary Gaussian process with $E\xi(t) = 0$, $E\xi^2(t) = 1$ and c.f.

$$B(t) = B_{0,\alpha}(t) = E\xi(0)\xi(t) = \frac{L(|t|)}{|t|^\alpha},\ 0 < \alpha < 1,\ t \in \mathbb{R}, \tag{9.1}$$

where $L(t), t \in [0, \infty)$, is a slowly varying function (s.v.f.) at infinity bounded over finite intervals.

Remark 1. Note that the class of c.f.'s $B(t), B(0) = 1$ of real-valued mean-square continuous stationary processes coincides with the class of even characteristic functions (ch.f.). From the theory of ch.f.'s we are able to present the following four examples:

$$B_{1,\alpha}(t) = \frac{1}{(1+t^2)^{\alpha/2}},\ B_{2,\alpha}(t) = \frac{1}{1+|t|^\alpha},\ B_{3,\alpha}(t) = \frac{1}{(1+|t|)^\alpha},\ 0 < \alpha < 1,$$

$$B_{4,\alpha,\gamma}(t) = \frac{1}{(1+|t|^\alpha)^\gamma}, 0 < \alpha\gamma < 1,\ t \in \mathbb{R}. \tag{9.2}$$

The c.f. $B_{1,\alpha}(t)$ is known as the ch.f. of the symmetric Bessel distribution. The function $B_{2,\alpha}(t)$ is the ch.f. of the Linnik distribution and the c.f.'s $B_{3,\alpha}(t)$ and $B_{4,\alpha,\gamma}(t)$ are known as the ch.f.'s of the symmetric generalized Linnik distribution. The analytic form of their spectral densities can be found in the papers of Ivanov and Leonenko (2004) and Anh et al. (2004). It is clear that all c.f.'s (9.2) satisfy condition A with the following s.v.f.'s:

$$L_1(|t|) = |t|^{\alpha}/(1+t^2)^{\alpha/2},\ L_2(t) = |t|^{\alpha}/(1+|t|^{\alpha}),$$
$$L_3(t) = |t|^{\alpha}/(1+|t|)^{\alpha},\ L_4(|t|) = |t|^{\alpha}/(1+|t|^{\alpha})^{\gamma}.$$

Note that $B_{1,\alpha}$ is a c.f. for any $\alpha > 0$, $B_{2,\alpha}$ is a c.f. for $\alpha \in (0,2]$, $B_{3,\alpha}$ is a c.f. for $\alpha > 0$, while $B_{4,\alpha,\gamma}(t)$ is a c.f. for $\alpha \in (0,2]$ and $\gamma > 0$.

A flexible class of c.f.'s has been introduced by Ma (2003) and Ivanov and Leonenko (2004). By randomizing the time-scale of the stationary process $\xi(t)$ with c.f. $B(t)$, $t \in \mathbb{R}$, one can introduce a new stochastic process $\eta(t) = \xi(tV)$, $t \in \mathbb{R}$, where V is a non-negative random variable (r.v.) with the distribution function (d.f.) $F_V(v)$ and is independent of the process ξ. The $\eta(t), t \in \mathbb{R}$, is a stationary process with the same marginal distribution as $\xi(t), t \in \mathbb{R}$, but not the same bivariate distribution, and c.f.

$$\varrho(\tau) = \int_0^{\infty} B(\tau v)\mathrm{d}F_V(v),\ \tau \in \mathbb{R}.$$

From the Tauberian theorem (see Feller (1971), p. 445) it follows that for the c.f. $B(t) = \mathrm{e}^{-|t|}$, $t \in \mathbb{R}$, of the Orstein–Uhlenbeck-type process (possibly, non-Gaussian, see Barndorff-Nielsen and Sheppard (2001) and Barndorff-Nielsen and Leonenko (2005) and the references therein)

$$\varrho(\tau) = \int_0^{\infty} \mathrm{e}^{-|\tau|v}\mathrm{d}F_V(v) \sim \tau^{-\beta}L(\tau^{-1}),\ \tau \to \infty,$$

if and only if

$$F_V(v) \sim \frac{1}{\Gamma(\beta+1)}v^{\beta}L(v),\ v \to 0^+,$$

where $L(\cdot)$ is an s.v.f. at the origin.

For instance, if the r.v. V has the Pareto distribution:

$$\mathrm{d}F_V(v) = \frac{\Gamma(\alpha+\beta)}{\Gamma(\alpha)\Gamma(\beta)}\frac{v^{\beta-1}}{(1+v)^{\alpha+\beta}}\mathrm{d}v,\ \alpha > 0,\ \beta > 0, v > 0,$$

then

$$\varrho_1(\tau) \sim \frac{\Gamma(\alpha+\beta)}{\Gamma(\alpha)}\tau^{-\beta},\ \tau \to \infty$$

and if $\beta \in (0,1)$ the process $\eta(t) = \xi(tV)$, $t > 0$ displays LRD. Some other examples of the d.f.'s $F_V(v)$ and corresponding covariance structures of stationary processes with LRD can be found in Barndorff-Nielsen and Leonenko (2005). In particular, the function

$$\varrho_2(\tau) = E_\gamma(-|\tau|^\alpha),\ \tau \in \mathbb{R},$$

is a c.f. of stationary processes with LRD for $0 < \alpha < 1, 0 < \gamma < 1$, where $E_\gamma(z) = \sum_{k=0}^{\infty} z^k/\Gamma(1+\gamma k)$, is the Mittag–Leffler function of the complex variable $z \in \mathbb{C}$. Note that

$$E_\gamma(-|\tau|^\alpha) \sim \frac{1}{|\tau|^\alpha\, \Gamma(1-\gamma)}, \tau \to \infty.$$

Moreover, one can prove that the function

$$\begin{aligned}
\varrho_3(\tau) &= [1 + \ln(1 + |\tau|)]^{-\beta}, \beta > 0;\\
\varrho_4(\tau) &= [1 + \ln\{1 + \ln(1 + |\tau|)\}]^{-\beta}, \beta > 0;\\
\varrho_5(\tau) &= \ln[1 + \ln(1 + |\tau|)] / \ln(1 + |\tau|);
\end{aligned}$$

are c.f.'s of stationary stochastic processes with LRD, that is for $i = 3, 4, 5$

$$\int_0^\infty \varrho_i(\tau)\mathrm{d}\tau = \infty, \tau \in \mathbb{R}, \text{ or } \sum_{\tau=0}^{\infty} \varrho_i(\tau) = \infty, \tau \in \mathbb{Z}.$$

It is clear that all the above covariance structures can be used as c.f.'s of stationary processes with discrete time if $t \in \mathbb{Z}$.

9.2.2 Non-linear Regression with LRD

We shall assume that:
A$_1$. The process $\varepsilon(t) = G(\xi(t)), t \in \mathbb{S}$, where stochastic process $\xi(t)$ satisfies condition A and $G : \mathbb{R} \to \mathbb{R}$ is a non-random Borel function such that $EG(\xi(0)) = 0$, $EG^2(\xi(0)) = 1$.

Consider now the regression model

$$x(t) = g(t, \theta) + \varepsilon(t), \tag{9.3}$$

where $g(t, \theta) : \mathbb{S} \times \Theta^c \to \mathbb{R}$ is a measurable function, Θ^c is the closure in $\mathbb{R}^q$ of the open bounded set $\Theta \subset \mathbb{R}^q$, and $\varepsilon(t), t \in \mathbb{S}$, is a random noise satisfying assumption A$_1$. We do not suppose that the function $g(t, \theta)$ is a linear form of the vector $\theta = (\theta^1, \ldots, \theta^q)$ coordinates. The problem is to estimate the unknown parameter θ from the observation of the random process $x(t), t \in \mathbb{Y} = \{0, \ldots, T-1\}$ for discrete time and $t \in \mathbb{Y} = [0, T]$, for continuous time, as $T \to \infty$.

Remark 2. Let us provide a few examples of condition A$_1$, in which $\varepsilon(t) = \tilde{\varepsilon}(t) - E\tilde{\varepsilon}(t)$; in this sequel, we denote by Φ the distribution function (d.f.) of a standard normal distribution, by U the uniform distribution on $[0, 1]$, by "$\overset{d}{=}$" equality in distribution, and we repeatedly use the well-known property of the Smirnov transformation, i.e. $\Phi(\xi(t)) \overset{d}{=} U$; then the process $\tilde{\varepsilon}(t) =$

$\Phi(\xi(t)), t \in \mathbb{S}$, has the uniform marginal distribution in $[0,1]$ and the process $\tilde{\varepsilon}(t) = \tan(\frac{\pi}{2}\Phi(\xi(t))), t \in \mathbb{S}$, has a Cauchy distribution. Similarly one can construct the regression models with the Student t_ν marginal density function

$$f_\nu(z) = \left\{\Gamma\left(\frac{\nu+1}{2}\right) / \left[\sqrt{\pi\nu}\,\Gamma\left(\frac{\nu}{2}\right)\right]\right\}\left(1+\frac{z^2}{2}\right)^{-\frac{\nu+1}{2}}, \quad z \in \mathbb{R},$$

where for the numbers of degrees of freedom $\nu = 1$ (Cauchy distribution) and $\nu = 2,3,4,5,6$ we have the following exact forms of d.f.'s:

$$F_3(z) = \frac{1}{2} + \frac{1}{\pi}\tan^{-1}\left(\frac{z}{\sqrt{3}}\right) + \frac{z\sqrt{3}}{\pi(3+z^2)}, \quad F_4(z) = \frac{1}{2} + \frac{z(6+z^2)}{2(4+z^2)^{3/2}},$$

$$F_5(z) = \frac{1}{2} + \frac{1}{\pi}\tan^{-1}\left(\frac{z}{\sqrt{5}}\right) + \frac{z\sqrt{5}}{\pi(5+z^2)} + \frac{10\sqrt{5}z}{3\pi(5+z^2)^2},$$

$$F_6(z) = \frac{1}{2} + \frac{z(135+30z^2+2z^4)}{4(6+z^2)^{5/2}}.$$

Thus, the process

$$\tilde{\varepsilon}(t) = G(\xi(t)) = F_\nu^{-1}(\Phi(\xi(t)), \quad t \in \mathbb{S}$$

has the Student t_ν marginal density function for $\nu = 2,3,4,5,6$. It is not difficult to see that any strictly increasing distribution function F can actually be obtained from the marginal d.f. of the process $\tilde{\varepsilon}(t), t \in \mathbb{S}$ by defining $\tilde{\varepsilon}(t) = F^{-1}(\Phi(\xi(t))$. For instance, the process $\tilde{\varepsilon}(t) = F_{\chi^2_\nu}^{-1}(\Phi(\xi(t)), t \in \mathbb{R}$, has a Chi-Square distribution with ν degrees of freedom, where $F_{\chi^2_\nu}$ denotes the d.f. of a χ^2_ν-distributed r.v.

To estimate θ we will use the M-estimator generated by the smooth function $\varrho(x) : \mathbb{R} \to \mathbb{R}$ (see, e.g., Huber (1981) or Hampel et al. (1986)).

The M-estimator of the unknown parameter $\theta \in \Theta$ obtained from the observations $x(t), t \in \mathbb{Y}$ is said to be any random vector $\hat{\theta}_T \in \Theta^c$ having the property:

$$Q_T(\hat{\theta}_T) = \inf_{\tau \in \Theta^c} Q_T(\tau), \quad Q_T(\tau) = \int_0^T \varrho(X(t) - g(t,\tau))\mu(\mathrm{d}\tau).$$

The function $\psi(x) = \varrho'(x)$ is called the score function or influence function.

Remark 3. Typical examples of functions $\varrho(x)$ and $\psi(x)$ are

$$\varrho(x) = x^2, \psi(x) = x,$$

the least squares estimator (LSE). More general estimates with

$$\varrho(x) = |x|^p, \psi(x) = p\ \mathrm{sign}(x)\,|x|^{p-1}, 1 < p \le 2$$

are known as L_p-estimates, while the functions

$$\varrho(x) = |x|, \psi(x) = \mathrm{sign}(x).$$

define the least absolute error estimator.

For positive ν,

$$\psi(x) = x \exp\{-\nu x^2\}$$

leads to the so-called alpha estimator. For $\nu, s > 0$,

$$\psi(x) = \nu \arctan(sx) \quad \text{and} \quad \psi(x) = \nu \tanh(sx)$$

determine the trigonometric and the hyperbolic estimators. The latter influence functions trimmed at $0 < r < \infty$,

$$\psi(x) = x 1_{(-r,r)}(x), \qquad \psi(x) = 1_{(-r,r)}(x)\ \mathrm{sign}(x),$$

specify the trimmed least squares and the trimmed absolute error estimators.

The function

$$\varrho_\gamma(x) = \begin{cases} \gamma x, & x \geq 0 \\ (1-\gamma)x, & x < 0 \end{cases}$$

defines the class of Koenker and Basset estimators $(0 < \gamma < 1)$. The influence function

$$\psi(x) = x\left(r^2 - x^2\right) 1_{(-r,r)}(x), r > 0,$$

defines the Tukey biweight estimator. The function

$$\psi(x) = x 1_{(-\gamma,\gamma)}(x) + \gamma 1_{(\gamma,\beta)}(|x|) + \frac{r-x}{r-\beta}\gamma 1_{(\beta,r)}(|x|)$$

for $\gamma < \beta \leq r$ defines the Hampel trapezoidal estimator. The function

$$\varrho_c(x) = \begin{cases} \frac{1}{2}x^2, & |x| < c \\ c|x| - \frac{1}{2}x^2, & |x| \geq c \end{cases}$$

and the score function

$$\psi_c(x) = \max\{\min(x, c), -c\}, c > 0,$$

define the Huber estimates, which has some optimal properties (Huber (1981)) and the function $\psi_c(x)$ can be approximated by a twice differentiable score function (see Braiman et al. (2001)). It is defined as follows

$$\psi_r(x) = \mathrm{sign}(x) \begin{cases} \left|\frac{x}{r}\right|, & |x| \leq 0.8r \\ s(\left|\frac{x}{r}\right|), & 0.8r < |x| \leq r \\ s(1), & |x| > r \end{cases}$$

where $r > 0$ is a tuning constant and $s(x) = 38.4 - 175x + 300x^2 + 62.5x^4$.

9.2.3 Assumptions on Regression Function and Noise

To proceed further we need some assumptions on the random noise $\varepsilon(t)$ and the regression function $g(t,\tau)$. The following conditions are valid for many much-used regression functions.

Suppose that $g(t,\tau)$ is twice differentiable with respect to $\tau \in \Theta^c$ and set

$$g_i(t,\tau) = \frac{\partial}{\partial \tau^i} g(t,\tau), \quad g_{il}(t,\tau) = \frac{\partial^2}{\partial \tau^i \partial \tau^l} g(t,\tau), \quad i,l = 1,\dots,q,$$
$$d_T(\tau) = \operatorname{diag}(d_{iT}(\tau))_{i=1}^q, \quad \tau \in \Theta^c,$$

where

$$d_{iT}^2(\theta) = \int_0^T g_i^2(t,\theta)\mu(\mathrm{d}t) \text{ and } \underline{\lim} T^{-1} d_{iT}^2(\theta) > 0, \quad T \to \infty, i = 1,\dots,q.$$

These limits can be equal to ∞ in particular. Let also

$$d_{il,T}^2(\tau) = \int_0^T g_{il}^2(t,\tau)\mu(\mathrm{d}t), \ \tau \in \Theta^c, \ i,l = 1,\dots,q.$$

We will denote by the letter k the positive constants. Suppose that for any $T > 0$:

B1.

$$\sup_{t\in\mathbb{Y}} \sup_{\tau\in\Theta^c} \frac{|g_i(t,\tau)|}{d_{iT}(\theta)} \leq k^i T^{-1/2}, \ i = 1,\dots,q. \tag{9.4}$$

B2.

$$\sup_{t\in\mathbb{Y}} \sup_{\tau\in\Theta^c} \frac{|g_{il}(t,\tau)|}{d_{il,T}(\theta)} \leq k^{il} T^{-1/2}, \ i,l = 1,\dots,q.$$

B3.

$$\sup_{\tau\in\Theta^c} \frac{d_{il,T}(\tau)}{d_{iT}(\theta) d_{lT}(\theta)} \leq \tilde{k}^{il} T^{-1/2}, \ i,l = 1,\dots,q.$$

Write

$$J_T(\theta) = (J_{il,T}(\theta))_{i,l=1}^q, \quad J_{il,T}(\theta) = d_{iT}^{-1} d_{lT}^{-1} \int_0^T g_i(t,\theta) g_l(t,\theta)\mu(\mathrm{d}t);$$
$$\Lambda_T(\theta) = \left(\Lambda_T^{il}(\theta)\right)_{i,l=1}^q = J_T^{-1}(\theta).$$

B4. There exists

$$\lim_{T\to\infty} J_T(\theta) = J_0(\theta),$$

where $J_0(\theta)$ is some positive definite matrix.

Under the conditions B1 and B4 there exists

$$\Lambda_0(\theta) = J_0^{-1}(\theta) = \left(\Lambda_0^{il}(\theta)\right)_{i,l=1}^q. \tag{9.5}$$

Assume further that

B5. The function $\varrho(x)$ is twice differentiable and its derivatives $\varrho'(x) = \psi(x)$ and $\varrho''(x) = \psi'(x)$ possess the following properties:

(i) $E\psi(G(\xi(0))) = 0$,
(ii) $E\psi'(G(\xi(0))) \neq 0$,
(iii) The function $\psi'(x)$ satisfies the Lipschitz condition: for any $x, h \in \mathbb{R}$ and some constant $K < \infty$

$$|\psi'(x+h) - \psi'(x)| \leq K\,|h|\,.$$

From A1, B5 (iii) we obtain: $E\psi^2(G(\xi(0))) < \infty$; $E(\psi'(G(\xi(0))))^2 < \infty$; and the functions $\psi(G(x)), \psi'(G(x)), x \in \mathbb{R}$ can be expanded into the series

$$\psi(G(x)) = \sum_{k=0}^{\infty} \frac{C_k(\psi)}{k!} H_k(x)\,, \quad C_k(\psi) = \int_{\mathbb{R}} \psi(G(x)) H_k(x) \varphi(x) \mathrm{d}x\,,$$

$$\psi'(G(x)) = \sum_{k=0}^{\infty} \frac{C_k(\psi')}{k!} H_k(x)\,, \quad C_k(\psi') = \int_{\mathbb{R}} \psi'(G(x)) H_k(x) \varphi(x) \mathrm{d}x\,,$$

by Chebyshev–Hermite polynomials

$$H_k(x) = (-1)^k \mathrm{e}^{\frac{x^2}{2}} \frac{\mathrm{d}^k}{\mathrm{d}x^k} \mathrm{e}^{-\frac{x^2}{2}}\,, k = 0, 1, \ldots$$

in Hilbert space $L_2(\mathbb{R}, \varphi(x)\mathrm{d}x)$, where

$$\varphi(x) = \frac{1}{\sqrt{2\pi}} \mathrm{e}^{-\frac{x^2}{2}}\,, x \in \mathbb{R}.$$

The conditions B5(i) and B5(ii) mean in particular that $C_0(\psi) = 0, C_0(\psi') \neq 0$. Suppose $C_1(\psi) \neq 0$ or $C_1(\psi') \neq 0$, or for some $m, m' \geq 2$

$$\begin{aligned} C_1(\psi) &= \cdots = C_{m-1}(\psi) = 0\,, \quad C_m(\psi) \neq 0\,; \\ C_1(\psi') &= \cdots = C_{m'-1}(\psi') = 0\,, \quad C_{m'}(\psi') \neq 0. \end{aligned} \tag{9.6}$$

Then the integers $m, m' \geq 1$ are said to be the Hermite ranks of $\psi(G), \psi'(G)$, respectively: $\mathrm{Hrank}(\psi(G)) = m$, $\mathrm{Hrank}(\psi'(G)) = m'$.

B6. $\alpha m < 1$, $\alpha m' < 1$ where $\alpha \in (0,1)$ is the c.f. B parameter.

Let $\nabla g(t,\theta)$ be the column vector-gradient of the function $g(t,\theta)$:

$$\nabla g(t,\theta) = \begin{bmatrix} g_1(t,\theta) \\ \vdots \\ g_q(t,\theta) \end{bmatrix}, \nabla^* g(t,\theta) = [g_1(t,\theta), \ldots, g_q(t,\theta)]\,.$$

Then

$$J_T(\theta) = d_T^{-1}(\theta)\left(\int_0^T \nabla g(t,\theta)\nabla^* g(t,\theta)\mu(\mathrm{d}t)\right) d_T^{-1}(\theta)$$
$$= D_T^{-1}(\theta)\left(\int_0^1 \nabla g(Tt,\theta)\nabla^* g(Tt,\theta)\mathrm{d}t\right) D_T^{-1}(\theta)$$

with

$$D_T^2(\theta) = \frac{1}{T} d_T^2(\theta) = \operatorname{diag}\left(\int_0^1 g_i^2(Tt,\theta)\mathrm{d}t\right)_{i=1}^q = \operatorname{diag}\left(D_{iT}^2(\theta)\right)_{i=1}^q$$

Introduce the matrix

$$\sigma_T(\theta) = D_T^{-1}(\theta)\left(\int_0^1\int_0^1 \frac{\nabla g(Tt,\theta)\nabla^* g(Ts,\theta)}{|t-s|^{\alpha m}}\mathrm{d}t\mathrm{d}s\right) D_T^{-1}(\theta)$$

with m taken from the assumption B6.

B7. $\sigma_T(\theta) \underset{T\to\infty}{\to} \sigma_m(\theta)$, where $\sigma_m(\theta)$ is some positive definite matrix.

Introduce the following notations

$$u = B^{-m/2}(T)T^{-1/2}d_T(\theta)(\tau-\theta)\,, \tag{9.7}$$
$$\gamma = \left(E\psi'(G(\xi(0)))\right)^{-1}, \tag{9.8}$$
$$h(t,u) = g(t,\theta + B^{m/2}(T)T^{1/2}d_T^{-1}(\theta)u)\,, \tag{9.9}$$
$$h_i(t,u) = g_i(t,\theta + B^{m/2}(T)T^{1/2}d_T^{-1}(\theta)u)\,. \tag{9.10}$$

Denote by

$$M_T(u) = \left(M_T^i(u)\right)_{i=1}^q,$$

where

$$M_T^i(u) = \frac{\gamma}{T^{1/2}B^{m/2}(T)}\int_0^T \psi(x(t)-h(t,u))\frac{h_i(t,u)}{d_{iT}(\theta)}\mu(\mathrm{d}t),\ i=1,\ldots,q; \tag{9.11}$$

B8. For some $m \geq 1$, the normalized M-estimator

$$\hat{u}_T = \hat{u}_T(\theta) = B^{-m/2}(T)T^{-1/2}d_T(\theta)(\hat{\theta}_T - \theta)$$

is the unique solution of the system of equations:

$$M_T(u) = 0 \tag{9.12}$$

at least for sufficiently large T (we will write $T > T_0$).

9.2.4 The Case of Hermite Rank $m = 1$

Now we are in a position to formulate one of the main result of this chapter.

Theorem 1. *Suppose assumptions A, B1–B8 are fulfilled with*

$$m = \mathrm{Hrank}(\psi(G)) = m' = \mathrm{Hrank}(\psi'(G)) = 1\,.$$

Then the normalized M-estimator

$$\hat{u}_T = \hat{u}_T(\theta) = B^{-1/2}(T)T^{-1/2}d_T(\theta)(\hat{\theta}_T - \theta)$$

converges in distribution as $T \to \infty$ *to the Gaussian vector*

$$u_{01}(\theta) = (u_{10}(\theta), \dots, u_{q0}(\theta))^*$$

with zero mean and covariance matrix

$$\frac{C_1^2(\psi)}{C_0^2(\psi')}\Lambda_0(\theta)\sigma_1(\theta)\Lambda_0(\theta), \tag{9.13}$$

where $\Lambda_0(\theta)$ *is defined in* (9.5), $\sigma_1(\theta)$ *is defined in condition B7 with* $m = 1$ *and* $C_1(\psi), C_0(\psi')$ *are defined in* (9.6).

The proof of Theorem 1 is given in Sect. 9.4.

Remark 4. All the conditions of Theorem 1 are valid, for example, for functions $g(t,\theta) = t\log(\theta + t), \theta \in \Theta \subset \mathbb{R}$; $g(t,\theta) = \sqrt{\theta_1 t^2 + \theta_2 t + 1}, \theta = (\theta_1, \theta_2) \in \Theta \subset \mathbb{R}^2$.

Suppose that in the definition of the random noise $\varepsilon(t)$ in the basic model (9.3) the function $G(x) = x$, that is $\varepsilon(t) = \xi(t), t \in \mathbb{S}$, and (9.3) is the Gaussian non-linear regression model with LRD property of the noise $\xi(t)$. Then taking into account the relation

$$H_k(u)\varphi(u) = (-1)^k \frac{\mathrm{d}^k}{\mathrm{d}u^k}\varphi(u), k \geq 1$$

we obtain, by integration by parts, for $k \geq 1$

$$\begin{aligned} C_k(\psi) &= \int\limits_{\mathbb{R}} \psi(u)H_k(u)\varphi(u)\mathrm{d}u = \int\limits_{\mathbb{R}} \psi(u)(-1)^k \frac{\mathrm{d}^k}{\mathrm{d}u^k}\varphi(u)\mathrm{d}u \\ &= (-1)^k\psi(u)\frac{\mathrm{d}^{k-1}}{\mathrm{d}u^{k-1}}\varphi(u)\,|_{-\infty}^{\infty} + \int\limits_{\mathbb{R}} \psi'(u)(-1)^{k-1}\frac{\mathrm{d}^{k-1}}{\mathrm{d}u^{k-1}}\varphi(u)\mathrm{d}u \\ &= \int\limits_{\mathbb{R}} \psi'(u)H_{k-1}(u)\varphi(u)\mathrm{d}u = C_{k-1}(\psi') \end{aligned} \tag{9.14}$$

provided that B5(iii), for example, holds.

We arrive at the following:

Corollary 1. *If* $G(x) = x$, *then* $C_1(\psi) = C_0(\psi') \neq 0$ *and the limiting distribution of the normed M-estimator* $\hat{u}_T$, *under the conditions of Theorem* 1, *does not depend on the choice of the loss function* ϱ, *that is in the case of Gaussian non-linear regression models with long-range dependent errors in most cases the M-estimates of the regression parameters are asymptotically equivalent in the first order to the least-squares estimator.*

Proof. Obviously $\mathrm{Hrank}(\psi) = 1$ and in the formula (9.13) the ratio

$$\frac{C_1(\psi)}{C_0(\psi')} = 1.$$

by (9.14) with $k = 1$. □

9.2.5 The Case of Hermite Rank $m = 2$

Consider the random vector (for any $\mathrm{Hrank}(\psi) = m \geq 1$)

$$\mu_T(\theta) = (\mu_{jT}(\theta))_{j=1}^{q},$$

where

$$\mu_{jT}(\theta) = \gamma \int_0^T \psi(G(\xi(t)))g_j(t,\theta)\mu(\mathrm{d}t) / \left[T^{1/2} d_{jT}(\theta) B^{m/2}(T)\right] \tag{9.15}$$

and γ is given by (9.8).

Consider the following assumption:

D. For $\alpha m \in (0,1)$ and $j \in \{1, \ldots, q\}$ there exists a square function $\bar{g}_j(t,\theta)$, $t \in [0,1]$, $\theta \in \Theta^c$, integrable with respect to t, such that

$$\lim_{T\to\infty} \left| \frac{g_j(tT,\theta)}{D_{iT}(\theta)} - g_j(t,\theta) \right| = 0$$

uniformly for $t \in [0,1]$ and $\theta \in \Theta^c$, and

$$\int_0^1 \int_0^1 \bar{g}_l(t,\theta)\bar{g}_j(s,\theta) \frac{\mathrm{d}t\,\mathrm{d}s}{|t-s|^{m\alpha}} < \infty, \ j,l \in \{1,\ldots,q\}, \theta \in \Theta^c.$$

Observe that assumption B_7 holds if D holds and $\sigma_m(\theta) = (\sigma_{lj}(\theta))_{l,j=1}^{q} > 0$, where

$$\sigma_{lj}(\theta) = \int_0^1 \int_0^1 \bar{g}_l(t,\theta)\bar{g}_j(s,\theta) \frac{\mathrm{d}t\,\mathrm{d}s}{|t-s|^{m\alpha}}.$$

Now we use a generalization of Rosenblatt distribution (see Rosenblatt (1961), Ibragimov (1963), Taqqu (1979), Albin (1998), Leonenko and Anh

(2001)) introduced in the papers by Ivanov and Leonenko (2004) and Leonenko and Taufer (2006).

Under condition D with $m = 2$, let us introduce the r.v.'s

$$\Xi_{j2}(\theta) = \frac{C_2(\psi)}{2C_0(\psi')}\bar{\varkappa}_{j2}(\theta), \tag{9.16}$$

where $\bar{\varkappa}_{j2}(\theta)$ are r.v. with the ch.f.

$$\phi_j(z) = E\exp\{iz\bar{\varkappa}_{j2}(\theta)\} = \exp\left\{\frac{1}{2}\sum_{k=2}^{\infty}\frac{(2iz)^k}{k}c_{kj}(\theta)\right\},$$

where $C_2(\psi), C_0(\psi')$ are defined in (9.6) and

$$c_{kj}(\theta) = \int_0^1 \overset{k}{\cdots} \int_0^1 \frac{\bar{g}_j(x_1,\theta)\bar{g}_j(x_2,\theta)\cdots\bar{g}_j(x_k,\theta)}{|x_1-x_2|^{\alpha}|x_2-x_3|^{\alpha}\cdots|x_k-x_1|^{\alpha}}\mathrm{d}x_1\cdots\mathrm{d}x_k,$$

$j \in \{0,1,\ldots,q\}$, $0 < \alpha < 1/2$, $\theta \in \Theta^c$.

It follows from the second condition in D with $m = 2$ that $c_{kj}(\theta) < \infty$ (see Ivanov and Leonenko (2004) or Leonenko and Taufer (2006) for details). We note that the ch.f. $\phi_j(z)$, $j \in \{1,\ldots,q\}$ equals to the ch.f. of a Rosenblatt type r.v.'s (see Rosenblatt (1961), Ibragimov (1963) and Taqqu (1979)), except for the definition of the c_k's which in our case depend on the function $\bar{g}_j(t,\theta)$, while in the Rosenblatt case this may be taken to be 1. The cumulants χ_p of these distributions may be computed (see Ivanov and Leonenko (2004) or Leonenko and Taufer (2006)):

$$\chi_{pj} = (p-1)!2^{p-1}c_{pj}(\theta), p \geq 2,\ j \in \{1,\ldots,q\}.$$

The mean and variance are given, respectively, by

$$\chi_{1j} = 0,$$

$$\chi_{2j} = 2\int_0^1\int_0^1 \frac{\bar{g}_j(x_1,\theta)\bar{g}_j(x_2,\theta)}{|x_1-x_2|^{2\alpha}}\mathrm{d}x_1\mathrm{d}x_2,\ 0 < \alpha < 1/2.$$

Note (see again Leonenko and Taufer (2006)) that the ch.f. $\phi_j(z)$ is analytic in some neighbourhood of zero.

Theorem 2. *Suppose assumptions A, B1–B8 and D are fulfilled for* $m = \mathrm{Hrank}\{\psi(G)\} = 2$. *Then for every* $j = 1,\ldots,q$

$$\lim_{T\to\infty} \operatorname{var}\left[\mu_{jT}(\theta) - \Xi_{j2}(\theta))\right]^2 = 0,$$

and the normalized M-estimator

$$\hat{u}_T = \hat{u}_T(\theta) = B^{-1}(T)T^{-1/2}d_T(\theta)(\hat{\theta}_T - \theta)$$

converges in distribution as $T \to \infty$ *to the random vector*

$$u_{02}(\theta) = \left(\sum_{j=1}^{q} \Lambda_0^{lj}(\theta)\Xi_{j2}(\theta) \right)_{l=1}^{q}, \tag{9.17}$$

where Λ_0^{lj} *are defined in* (9.17) *and* $\Xi_{j2}(\theta)$ *are defined in* (9.16).

The proof of Theorem 2 is given in Sect. 9.4.

Remark 5. Theorem 2 should be compared with results of Taqqu (1979) and Dobrushin and Major (1979). They obtained non-Gaussian limiting distributions of normalized averages of local functionals of Gaussian random processes and fields. Observe, that in particular for $q = 1, g(t, \theta) = \theta, \varrho(x) = x^2$, our limiting distributions as well as normalized factors coincide with limiting distributions and normalized factors of the papers of Taqqu (1979) and Dobrushin and Major (1979).

Remark 6. All the conditions of Theorem 2 are valid, for example, for functions $g_1(t, \theta) = t \log(\theta + t)$, $g_2(t, \theta) = \log(\theta + t)$, $g_3(t, \theta) = \sqrt{\theta t + 1}, \theta \in \Theta \subset \mathbb{R}$; (in this case $\bar{g}_1(t, \theta) = \bar{g}_2(t, \theta) = 1$, $\bar{g}_3(t, \theta) = \sqrt{t}$), or $g_4(t, \theta) = \sqrt{\theta_1 t^2 + \theta_2 t + 1}, \theta = (\theta_1, \theta_2) \in \Theta \subset \mathbb{R}^2$.

To describe the limiting distributions for the Hermite rank $m \geq 3$ we need more restrictive assumptions and these distributions can be given in the form of Wiener–Ito integrals (see Ivanov and Leonenko (2000, 2002, 2004)).

9.3 Auxiliary Assertions

The following result on the s.v.f.'s will be used below.

Lemma 1. *Let* $\eta \geq 0$ *be a given real number and function* $f(t, s)$ *be defined in* $(0, \infty) \times (0, \infty)$ *such that the integral*

$$\int_0^\beta \int_0^\beta f(t, s) \frac{\mathrm{d}t\,\mathrm{d}s}{|t - s|^\eta}$$

converges for some β *from* $(0, \infty)$. *Let* L *be a s.v.f. at infinity bounded on every finite interval from* $(0, \infty)$. *Then for* $\eta > 0$

$$\int_0^\beta \int_0^\beta f(t, s) L(T\,|t - s|) \frac{\mathrm{d}t\,\mathrm{d}s}{|t - s|^\eta} \underset{T \to \infty}{\sim} L(T) \int_0^\beta \int_0^\beta f(t, s) \frac{\mathrm{d}t\,\mathrm{d}s}{|t - s|^\eta}.$$

When $\eta = 0$ *this relation is valid when the function* L *is non-decreasing on the semi-axis* $(0, \infty)$.

The proof of Lemma 1 is similar to the proof of Theorem 2.7 in the book of Seneta (1976) (see also Ivanov and Orlovsky (2001)).

Let us perform the change of variables

$$u = B^{-m/2}(T)T^{-1/2}d_T(\theta)(\tau - \theta)$$

in the regression function and its derivatives corresponding to the normalization (9.7). Together with (9.7–9.12) we will also use the notation

$$H(t;u_1,u_2) = h(t,u_1) - h(t,u_2)\,,$$
$$H_i(t;u_1,u_2) = h_i(t,u_1) - h_i(t,u_2)\,,\ i = 1,\dots,q\,.$$

Introduce the vectors

$$\Psi_T(u) = \left(\Psi_T^i(u)\right)_{i=1}^q,$$

$$\Psi_T^i(u) = \frac{1}{T^{1/2}B^{m/2}(T)}\left(\int_0^T \gamma\psi(G(\xi(t)))\frac{h_i(t,u)}{d_{iT}(\theta)}\mu(\mathrm{d}t) + \int_0^T H(t;0,u)\frac{h_i(t,u)}{d_{iT}(\theta)}\mu(\mathrm{d}t)\right)$$

where $i = 1,\dots,q$.

The vectors $M_T(u)$ (see condition B8) and $\Psi_T(u)$ are defined for $u \in U_T^c(\theta)$,

$$U_T(\theta) = B^{-m/2}(T)T^{-1/2}d_T(\theta)(\Theta - \theta).$$

Note that under our assumptions the sets $U_T(\theta)$ expand to $\mathbb{R}^q$ as $T \to \infty$. Thus for any $R > 0$ $v(R) = \{u \in R^q : \|u\| \le R\} \subset U_T(\theta)$ for $T > T_0(R)$.

It is easy to understand the statistical meaning of the vectors $M_T(u)$ and $\Psi_T(u)$. Consider the functional

$$\frac{\gamma}{TB^m(T)}Q_T\left(\theta + B^{m/2}(T)T^{1/2}d_T^{-1}(\theta)u\right);$$

then the normed M-estimator $\hat{u}_T$ satisfies the system of equations (see condition B8):

$$M_T(u) = 0.$$

Let $\eta(t) = \gamma\psi(G(\xi(t))), t \in \mathbb{S}$, and the observations have the form

$$Y(t) = g(t,\theta) + \eta(t),\ t \in \mathbb{Y}, \tag{9.18}$$

then

$$\Psi_T(u) = 0$$

is the system of normal equations for determining the normed LSE

$$\check{u}_T = \check{u}_T(\theta) = B^{-m/2}(T)T^{-1/2}d_T(\theta)(\check{\theta}_T - \theta)$$

of unknown parameter θ of an auxiliary non-linear regression model (9.18): set $\varrho(x) = x^2$.

Lemma 2. *Under assumptions B1, B2, B3, B5 and B6 for any $R > 0, r > 0$*

$$P\left\{\sup_{u \in v(R)} \|M_T(u) - \Psi_T(u)\| > r\right\} \underset{T\to\infty}{\to} 0. \tag{9.19}$$

Proof. For fixed i write down the difference $M_T^i(u) - \Psi_T^i(u)$, which is equal to

$$\frac{\gamma}{T^{1/2}B^{m/2}(T)} \int_0^T \frac{h_i(t,u)}{d_{iT}(\theta)} \left[\psi\left(G(\xi(t)) + H(t;0,u)\right) - \psi(G(\xi(t))) -\right.$$

$$\left.\psi'(G(\xi(t)))H(t;0,u)\right] \mu(\mathrm{d}t) + \frac{\gamma}{T^{1/2}B^{m/2}(T)} \int_0^T H(t;0,u) \frac{h_i(t,u)}{d_{iT}(\theta)} \zeta(t)\, \mu(\mathrm{d}t)$$

$$= I_1(u) + I_2(u),$$

where $\zeta(t) = \psi'(G(\xi(t))) - E\psi'(G(\xi(0)))$, $t \in \mathbb{S}$. We have to prove that $I_1(u)$ and $I_2(u)$ tend to zero in probability uniformly in $u \in v(R)$.

Let $u \in v(R)$ be fixed. Then

$$EI_2^2(u) = \frac{\gamma^2}{TB^m(T)} \int_0^T\int_0^T H(t;0,u)H(s;0,u) \frac{h_i(t,u)}{d_{iT}(\theta)} \frac{h_i(s,u)}{d_{iT}(\theta)}$$
$$\times \mathrm{cov}(\zeta(t),\zeta(s))\, \mu(\mathrm{d}t)\, \mu(\mathrm{d}s)\,. \tag{9.20}$$

By the condition B1

$$\sup_{t\in\mathbb{Y}} |H(t;0,u)| = T^{1/2}B^{m/2}(T) \sup_{t\in\mathbb{Y}} \left|\sum_{i=1}^q \frac{h_i(t,u_t^*)u_i}{d_{iT}(\theta)}\right| \le B^{m/2}(T)\, \|k\|\, \|u\|\,, \tag{9.21}$$

where $\|u_t^*\| \le \|u\|, k = (k^1, \ldots, k^q)$ are constants from the inequalities (9.4). Utilizing the bound (9.21) and the condition B1 once more to the integral (9.20) we obtain

$$EI_2^2(u) \le \gamma^2 \|k\|^2 (k^i)^2 R^2 \frac{1}{T^2} \int_0^T\int_0^T |\mathrm{cov}(\zeta(t),\zeta(s))|\, \mu(\mathrm{d}t)\mu(\mathrm{d}s).$$

Due to the conditions B5(ii) and B6

$$\frac{1}{T^2} \int_0^T\int_0^T |\mathrm{cov}(\zeta(t),\zeta(s))|\, \mu(\mathrm{d}t)\mu(\mathrm{d}s) = \mathcal{O}(B^{m'}(T)) \tag{9.22}$$

as $T \to \infty$ and $I_2(u) \overset{P}{\to} 0$, as $T \to \infty$ pointwise.

For $u_1, u_2 \in v(R)$ consider the difference

$$I_2(u_1) - I_2(u_2) = \frac{\gamma}{T^{1/2}B^{m/2}(T)} \int_0^T H(t;0,u_1)\frac{H_i(t,u_1,u_2)}{d_{iT}(\theta)}\zeta(t)\mu(\mathrm{d}t)$$

$$+\frac{\gamma}{T^{1/2}B^{m/2}(T)} \int_0^T H(t;u_2,u_1)\frac{h_i(t,u_1)}{d_{iT}(\theta)}\zeta(t)\mu(\mathrm{d}t) = I_3(u_1,u_2) + I_4(u_1,u_2)\,.$$

For arbitrary $h > 0, r > 0$ we shall examine the probability

$$\begin{aligned} P\left\{ \sup_{\|u_1-u_2\|\le h} |I_3(u_1,u_2)| > r \right\} &\le r^{-1}E \sup_{\|u_1-u_2\|\le h} |I_3(u_1,u_2)| \\ &\le 2r^{-1}\,|\gamma|\,E\,|\psi'(G(\xi(0)))|\,T^{1/2}B^{-m/2}(T) \sup_{\substack{u\in v(r)\\ t\in\mathbb{Y}}} |H(t;0,u)| \\ &\quad \times \sup_{\|u_1-u_2\|\le h} \sup_{t\in\mathbb{Y}} \frac{|H_i(t;u_1,u_2)|}{d_{iT}(\theta)}\,. \end{aligned} \tag{9.23}$$

Now,

$$\begin{aligned} &\sup_{\|u_1-u_2\|\le h} \sup_{t\in\mathbb{Y}} \left| \frac{H_i(t;u_1,u_2)}{d_{iT}(\theta)} \right| \\ &\quad\le h \sup_{t\in\mathbb{Y}} \left[\sum_{l=1}^q \left(\sup_{u\in v(r)} \frac{|h_{il}(t,u)|}{d_{il,T}(\theta)} \right) \frac{d_{il,T}(\theta)}{d_{iT}(\theta)d_{lT}(\theta)} \right] T^{1/2}B^{m/2}(T) \\ &\quad\le \sum_{l=1}^q k^{il}\tilde{k}^{il} h T^{-1/2}B^{m/2}(T) \end{aligned} \tag{9.24}$$

thanks to the conditions B2 and B3. From (9.21), (9.23) and (9.24) it follows that

$$P\left\{ \sup_{\|u_1-u_2\|\le h} |I_3(u_1,u_2)| > r \right\} \le k_1 r^{-1} B^{m/2}(t) h \tag{9.25}$$

with $k_1 = 2\,|\gamma|\,E\,|\psi'(G(\xi(0)))|\,R\,\|k\|\left(\sum_{l=1}^q k^{il}\tilde{k}^{il}\right)$.

Similarly, from (9.4) and (9.21) we deduce

$$\begin{aligned} P\left\{ \sup_{\|u_1-u_2\|\le h} |I_4(u_1,u_2)| > r \right\} &\le r^{-1}E \sup_{\|u_1-u_2\|\le h} |I_4(u_1,u_2)| \\ \le 2r^{-1}\,|\gamma|\,E\,|\psi'(G(\xi(0)))|\,T^{1/2}B^{-m/2}(T) & \\ \times \sup_{\substack{u\in v(r)\\ t\in[0,T]}} \frac{|h_i(t,u)|}{d_{iT}(\theta)} \sup_{\|u_1-u_2\|\le h} \sup_{t\in\mathbb{Y}} |H(t,u_1,u_2)| &\le k_2 r^{-1} h \end{aligned} \tag{9.26}$$

with $k_2 = 2\,|\gamma|\, E\,|\psi'(G(\xi(0)))|\, k^i\, \|k\|$. From (9.25) and (9.26) we derive

$$P\left\{\sup_{\|u_1-u_2\|\le h} |I_2(u_1) - I_2(u_2)| > r\right\} \le 2r^{-1}h\left(k_1 B^{m/2}(T) + k_2\right) \le k_3 r^{-1}h. \tag{9.27}$$

Let N_h be a finite h-net of the ball $v(R)$. Then

$$\sup_{u\in v(R)} |I_2(u)| \le \sup_{\|u_1-u_2\|\le h} |I_2(u_1) - I_2(u_2)| + \max_{u\in N_h} |I_2(u)|\,. \tag{9.28}$$

From (9.27) and (9.28) it follows for any $r > 0$

$$P\left\{\sup_{u\in v(R)} |I_2(u)| > r\right\} \le 2k_3 r^{-1}h + P\left\{\max_{u\in N_h} |I_2(u)| > r/2\right\}.$$

Let $\varepsilon > 0$ be an arbitrarily small number. Set $h = \varepsilon r/4k_3$. Then for $T > T_0$

$$P\left\{\sup_{u\in v(R)} |I_2(u)| > r\right\} \le \varepsilon,$$

because for sufficiently large T

$$P\left\{\max_{u\in N_{\varepsilon r/4k_3}} |I_2(u)| > r/2\right\} \le \varepsilon/2$$

due to the pointwise convergence of $I_2(u)$ to zero in probability.

On the other hand for some random variable $u_T^* \in v(R)$

$$\sup_{u\in v(R)} |I_1(u)| \le \frac{|\gamma|}{T^{1/2}B^{m/2}(T)} \int_0^T \left|\frac{h_i(t,u_T^*)}{d_{iT}(\theta)}\right| |[\psi\left(G(\xi(t)) + H(t;0,u_T^*)\right)$$
$$-\psi(G(\xi(t))) - \psi'(G(\xi(t)))H(t;0,u_T^*)]|\,\mu(\mathrm{d}t) \tag{9.29}$$

According to the assumption B5 (iii) and (9.22)

$$\begin{aligned}&|\psi\left(G(\xi(t)) + H(t;0,u_T^*)\right) - \psi(G(\xi(t))) - \psi'(G(\xi(t)))H(t;0,u_T^*)|\\ &\quad = |\psi'\left(G(\xi(t)) + \delta H(t;0,u_T^*)\right) - \psi'(G(\xi(t)))|\,|H(t;0,u_T^*)|\\ &\quad \le K H^2(t;0,u_T^*) \le K\,\|k\|^2\,R^2 B^m(T),\end{aligned}$$

$\delta \in (0,1)$, a.s. Due to (9.28), (9.29) and (9.4)

$$\sup_{u\in v(R)} |I_1(u)| \le K\,|\gamma|\,\|k\|^2\,k^i R^2 B^{m/2}(T)$$

a.s., and Lemma 2 is proved. □

Define further a random vector

$$L_T(u) =$$

$$\left(\frac{1}{TB^m(T)} \int_0^T \left(\eta(t) - \sum_{l=1}^q \frac{g_l(t,\theta)}{d_{lT}(\theta)} T^{1/2} B^{m/2}(T) u^l \right) \frac{g_i(t,\theta)}{d_{iT}(\theta)} T^{1/2} B^{m/2}(T) \mu(\mathrm{d}t) \right)_{i=1}^q$$

$$= \left(L_T^i(u) \right)_{i=1}^q \tag{9.30}$$

corresponding to the auxiliary linear regression model

$$Z(t) = \sum_{i=1}^q g_i(t,\theta)\beta^i + \eta(t), t \in \mathbb{Y} \tag{9.31}$$

with $\eta(t) = \psi(G(\xi(t)))\gamma, t \in \mathbb{S}$. The system of normal equations

$$L_T(u) = 0 \tag{9.32}$$

determines the normed linear LSE $\tilde{\beta}_T$ of the parameter $\beta \in \mathbb{R}^q$

$$\widetilde{u}_T = \widetilde{u}_T(\theta) = T^{-1/2} B^{-m/2}(T) d_T(\theta)(\tilde{\beta}_T - \beta). \tag{9.33}$$

Lemma 3. *Under the assumptions B1, B2, B3, B5 and B6, for any $R > 0$, $r > 0$*

$$P\left\{ \sup_{u \in v(R)} \|\Psi_T(u) - L_T(u)\| > r \right\} \underset{T\to\infty}{\longrightarrow} 0 \tag{9.34}$$

Proof. For any $i = 1, \ldots, q$

$$\Psi_T^i(u) - L_T^i(u) =$$

$$\frac{1}{T^{1/2}B^{m/2}(T)} \int_0^T \eta(t) \frac{h_i(t,u)}{d_{iT}(\theta)} \mu(\mathrm{d}t) + \frac{1}{T^{1/2}B^{m/2}(T)} \int_0^T H(t;0,u) \frac{h_i(t,u)}{d_{iT}(\theta)} \mu(\mathrm{d}t)$$

$$- \frac{1}{T^{1/2}B^{m/2}(T)} \int_0^T \eta(t) \frac{g_i(t,\theta)}{d_{iT}(\theta)} \mu(\mathrm{d}t)$$

$$+ \frac{1}{T^{1/2}B^{m/2}(T)} \int_0^T \frac{g_i(t,\theta)}{d_{iT}(\theta)} \sum_{l=1}^q \frac{g_l(t,\theta)}{d_{lT}(\theta)} u^l T^{1/2} B^{m/2}(T) \mu(\mathrm{d}t)$$

$$+ \frac{1}{T^{1/2}B^{m/2}(T)} \int_0^T \eta(t) \frac{H_i(t;u,0)}{d_{iT}(\theta)} \mu(\mathrm{d}t) + \frac{1}{T^{1/2}B^{m/2}(T)} \int_0^T H(t;0,u) \frac{H_i(t;u,0)}{d_{iT}(\theta)} \mu(\mathrm{d}t)$$

$$+ \frac{1}{T^{1/2}B^{m/2}(T)} \int_0^T \frac{g_i(t,\theta)}{d_{iT}(\theta)} \left[H(t;0,u) + \sum_{l=1}^q \frac{g_l(t,\theta)}{d_{lT}(\theta)} u^l T^{1/2} B^{m/2}(T) \right] \mu(\mathrm{d}t)$$

$$= I_5(u) + I_6(u) + I_7(u).$$

For fixed $u \in v(R)$ using the bound (9.24) we get

$$EI_5^2(u) = \frac{1}{TB^m(T)} \int_0^T \int_0^T \operatorname{cov}(\eta(t), \eta(s)) \frac{H_i(t;u,0)}{d_{iT}(\theta)} \frac{H_i(s;u,0)}{d_{iT}(\theta)} \mu(\mathrm{d}t)\mu(\mathrm{d}s)$$

$$\leq \left(\sum_{l=1}^{q} k^{il} \tilde{k}^{il} \right) R^2 \frac{1}{T^2} \int_0^T \int_0^T \left| \operatorname{cov}(\eta(t), \eta(s)) \right| \mu(\mathrm{d}t)\mu(\mathrm{d}s) .$$

Thanks to the conditions B5 and B6 as $T \to \infty$

$$\frac{1}{T^2} \int_0^T \int_0^T \left| \operatorname{cov}(\eta(t), \eta(s)) \right| \mu(\mathrm{d}t)\mu(\mathrm{d}s) = \mathcal{O}(B^m(T)),$$

and therefore $I_5(u) \xrightarrow{P} 0$ pointwise. On the other hand due to (9.24)

$$E \sup_{\|u_1 - u_2\| \leq h} |I_5(u_1) - I_5(u_2)| \leq |\gamma| \, E \left| \psi(G(\xi(0))) \right| \left(\sum_{l=1}^{q} k^{il} \tilde{k}^{il} \right) h,$$

and one can prove the uniform (in $u \in v(R)$) convergence of $I_5(u)$ to zero in probability similarly to $I_2(u)$ of Lemma 2.

Taking into account the bounds (9.21) and (9.24) we obtain

$$\sup_{u \in v(R)} |I_6(u)| = \|k\| \left(\sum_{l=1}^{q} k^{il} \tilde{k}^{il} \right) R^2 B^{m/2}(T) \underset{T \to \infty}{\longrightarrow} 0.$$

Note that $I_7(u)$ can be rewritten in the form

$$I_7(u) = -\frac{1}{2} \sum_{j,l=1}^{q} \left(\int_0^T \frac{h_{jl}(t, \widetilde{u}_T)}{d_{jT}(\theta) d_{lT}(\theta)} \frac{g_i(t,\theta)}{d_{iT}(\theta)} \mu(\mathrm{d}t) \right) u^j u^l T^{1/2} B^{m/2}(T),$$

for some $\widetilde{u}_T \in v(R)$. Then by the Cauchy–Schwarz inequality and the condition B3

$$|I_7(u)| \leq \frac{1}{2} \sum_{j,l=1}^{q} \tilde{k}^{jl} k^{il} \left| u^j \right| \left| u^l \right| B^{m/2}(T) \leq \frac{q}{2} \max_{i,l=1,\dots,q} \tilde{k}^{jl} k^{il} \, \|u\|^2 \, B^{m/2}(T).$$

Therefore as $T \to \infty$

$$\sup_{u \in v(R)} |I_7(u)| = \mathcal{O}(B^{m/2}(T))$$

and Lemma 3 is proved. □

From (9.19) and (9.34) it follows

Corollary 2. *Under the conditions of Lemmas 2, 3 for any $R>0$ and $r>0$*

$$P\left\{\sup_{u\in v(R)}\|M_T(u)-L_T(u)\|>r\right\}\underset{T\to\infty}{\to}0.$$

From (9.31) and (9.32) we find (see, (9.33))

$$\widetilde{u}_T=\Lambda_T(\theta)T^{-1/2}B^{-m/2}(T)\int_0^T\eta(t)d_T^{-1}(\theta)\nabla g(t,\theta)\mu(\mathrm{d}t). \tag{9.35}$$

Lemma 4. *If the conditions B1, B4, B5(i), B5(ii), B6 and B7 hold, then*

$$E\widetilde{u}_T\widetilde{u}_T^*\underset{T\to\infty}{\to}\frac{C_m^2(\psi)}{m!C_0^2(\psi')}\Lambda_0(\theta)\sigma_m(\theta)\Lambda_0(\theta) \tag{9.36}$$

with $\Lambda(\theta)=J^{-1}(\theta)$.

Proof. Let $L_2(\Omega)$ be a Hilbert space of r.v.'s with finite second moments. Consider the following Chebyshev–Hermite expansion in $L_2(\Omega)$;

$$\eta(t)=\gamma\psi(G(\xi(t))=\gamma\sum_{k=m}^{\infty}(C_k(\psi)/k!)H_k(\xi(t)) \tag{9.37}$$

where $C_k(\psi)$ is defined in (9.6) and γ is defined in (9.9).

It is well known that

$$EH_k(\xi(s))H_l(\xi(t))=\delta_k^l k!B^k(t-s),l,k\geq 1,$$

where δ_k^l is the Kronecker delta and B is defined in (9.1). Hence

$$E\widetilde{u}_T\widetilde{u}_T^*=\Lambda_T(\theta)\sigma_{m,T}(\theta)\Lambda_T(\theta)+\Lambda_T(\theta)R_T\Lambda_T(\theta) \tag{9.38}$$

where $\Lambda_T(\theta)$ is defined in B3 and by B7,

$$\begin{aligned}\lim_{T\to\infty}\sigma_{m,T}(\theta)&=\lim_{T\to\infty}\frac{C_m^2(\psi)}{m!C_0^2(\psi')}T^{-1}B^{-m}(T)d_T^{-1}(\theta)\\&\quad\times\left[\int_0^T\int_0^T B^m(t-s)\nabla g(t,\theta)\nabla^*g(s,\theta)\mu(\mathrm{d}t)\mu(\mathrm{d}s)\right]d_T^{-1}(\theta)\\&=\frac{C_m^2(\psi)}{m!C_0^2(\psi')}\lim_{T\to\infty}\left(\int_0^1\int_0^1\frac{g_l(tT,\theta)g_j(sT,\theta)}{D_{l\theta}(T)D_{j\theta}(T)}\frac{\mathrm{d}t\mathrm{d}s}{|t-s|^{m\alpha}}\right)_{j,l=1}^q\\&=\frac{C_m^2(\psi)}{m!C_0^2(\psi')}\sigma_m(\theta).\end{aligned} \tag{9.39}$$

In (9.38)

$$R_T = \gamma^2 T^{-1} B^{-m}(T) \sum_{k=m+1}^{\infty} (C_k^2(\psi)/k!) d_T^{-1}(\theta) \tilde{\sigma}_{k,T}(\theta) d_T^{-1}(\theta) \quad (9.40)$$

and

$$\tilde{\sigma}_{k,T}(\theta) = \int_0^T \int_0^T B^k(t-s) \nabla g(t,\theta) \nabla^* g(s,\theta) \mu(\mathrm{d}t)\mu(\mathrm{d}s) = \iint_{D_1} \cdots + \iint_{D_2} \cdots$$

where D_1 and D_2 are defined as follows:

$$D_1 = \{(u,v) \in [0,T] \times [0,T] : |u-v| \le A_0\},$$
$$D_2 = \{(u,v) \in [0,T] \times [0,T] : |u-v| > A_0\},$$

and A_0 is such that for given $\varepsilon > 0, B(t) < \varepsilon$, for $t > A_0$.

Now, for $(t,s) \in D_2 : B^{k+1}(t-s) \le \varepsilon B^k(t-s), k \ge m$, and

$$\begin{aligned}
\tilde{\sigma}_{k,T}(\theta) &\le' \sum_{k=m+1}^{\infty} (C_k^2(\psi)/k!) \int_0^T \int_0^T B^{m+1}(t-s) \left|\nabla g(t,\theta) \nabla^* g(s,\theta)\right| \mu(\mathrm{d}t)\mu(\mathrm{d}s) \\
&\le' k_0 \left[\iint_{D_1} + \iint_{D_2} \right] B^{m+1}(t-s) \left|\nabla g(t,\theta) \nabla^* g(s,\theta)\right| \mu(\mathrm{d}t)\mu(\mathrm{d}s) \\
&\le' k_0 \left[\iint_{D_1} \left|\nabla g(t,\theta) \nabla^* g(s,\theta)\right| \mu(\mathrm{d}t)\mu(\mathrm{d}s) \right. \\
&\qquad \left. + \varepsilon \iint_{D_2} B^m(t-s) \left|\nabla g(t,\theta) \nabla^* g(s,\theta)\right| \mu(\mathrm{d}t)\mu(\mathrm{d}s) \right] \quad (9.41)
\end{aligned}$$

where the sign $\le'$ means the relationship $\le$ between respective entries of matrices. From (9.40, 9.41) we obtain that every entry of matrix $\Lambda_T(\theta) R_T \Lambda_T(\theta)$ tends to zero as $T \to \infty$ and (9.36) follows from (9.39) and B3–B4. □

9.4 Proofs

Proof of Theorem 1. We use the ideas of the papers of Ivanov and Leonenko (1989, 2002, 2004). From (9.15) and (9.37) we obtain the following expansion in $L_2(\Omega)$;

$$\mu_{jT}(\theta) = \gamma \frac{C_m(\psi)}{m!} \int_0^T H_m(\xi(t)) g_j(t,\theta) \mu(\mathrm{d}t) \left[T^{1/2} B^{m/2}(T) d_{jT}(\theta) \right]^{-1} + \tilde{R}_T.$$

where similar to the proof of Lemma 4 $\tilde{R}_T \to 0$ in probability as $T \to \infty$.

From the Slutsky Lemma and the Cramer–Wald argument we conclude that random vector $\widetilde{u}_T$ defined in (9.35) converges weakly as $T \to \infty$ to the random vector $u_{01}(\theta)$ described in Theorem 1.

Let us show that for any $r > 0, \varepsilon > 0, T > T_0$

$$P\left\{\|\hat{u}_T - \widetilde{u}_T\| > r\right\} \leq \varepsilon. \tag{9.42}$$

Define the event $A_T = \{\widetilde{u}_T \in v(R-r)\}$ with $\mathbb{R}^q$ chosen so that for $T > T_0$ $P\{\hat{A}_T\} \leq \varepsilon/2$. Introduce one more event,

$$B_T = \left\{ \sup_{u \in v(R)} \|\Lambda_T(\theta)\,(M_T(u) - L_T(u)\| \leq r \right\}.$$

From the condition B4 and Corollary 2 it follows that for some $\underline{\lambda} > 0$ and $T > T_0$

$$P\{\bar{B}_T\} \leq P\left\{ \sup_{u \in v(R)} \|M_T(u) - L_T(u)\| > \underline{\lambda} r \right\} \leq \varepsilon/2, \tag{9.43}$$

and hence for $T > T_0$

$$P\{A_T \cap B_T\} > 1 - \varepsilon.$$

Taking into account the formulae (9.30) and (9.35) one can observe that

$$\Lambda_T(\theta)\, L_T(u) = \widetilde{u}_T - u.$$

If the event $A_T \cap B_T$ occurs then for $u \in v(R)$

$$\|u + \Lambda_T(\theta)\, M_T(u)\| \leq \|\Lambda_T(\theta)\,(M_T(u) - L_T(u)\| + \|\widetilde{u}_T\| \leq r + (R - r) = R,$$

and $F_T(u) = u + \Lambda_T(\theta)\, M_T(u)$ is a continuous map of $v(R)$ into $v(R)$. Thanks to the Brouwer fixed point theorem: any continuous function $F : v(R) \to v(R)$ has a fixed point (see Milnor (1965), p. 14), there exists $u_T^0 \in v(R)$ such that $F_T(u_T^0) = u_T^0$, or, because the matrix $\Lambda_T(\theta)$ is non-singular, $M_T(u_T^0) = 0$. According to our assumption B8 the unique solution of the system of the equations $M_T(u) = 0$ is the normed M-estimator $\hat{u}_T$. Thus $A_T \cap B_T \subset \{\hat{u}_T \in v(R)\}$, and

$$P\{\hat{u}_T \in v(R)\} > 1 - \varepsilon.$$

Note also that by (9.43)

$$\begin{aligned} 1 - \varepsilon < P\left\{\{\hat{u}_T \in v(R)\} \cap B_T\right\} &\leq P\left\{\|\Lambda_T(\theta)\,(M_T(\hat{u}_T) - L_T(\hat{u}_T)\| \leq r\right\} \\ &= P\{\|\hat{u}_T - \widetilde{u}_T\| \leq r\}, \end{aligned} \tag{9.44}$$

and (9.44) is equivalent to (9.42). However (9.44) obviously means that the normed M-estimator $\hat{u}_T$ has the same limiting distribution as $\widetilde{u}_T$.

Thus for $m = 1$ we arrive at the Gaussian distribution described in Theorem 1. □

Proof of Theorem 2. Similarly to the proof of Theorem 1 we obtain the following expansion in $L_2(\Omega)$;

$$\mu_{jT}(\theta) = \gamma \frac{C_2(\psi)}{2!} \int_0^T H_2(\xi(t)) g_j(t,\theta)\mu(\mathrm{d}t) \left[T^{1/2} B(T) d_{jT}(\theta)\right]^{-1} + \tilde{R}_T,$$

where similar to the proof of Lemma 4 $\tilde{R}_T \to 0$ in probability as $T \to \infty$.

From Slutsky Lemma and Cramer–Wald argument we conclude that random vector $\widetilde{u}_T$ defined in (9.35) converges weakly as $T \to \infty$ to the random vector $u_{02}(\theta)$ described in Theorem 2 (see Ivanov and Leonenko (2004) or Leonenko and Taufer (2006) for details). The rest of the proof is similar to the proof of Theorem 1. □

Acknowledgements

The second author was supported, in part, by the EPSRC grant RCMT 119 and the Australian Research Council Discovery grant DP0559807.

References

Adenstedt, R.K. (1974). On large-sample estimation of the mean of a stationary random sequence. *The Annals of Statistics,* **2**, 1095–1107.

Albin, J.M.P. (1998). A note on Rosenblatt distribution. *Statistics & Probability Letters*, **40**, 83–91.

Anh, V.V., Leonenko, N.N., and McVinish, R. (2001). Models for fractional Riesz-Bessel motion and related processes. *Fractals*, **9**, N3, 329–346.

Anh, V.V., Knopova, V.P., and Leonenko, N.N. (2004). Continuous-time stochastic processes with cyclical long-range dependence. *Australian and New Zealand Journal of Statistics*, **46**, N2, 275–296.

Arcones, M.A. (2001). Asymptotic distribution of regression M estimators. *Journal of Statistical Planning and Inference*, **97**, 235–261.

Bardadym, T.A. and Ivanov, A.V. (1999). On asymptotic normality of l_α estimates of parameter of non-linear regression. *Theory of Probability and Mathematical Statistics*, **60**, 1–10.

Barndorff-Nielsen, O.E. and Shephard, N. (2001). Non-Gaussian Ornstein-Uhlenbeck-based models and some of their uses in financial econometrics (with discussion). *Journal of the Royal Statistical Society B,* **63**, 167–241.

Barndorff-Nielsen, O.E. and Leonenko, N.N. (2005). Spectral properties of superpositions of Ornstein-Uhlenbeck type processes. *Methodology and Computing in Applied Probability,* **7**, 335–352.

Beran, J. (1991). M-estimators of location for data with slowly decaying correlations. *Journal of the American Statistical Association,* **86**, 704–708.

Beran J. (1994). *Statistics for Long-Memory Processes.* Chapman and Hall, New York.

Beran, J. and Ghosh, S. (1998). Root-n-consistent estimation in partial linear models with long-memory errors. *Scandinavian Journal of Statistics*, **25**, 345–357.

Braiman R., Yohai, V.J., and Zamar, R.H. (2001). Optimal robust M-estimates of location. *The Annals of Statistics,* **29**, 194–223.

Cantoni, E. and Ronchetti, E. (2001). Robust inference for generalized linear models. *Journal of the American Statistical Association,* **96**, 1022–1030.

Choy, K. and Taniguchi, M. (2001). Stochastic regression model with dependent disturbances. *Journal of Time Series Analysis*, **22**, 175–196.

Dahlhaus, R. (1995). Efficient location and regression estimation for long-range dependent regression models. *The Annals of Statistics,* **23**, 1029–1047.

Deo, R.S. (1997). Asymptotic theory for certain regression models with long memory errors. *Journal of Time Series Analysis*, **18**, 385–393.

Deo, R.S. and Hurvich, C.M. (1998). Linear trend with fractional integrated errors. *Journal of Time Series Analysis*, **19**, 379–397.

Dobrushin, R.L. and Major, P. (1979). Non-central limit theorems for non-linear functionals of Gaussian fields. *Z. Wahrsch. verw. Gebiete.*, **50**, 27–52.

Doukhan, P., Oppenheim, G., and Taqqu, M.S. (2003). *Theory and Applications of Long-Range Dependence.* Birkhauser, Boston.

Feller, W. (1971). *An Introduction to Probability Theory and Its Applications,* v. **2**. John Willey, New York.

Giraitis L. and Koul H. (1997). Estimation of the dependence parameter in linear regression with long-range dependent errors. *Stochastic Processes and their Applications*, **71**, 207–224.

Giraitis L., Koul H., and Surgailis, D. (1996). Asymptotic normality of regression estimators with long memory errors. *Statistics & Probability Letters,* **29**, 317–335.

Grenander, U. and Rosenblatt, M. (1984). *Statistical Analysis of Stationary Time Series.* Chelsea Publ. Company, New York.

Hall, P., Koul, H.L., and Turlach, A. (1997). Note on the convergence rates of semiparametric estimators of dependence index. *The Annals of Statistics,* **25**, 1725–1739.

Hannan, E. (1973). The asymptotic theory of linear time-series models. *Journal of Applied Probability*, **10**, 130–145.

Hampel, F.R., Ronchetti, E.M., Rousseeuw, P.J., and Stahel, W.A. (1986). *Robust Statistics. The Approach Based on Influence Functions.* Wiley, New York.

Heyde, C.C. and Leonenko, N.N. (2005). Student processes. *Advances in Applied Probability*, **37**, 342–365.

Holevo, A.S. (1976). On the asymptotic efficient regression estimates in the case of degenerate spectrum. *Theory of Probability and Its Applications*, **21**, 324–333.

Huber, P.J. (1981). *Robust Statistics.* Wiley, New York.

Ibragimov, I.A. (1963). On estimation of the spectral function of a stationary Gaussian process. *Theory of Probability and Its Applications*, **8**, 366–401.

Ibragimov, I.A. and Rozanov, Yu.A. (1970). *Gaussian Random Processes.* Nauka. Moscow (in Russian).

Ivanov, A.V. (1997). *Asymptotic Theory of Nonlinear Regression.* Kluwer Academic Publishers. Dordrecht.

Ivanov, A.V. and Leonenko, N.N. (1989). *Statistical Analysis of Random Fields.* Kluwer Academic Publishers. Dordrecht.

Ivanov, A.V. and Leonenko, N.N. (2000). Asymptotic inference for non-linear regression with long-range dependence. *Theory of Probability and Mathematical Statistics*, **63** 61–79.

Ivanov, A.V. and Leonenko, N.N. (2002). Asymptotic behavior of M-estimators in continuous-time non-linear regression with long-range dependent errors. *Random Operators and Stochastic Equations*, **10**, N 3, 201–222.

Ivanov, A.V. and Leonenko, N.N. (2004). Asymptotic theory for non-linear regression with long-range dependence. *Mathematical Methods of Statistics*, **13**, 153–178.

Ivanov, A.V. and Leonenko, N.N. (2008). Semiparametric analysis of long-range dependence in nonlinear regression. *Journal of Statistical Planning and Inference*, **138**, 1733–1753.

Ivanov, A.V. and Orlovsky, I.V. (2001). L_p-estimates in nonlinear regression with long-range dependence, *Theory of Stochastic Processes*, **7(23)**, N 3–4, 38–49.

Kato, T. and Masry, E. (2003). A time-domain semi-parametric estimate for strong dependent continuous-time stationary processes. *Journal of Time Series Analysis*, **24**, N 6, 679–703.

Koul, H.L. (1992). M-estimators in linear models with long range dependent errors. *Statistics & Probability Letters*, **14**, 153–164.

Koul, H.L. (1996). Asymptotics of M-estimations in non-linear regression with long-range dependence errors. In: *Proceedings of the Athens Conference on Applied Probability and Time Series Analysis* (P.M. Robinson and M. Rosenblatt, Eds.). *Springer Verlag Lecture Notes in Statistics*, II, 272–291.

Koul, H.L. (2002). Asymptotic distributions of some scale estimators in nonlinear regression models. *Metrika*, **55**, 75–90.

Koul, H.L. and Baillie, R. T. (2003). Asymptotics of M-estimators in nonlinear regression models with long-memory designs. *Statistics & Probability Letters*, **61**, 237–252.

Koul, H.L. and Mukherjee, K. (1993). Asymptotics of R-, MD- and LAD-estimators in linear regression models with long-range dependent errors. *Probability Theory and Related Fields*, **95**, 535–553.

Koul, H.L. and Mukherjee, K. (1994). Regression quantiles and related processes under long range dependent errors. *Journal of Multivariate Analysis*, **51**, 318–337.

Koul, H.L. and Surgailis, D. (2000a). Second-order behavior of M-estimators in linear regression with long-memory errors. *Journal of Statistical Planning and Inference*, **91**, 399–412.

Koul, H.L. and Surgailis, D. (2000b). Asymptotic normality of the Whittle estimator in linear regression models with long-memory errors. *Statistical Inference for Stochastic Processes*, **3**, 129–147.

Koul, H.L. and Surgailis, D. (2003). Robust estimators in regression models with long memory errors. In: *Theory and Applications of Long-Range Dependence* (P. Doukhan, G. Oppenheim and M.S, Taqqu, Eds.). Birkhäuser, Boston, 339–353.

Koul, H.L., Baillie, R.T., and Surgailis, D. (2004). Regression model fitting with a long memory covariance process. *Economic Theory*, **20**, 485–512.

Künsch, H.R., Beran, J., and Hampel, F. (1993). Contrasts under long-range correlations. *The Annals of Statistics,* **21**, 943–964.

Leonenko, N.N. (1999). *Limit Theorems for Random Fields with Singular Spectrum.* Kluwer Academic Publishers. Dordrecht.

Leonenko, N.N. and Anh, V.V. (2001). Rate of convergence to the Rosenblatt distribution for additive functionals of stochastic processes with long-range dependence. *Journal of Applied Mathematics and Stochastic Analysis*, **14**, 27–46.

Leonenko, N.N. and Taufer, E. (2006). Weak convergence of weighted quadratic functionals of stationary long memory processes to Rosenblatt-type distributions. *Journal of Statistical Planning and Inference*, **136**, 1220–1236.

Leonenko, N.N. and Sakhno, L.M. (2006). On the Whittle estimators for some classes of continuos-parameter random processes and fields. *Statistics & Probability Letters*, **76**, 781–795.

Leonenko, N.N., Sharapov, M.M., and El-Bassiouny, A.H. (2000). On the exactness of normal approximation of LSE of regression coefficient of long-memory random fields. *Statistics & Probability Letters*, **48**, 121–130.

Leonenko, N.N. and Woyczynski, W.A. (1999). Parameter identification for singular random fields arising in Burgers' turbulence. *Journal of Statistical Planning and Inference*, **80**, 1–13.

Lobato, I. and Robinson, P.M. (1996). Average periodogramm estimation of long memory. *Journal of Econometrics*, **73**, 303–324.

Ma, C. (2003). Long-memory continuous-time correlation models. *Journal of Applied Probability*, **40**, 113–1146.

Milnor, J.W. (1965). *Topology from the Differentiable Viewpoint.* Princeton University Press, Princeton, NJ.

Mukherjee, K. (2000). Linearization of randomly weighted empiricals under long range dependence with applications to nonlinear regression quantiles. *Economic Theory*, **16**, 301–323.

Pázman, A. and Pronzato, L. (2006). On the irregular behavior of LS estimates for asymptotially singular designs. *Statistics & Probability Letters*, **76**, 1089–1096.

Pollard, D. and Radchenko, P. (2006). Nonlinear least-squares estimation. *Journal of Multivariate Analysis*, **97**, 548–562.

Qian, G. and Künsch, H.R. (1998). On model selection via stochastic complexity in robust linear regression. *Journal of Statistical Planning and Inference*, **75**, 91–116.

Robinson, P.M. (1994a). Rate of convergence and optimal spectral bandwidth for long range dependence. *Probability Theory and Related Fields*, **99**, 443-473.

Robinson, P.M. (1994b). Semiparametric analysis of long-memory time series. *The Annals of Statistics,* **22**, 515–539.

Robinson, P.M. (1995a). Long-periodogram regression of time series with long-range dependence. *The Annals of Statistics,* **23**, 1048–1072.

Robinson, P.M. (1995b). Gaussian semiparametric estimation of long-range dependence. *The Annals of Statistics,* **23**, 1630–1661.

Robinson, P.M. and Hidalgo, F.J. (1997). Time series regression with long-range dependence. *The Annals of Statistics,* **25**, 77–104.

Rosenblatt, M. (1961). Independence and dependence. *Proceedings of the Fourth Berkeley Symposium on Mathematics Statistics and Probability.* University of California Press, California, 431–443.

Ronner, A.E. (1984). Asymptotic normality of p-norm estimators in multiple regression. *Z. Wahrsch.verw. Gebiete*, **66**, 613–620.

Samarov, A. and Taqqu, M.S. (1988). On the efficiency of the sample mean in long memory noise. *Journal of Time Series Analysis*, **9**, 191–200.

Seneta, E. (1976). *Regularly Varying Functions.* Lecture Notes in Mathematics, 508. Springer-Verlag, Berlin-New York.

Sibbersten, P. (2003). Log-periodogram estimation of the memory parameter of a long-memory process under trend. *Statistics & Probability Letters*, **61**, 261–268.

Skouras, K. (2000). Strong consistency in nonlinear regression models. *The Annals of Statistics,* **28**, 871–879.

Taqqu, M.S. (1979). Convergence of iterated processes of arbitrary Hermite rank. *Z. Wahrsch. verw. Gebiete.*, **50**, 53–58.

Yajima, Y. (1988). On estimation of a regression model with long-memory stationary errors. *The Annals of Statistics,* **16**, 791–807.

Yajima, Y. (1991). Asymptotic properties of the LSE in a regression model with long-memory stationary errors. *The Annals of Statistics,* **19**, 158–177.

Wu, Y. and Zen, M.M. (1999). A strongly consistent information criterion for linear model selection based on M-estimation. *Probability Theory and Related Fields*, **113**, 599–625.

Index

Printed in the United States of America